KV-106-954

MITOCHONDRIA CHLOROPLASTS AND BACTERIAL MEMBRANES

J. N. Prebble

Department of Biochemistry
Bedford College
University of London

THAMES POLYTECHNIC LIBRARY
574·
8734
2
PRE

Longman
London and New York

574.8734
229726
S. 82

THAMES POLYTECHNIC
10 JUN 1992
LIBRARY

Longman Group Limited

Longman House
Burnt Mill, Harlow, Essex, UK

Published in the United States of America
by Longman Inc., New York

© Longman Group Limited 1981

All rights reserved. No part of this publication may be reproduced, stored in a retrieval system, or transmitted in any form or by any means, electronic, mechanical, photocopying, recording, or otherwise, without the prior permission of the Copyright owner.

First published 1981

British Library Cataloguing in Publication Data

Prebble, J N
    Mitochondria, chloroplasts and bacterial
    membranes.
    1. Mitochondria
    2. Chloroplasts
    I. Title
    574.8'734   QH603.M5   80–40777

    ISBN 0–582–44133–1

Printed in Great Britain by
Butler & Tanner Ltd
Frome and London

MITOCHONDRIA, CHLOROPLASTS AND BACTERIAL MEMBRANES

PREBBLE: Mitochondria, chloroplasts & bacterial membranes.

Rachel McMillan Library

574.8734

229726

Contents

Preface

Books on mitochondria and on photosynthesis abound. Thus the addition of yet another book on these subjects requires some justification. Most of the books on mitochondria fall into two categories, those which are research monographs, many of them being the proceedings of a scientific meeting, and those which form an introduction to the subject at about the first- or second-year University level. A not dissimilar situation applies to photosynthesis. The chloroplast as an organelle receives rather less emphasis. Although the cross-fertilisation of ideas between the fields of chloroplast and mitochondrial biochemistry is widely acknowledged, treatment of both subjects together is less common. In the present work, I have endeavoured to bridge the gap between the introductory and the advanced research-orientated works by attempting to provide a broad coverage of the main aspects of mitochondria and chloroplasts with a view to providing a jumping-off point for the extensive literature of reviews. Each chapter terminates with a short list of recent reviews.

The book is written as a textbook with the intention of conveying concepts and information and is aimed primarily at third-year undergraduate and graduate students. The background knowledge of students tends to be very variable, even among those of the same institution, and an attempt has been made to limit the knowledge required to understand the major concepts discussed. The more elementary material has been presented in a historical manner in order to demonstrate the development of the subject. The number of references has been kept to a minimum not only for economy of space, but also because references can create a forbidding distraction to the text. A major objective is to present a balanced account of the organelles although the notion of balance is inevitably subjective. Thus, the mitochondrion has been viewed both as the organelle responsible for oxidative phosphorylation and as a particle possessing a complex pattern of metabolic pathways and permeation systems some of which are not directly related to its energetic function. In recent years, the study of bacterial biochemistry has contributed significantly to an under-standing of chloroplasts and mitochondria; two chapters summarising energetics of bacterial membranes have therefore been included.

I would like to acknowledge my indebtedness to many friends at Bedford College for their advice, help and encouragement. In particular I wish to thank Professor D. F. Cheesman who earlier encouraged my interest in biochemistry and who has kindly read the manuscript and made many valuable suggestions. My thanks go to Dr K. E. Howlett for advice on Chapter 1, Sarah Chapman and Sonia Copeland who typed the manuscript and to the staff of Longman for their help. Finally, I would also like to add my appreciation of the patience and support of my wife and family.

Department of Biochemistry   **J. N. Prebble**
Bedford College
(University of London)
Regents' Park
London NW1 4NS
October 1979

Abbreviations

$\Delta\psi$	membrane potential
Δp	proton motive force (PMF)
λ	wavelength
ADP	adenosine diphosphate
ALA	γ-aminolaevulinic acid
Ala	alanine
AMP	adenosine monophosphate
Asp	aspartic acid
ATP	adenosine triphosphate
CAM	crassulacean acid metabolism
CCCP	carbonylcyanide-m-chlorophenylhydrazone
Chl	chlorophyll
CoA	coenzyme A
CP	chlorophyll–protein complex
ctDNA	chloroplast DNA
cyt	cytochrome
DABS	p-(diazonium)-benzenesulphonate
DBTQ	dibromothymoquinone
DCCD	N,N-dicyclohexylcarbodiimide
DCMU	3(3,4-dichlorophenyl)-1,1-dimethylurea
DCPIP	dichlorophenol indophenol
DNA	deoxyribose nucleic acid
DNP	dinitrophenol
E_m	mid-point potential (pH included as subscript: $E_{m\,7.0}$)
EDTA	ethylenediaminetetra-acetic acid
EPR	electron paramagnetic resonance
F	Faraday
FAD	flavin adenine dinucleotide
FCCP	carbonylcyanide-p-trifluoromethoxyphenylhydrazone
Fd	ferredoxin
Fe–S	iron–sulphur centre or protein
FMN	flavin mononucleotide
Fp	flavoprotein
GDP	guanosine diphosphate
GTP	guanosine triphosphate
His	histidine
Ile	isoleucine
kb	kilobase (= 1000 bases)
Lys	lysine
Met	methionine

MK	menaquinone
mRNA	messenger RNA
mtDNA	mitochondrial DNA
NAD, $NADH_2$	nicotinamide adenine dinucleotide (oxidised and reduced respectively)
NADP, $NADPH_2$	nicotinamide adenine dinucleotide phosphate (oxidised and reduced respectively)
OSCP	oligomycin sensitivity conferring protein
PGA	3-phosphoglyceric acid
Phe	phenylalanine
Pi	inorganic phosphate
PP	pyrophosphate
PQ	plastoquinone
Pro	proline
PS I, PS II	pigment system or photosystem I and II respectively
Q	ubiquinone (also used for the primary acceptor of PS II)
RNA	ribonucleic acid
Ser	serine
SMP	submitochondrial particle
Thr	threonine
TMPD	tetramethylphenylenediamine
TPP	thiamin pyrophosphate
Try	tryptophan
Tyr	tyrosine
UDP	uridine diphosphate
UTP	uridine triphosphate
Val	valine

Chapter 1

Mitochondria and chloroplasts: basic concepts

1.1 The nature of the subcellular particles

(a) Introduction

The primary source of energy available for life on our planet is solar radiation. This may be converted to chemical energy by organisms possessing suitable photochemical pigments and associated systems. In green plants the process, photosynthesis, is concerned with the synthesis of organic compounds (e.g. hexose sugars) from carbon dioxide and water and may be represented as:

$$CO_2 + H_2O + \text{light energy} \longrightarrow C_6H_{12}O_6 + O_2$$

The fossil fuels, coal, oil and natural gas which are now being rapidly consumed, are ultimate products of this process and represent the photosynthetic activity of plants in earlier periods of geological time.

Just as technology harnesses the energy of organic compounds by their combustion in the presence of oxygen, so living cells are able to use the energy in organic compounds for life processes such as the synthesis of fresh cellular material and muscular movement. For hexoses, this oxidative process may be represented:

$$C_6H_{12}O_6 + O_2 \longrightarrow CO_2 + H_2O + \text{energy}$$

The two processes above are thermodynamically but not mechanistically the reverse of one another. However, as we shall see, they have similarities, for example in regard to ATP synthesis. Except in simple organisms (prokaryotes) the major part of both processes takes place in discrete organelles within the cell. The photosynthetic system is housed within the chloroplast, while the essential oxidative systems are the property of the mitochondrion. Both organelles are surrounded by a double membrane which is selectively permeable to cell constituents, so that the organelles constitute metabolic compartments separate from the rest of the cell but connected to it in a manner controlled by the permeation systems.

The reason for discussing both the chloroplast and the mitochondrion within the compass of a single book lies in the fact that both possess similar chains of oxidation – reduction reactions which are linked to ATP synthesis. In addition, the oxidation – reduction systems are also linked to complex metabolic

1

cycles. Further, both are semi-autonomous in possessing a partial hereditary system, a small 'chromosome' coding for some of the particle's proteins which can be totally synthesised within the particles.

(b) The mitochondrion – a metabolic system for ATP synthesis

The major functions of the mitochondrion can be conveniently listed as:

- (i) the oxidation of pyruvate to acetyl coenzyme A
- (ii) the oxidation of fatty acids to acetyl coenzyme A
- (iii) the oxidation of acetyl coenzyme A to CO_2 and reduced cofactors (e.g. $NADH_2$)
- (iv) the oxidation of reduced cofactors ($NADH_2$) by oxygen forming water
- (v) synthesis of ATP coupled to $NADH_2$ oxidation, oxidative phosphorylation.

All five of these processes are more or less intimately linked (see Fig. 1.1) and may be considered as part of the oxidative system for ATP synthesis. Thus the *raison d'être* of the mitochondrion is to be the major supplier of ATP for the cell. However, the mitochondrion also has a significant role in other processes such as nitrogen metabolism and various biosynthetic systems, for example in porphyrin synthesis and in steroid hormone synthesis in the adrenal cortex.

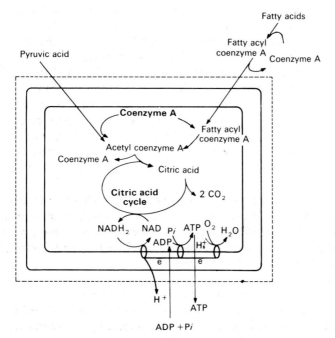

Fig. 1.1 Outline of the major metabolic pathways of the mitochondrion.

The study of the biochemistry of the mitochondrion has proceeded by several almost separate lines which have slowly converged on the particle as the main focus of investigation. Workers such as Szent-Györgyi who were interested in the chemistry of respiration demonstrated the catalytic effect of four-carbon dicarboxylic acids on oxygen uptake. This work was complemented by earlier studies on dehydrogenases by Wieland and others and was brought to fruition by Krebs with the formulation of the citric acid cycle. Later Lehninger showed the cycle to be a property of the mitochondrion.

A second line of investigation was initiated by Warburg, who showed, in his studies on carbon monoxide inhibition of respiration, that an iron porphyrin was involved in oxygen metabolism. Subsequently Keilin rediscovered the cytochromes and isolated cytochrome *c*. From observations of the absorption spectra of tissues, he was able to construct a scheme for the respiratory chain.

A third line of study was developed by Kalckar and by Belitzer and Tsybakova leading to the concept of oxidative phosphorylation in which ATP synthesis is coupled to respiratory activity.

Starting with the observations of Knoop, the system of fatty acid oxidation was investigated and this also was eventually seen as a property of the mitochondrion.

Cytologists in the late nineteenth century had become aware of the mitochondrial particles in cells. A number of attempts were made to study these particles either *in situ* with dyes or by isolating them. However, it was not until 1940, when Claude began a systematic investigation of large granules (mitochondria) and small granules (microsomal fraction) isolated from liver homogenates by differential centrifugation, that a biochemical understanding of mitochondria became possible. By 1950, it had become clear that the fatty acid oxidising system, the citric acid cycle and the respiratory chain with its associated phosphorylation system were all located in the mitochondrion. With the advent of the electron microscope, the detailed structure of the mitochondrion was formulated, particularly by Palade. This led to a study of enzyme distribution within the structure and to the problem of transport of metabolites across the membranes. A further area of investigation developed as workers became aware of the questions associated with the biogenesis of the mitochondrion leading to studies of the synthesis of mitochondrial protein.

Although bacteria are too small to possess mitochondria (*Escherichia coli*, a short rod 0.8×1.2 μm has dimensions similar to mitochondria) many are capable of similar oxidative metabolism and possess comparable systems of respiration and phosphorylation. Investigations of these systems in bacteria have augmented the studies on the mitochondrion.

(c) The chloroplast – an energy-transducing system

The prime function of the chloroplast is photosynthesis. The earliest ideas on this subject were developed in the eighteenth century by a colourful group of amateur naturalists. Stephen Hales (1677–1761) was the first of several workers to study photosynthesis as a process of gas exchange. In his essay *Vegetable Staticks*, he wrote 'Plants very probably draw through their leaves some part of their nourishment from the air . . . may not light also, by freely entering surfaces of leaves and flowers, contribute much to the ennobling principles of vegetables?'

The eighteenth century was also the era of development of ideas on the chemistry of gases associated with the names of Black, Scheele, Priestley, Cavendish, Lavoisier and others. The new insights had an impact on the understanding of respiration and particularly of photosynthesis. Joseph Priestley demonstrated the purification of air in the light in the presence of a photo-synthesising system – a piece of mint. Air in a glass jar in which a candle had burned out, was restored so that the candle would again burn in it. The full significance of these experiments was not realised by Priestley who interpreted them in terms of the current phlogiston theory. However, later work showed convincingly the evolution of oxygen from photosynthesising algae. Priestley's experiments, published in the Proceedings of the Royal Society in 1772, attracted the attention of a Dutch physician, Ingen-Housz, who became physician to the Empress Maria Theresa. The experiments of Ingen-Housz convinced him of the importance of light in the air-restoration process and the active production of oxygen in the light. In his book *Food of Plants and Renovation of Soil*, he claimed that plants obtained their carbon by decomposition of the carbonic acid of air and that they evolved oxygen. Thus towards the end of the eighteenth century a coherent concept of photosynthesis began to develop, namely the synthesis of organic matter from carbon dioxide in the light with the evolution of oxygen.

These views were supported by the careful weighing experiments on plants by de Saussure (1767–1845). He showed that plants obtain most of their elements from the soil, but all the carbon from the CO_2 of the atmosphere. He also found that the increase in dry weight of plants was considerably greater than the increase in carbon and concluded that water was also assimilated. His data showed that the assimilation of water was linked to the photosynthetic process although he did not realise this. An understanding of the role of water in photosynthesis had to wait until later.

The later part of the nineteenth century saw many developments in ideas on photosynthesis. Three only will be mentioned here. First, Robert Mayer, who formulated the law of conservation of energy, drew attention to the energetic aspects of photosynthesis as a process in which light energy is converted into chemical energy. Secondly, J. Sachs observed the development of starch grains in photosynthesising chloroplasts. Thirdly, Engelmann carried out the first action spectrum for photosynthesis. Engelmann illuminated a filamentous alga with light passed through a prism. Motile bacteria sensitive to the level of dissolved oxygen moved to those parts of the filament where the greatest oxygen evolution occurred. This was in the parts illuminated with the blue and red regions of the spectrum, providing confirmation for the role of chlorophyll in photosynthesis.

Thus at the beginning of the twentieth century it was possible to write a simple equation for photosynthesis:

$$\text{Carbon dioxide} + \text{water} + \text{light energy} \xrightarrow{\text{chlorophyll}} \text{carbohydrate} + \text{oxygen}$$

The first step in analysis of this equation was the demonstration that the photosynthetic process consisted of separable light and dark reactions. A variety of experiments led to this conclusion. At the beginning of the century Blackman showed that the rate of photosynthesis was temperature-dependent at high light intensity and low CO_2 concentration but temperature-independent at low light and high CO_2. These results were eventually interpreted as showing normal temperature-dependence when CO_2 fixation was the rate-limiting step (dark reaction) but a temperature-independent process when light was rate-limiting (light reaction). This view was given strong support by the work of Emerson and Arnold (1932a) using short flashes of light (3×10^{-3} s) with a varied dark period between them. They found that for the maximum yield of oxygen per flash, an intervening dark period of a few tenths of a second was necessary to allow the slower 'dark reaction' to utilise fully the products of the light reaction.

The implication of these studies is that the 'dark reaction', CO_2 fixation, is a metabolic system only loosely coupled to the light reaction. In 1954 Calvin published the first version of his cycle which showed the metabolic reactions involved in fixing CO_2 into organic carbon and the subsequent formation of hexose sugars (Bassham *et al.*, 1954). The process was dependent on a supply of reduced pyridine nucleotides ($NADPH_2$) and ATP.

The events associated with oxygen evolution proved to be capable of separate investigation and led to an appreciation of the light reaction. Molisch (1925) had extended earlier observations that preparations of dried leaves, when subsequently wetted, would evolve oxygen in the light. These experiments attracted the attention of R. Hill (1937, 1939) who isolated chloroplasts from green leaves and demonstrated that they would evolve oxygen in the light and in the presence of suitable electron acceptors such as oxaloferrate, and later ferricyanide. The fundamental reaction observed here, the Hill reaction, is a photolytic splitting of water:

$$2H_2O \xrightarrow{\text{light, chloroplasts}} O_2 + 4H^+ + 4e$$

There was, rather surprisingly, no apparent requirement for CO_2 in this reaction. However, a requirement for low levels of CO_2 was later demonstrated, although this CO_2 is not involved in fixation and the Calvin cycle.

What is the natural electron acceptor in the chloroplast when the photolysis of water takes place? Vishniac and Ochoa (1951) showed that chloroplasts would carry out a Hill reaction with pyridine nucleotides as electron acceptors:

$$2NADP + 2H_2O \longrightarrow 2NADPH_2 + O_2$$

The path by which reducing equivalents are transferred from water to NADP is complex and involves the chloroplast electron transport chain and two separate light reactions.

In 1954, the same year as the first formulation of the Calvin cycle, Arnon and coworkers provided a convincing demonstration of photophosphorylation, which was eventually seen as the synthesis of ATP coupled to the light-driven electron transport chain.

The major reactions of the chloroplast (see Fig. 1.2) can now be summarised:

1. Light energy is absorbed by a pigment complex (principally but not solely chlorophyll).
2. The light energy drives an electron transport chain in which water is oxidised to oxygen and NADP is reduced.
3. The synthesis of ATP (like that in the mitochondrion) is coupled to electron transport.
4. The fixation of carbon dioxide and its conversion to hexose. This process requires the ATP and reduced NADP formed in the reactions above.

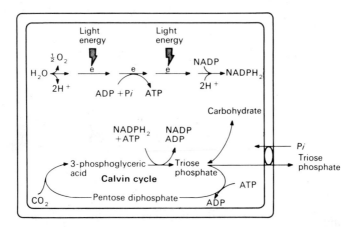

Fig. 1.2 Outline of chloroplast biochemistry.

A rather different form of photosynthesis with different chlorophylls is seen in a small group of bacteria. The organisms photosynthesise under anaerobic conditions in the light and do not evolve oxygen, although they possess an electron transport chain and carry out photophosphorylation. As in other fields, studies of a simpler bacterial system have contributed to our general understanding of photosynthesis.

(d) Distribution of particles
In higher plants, each cell in the green tissues normally contains several chloroplasts (Fig. 1.3). Cells near the upper surface of leaves frequently have a higher density of plastids than those near the lower; for example in *Ricinus communis* (castor oil plant) the palisade cells near the upper surface contain about 36 chloroplasts per cell, while mesophyll cells near the lower surface contain only about 20. In simpler organisms such as algae there may be several or only one relatively large chloroplast in each cell. Primitive algae such as the blue-greens (Cyanobacteria) are really prokaryotes and lack a properly defined chloroplast to house their photosynthetic apparatus. In general, plants tend to show an inverse relationship between the number of mitochondria and the number of chloroplasts per cell.

Mitochondria occur in considerable numbers in cells (Fig. 1.4). For

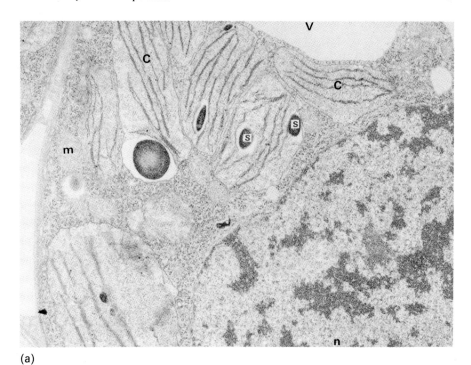

(a)

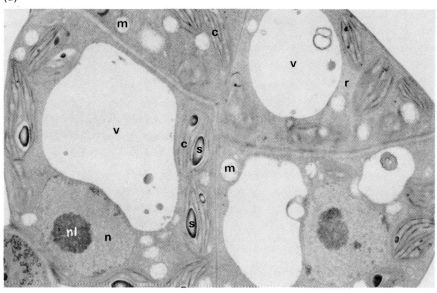

(b)

Fig. 1.3 Electron micrograph of plant cells showing chloroplasts. (a) part of a leaf cell from *Avena sativa* (cultivated oat), ×9900. (Courtesy of Dr. M. Steer, University of Belfast). (b) part of a mesophyll cell from a young leaf of *Phaseolus vulgaris* × 2400. (Courtesy of Dr. Jean Whatley, University of Oxford).
c = chloroplast, m = mitochondrion, n = nucleus, nl = nucleolus, r = ribosomes, s = starch grains, v = vacuole.

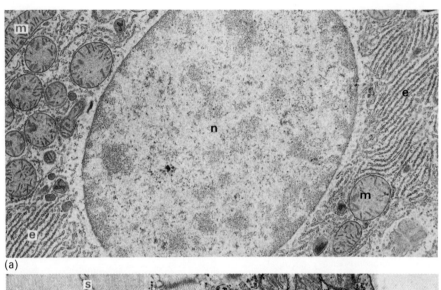

(a)

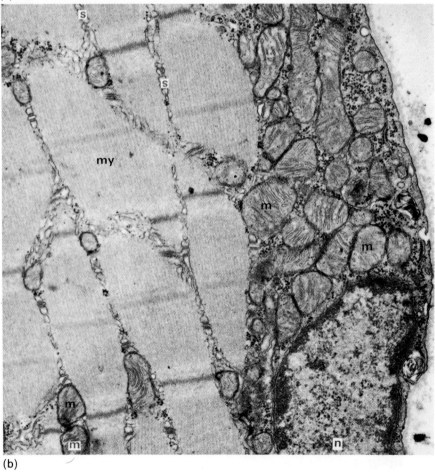

(b)

mammalian cells it is estimated that mitochondria occupy about 20 per cent of the cell volume. A relationship between the number of particles per cell and function can be seen in breast muscle of birds. A much higher density of mitochondria is found in birds which are good flyers than in those which are poor flyers. In liver tissue, estimates vary but show a complement of about 1000 to 1500 mitochondria per cell. Dividing undifferentiated liver cells possess fewer mitochondria (Table 1.1). By contrast the protozoan *Paramecium* possesses about 1000 mitochondria and the flagellate *Euglena* may have as few as 15 to 20 although it possesses a chloroplast.

Table 1.1 Mitochondrial content of cells

		Particles/cell
Liver cell (rat)		1000–1600
Yeast (*Saccharomyces*)*	aerobically grown	95
	anaerobically grown	5
Euglena		18
Paramecium		1000

* Recently, it has been suggested that some yeasts may possess one or very few much branched mitochondria rather than a number of separate particles.

(e) Isolation of mitochondrial and chloroplast fractions

Most of our knowledge of the structure and function of mitochondria and chloroplasts has been derived from studies of isolated particle preparations. Such preparations involve essentially three steps: homogenisation, separation and characterisation of the fractions separated (see Figs. 1.5–1.7).

Homogenisation is the disruption of the tissue, breakage of the cells and release of their contents into a suitable medium. This is achieved by a mechanical process such as grinding, blending or by forcing the tissue through a narrow space between a piston and cylinder wall. All these methods disrupt the tissue, break the cell envelope and release the cell contents but may also cause some damage to the subcellular particles. A further limitation to be borne in mind is that the particles are normally derived from a tissue which contains more than one type of cell. The medium in which the homogenisation is performed is important. Both mitochondria and chloroplasts are osmotically sensitive. Hogeboom and others (1948) were the first to recommend sucrose solutions (at a concentration of 0.25 M or greater) for mitochondria. Recipes for homogenisation media vary from

Fig. 1.4 Electron micrograph of animal tissues showing mitochondria. (a) part of a liver cell from mouse, × 7 300. (b) part of a skeletal muscle cell from mouse psoas, × 29 500. (Courtesy of Prof. S. Bullivant, University of Auckland, N.Z.).
e = endoplasmic reticulum, m = mitochondrion, my = myofibrils, n = nucleus, s = sarcoplasmic reticulum.

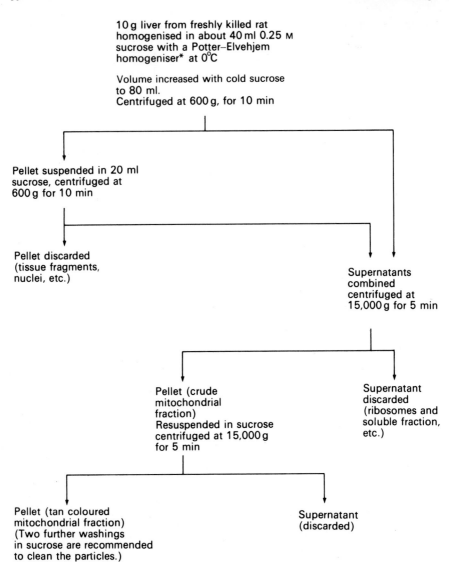

10 g liver from freshly killed rat
homogenised in about 40 ml 0.25 M
sucrose with a Potter–Elvehjem
homogeniser* at 0°C

Volume increased with cold sucrose
to 80 ml.
Centrifuged at 600 g, for 10 min

Pellet suspended in 20 ml
sucrose, centrifuged at
600 g for 10 min

Pellet discarded
(tissue fragments,
nuclei, etc.)

Supernatants
combined
centrifuged at
15,000 g for 5 min

Pellet (crude
mitochondrial
fraction)
Resuspended in sucrose
centrifuged at 15,000 g
for 5 min

Supernatant
discarded
(ribosomes and
soluble fraction,
etc.)

Pellet (tan coloured
mitochondrial fraction)
(Two further washings
in sucrose are recommended
to clean the particles.)

Supernatant
(discarded)

Fig. 1.5 Method for the preparation of mitochondria from rat liver. *Glass tube with a rotating glass or teflon pestle which is moved up and down by hand.

laboratory to laboratory and also depend on the final use to which the preparation is to be put. With chloroplasts 0.35 M NaCl (buffered) or 0.33 M sorbitol have been recommended as alternatives to 0.3–0.5 M sucrose. A problem with these particles is the loss of water-soluble enzymes when aqueous media are used. To overcome this difficulty, non-aqueous media (hexane and carbon tetrachloride) were used with freeze-dried plant tissue (Stocking, 1971). However, rapid methods such as that in Fig. 1.7 also overcome this problem (Nakatani and Barber, 1977).

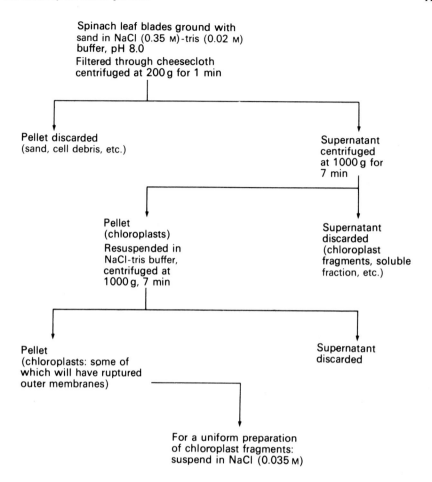

Fig. 1.6 Method for the preparation of chloroplasts and chloroplast fragments from spinach leaves. This method was used in Arnon's laboratory for many of the classical experiments on photophosphorylation. Although NaCl is used here to stabilise the particles osmotically in preference to the earlier sucrose, 0.33 M sorbitol is now preferred by many workers (see Fig. 1.7).

Separation of particles in homogenates is achieved by centrifugation. The rate at which particles in a dilute homogenate sediment will depend primarily on their size, shape and density and on the magnitude of the centrifugal field. Chloroplasts sediment more rapidly than mitochondria. Separation of either particle normally proceeds by way of an initial low-speed spin to remove debris and large particles but leaving the mitochondria or chloroplasts in suspension. A second spin at higher speed (greater centrifugal force) sediments the required particles while leaving ribosomal and fine material in suspension together with the soluble fraction of the cell. Further spins are included to purify the particle fraction.

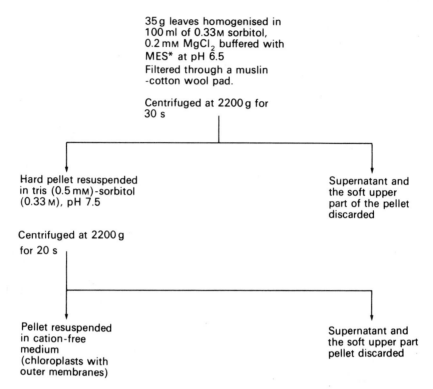

Fig. 1.7 Preparation of intact chloroplasts from plant leaves (Nakatani and Barber, 1977). MES buffer: 20 mM 2(*N*-morpholino) ethanesulphonic acid adjusted to pH 6.5 with tris (tris(hydroxymethyl)aminomethane).

A further refinement involves placing the homogenate, or more usually the crude particle preparation, on a density gradient frequently of sucrose. The particle preparation is carefully layered on top of the sucrose and the tube centrifuged long enough for the particles to settle to their own density level in a band. The bands can be removed either by piercing the bottom of the tube or by removal from the top with a pipette. Various forms of centrifugation provide the main means of particle separation although other methods have been used.

Particle fractions may be characterised and examined for purity by viewing samples with an electron microscope. Negatively stained preparations[1] or preferably sections of centrifugal pellets may be used. Alternatively, characterisation may be achieved by chemical means such as assay for substances or enzymes known to be located specifically in the particles. In addition one may wish to assay for contaminating material such as lysosomes in mitochondrial fractions. It must be remembered that what is usually isolated by these methods is a chloroplast or mitochondrial fraction rich in the desired particle rather than a pure preparation of the particle itself. However, current methods can provide a very high degree of purity.

[1] Drops of particle suspension are treated with a negative stain which renders the background electron-dense but leaves much of the particle unstained.

1.2   Energetic considerations – ATP

Both the chloroplast and the mitochondrion bring about the synthesis of adenosine triphosphate (ATP) coupled to their respiratory chains as an essential part of their energetic function. It is now necessary to examine the ideas associated with this key substance.

Interest in the purines of tissues was aroused when Embden and Zimmermann (1927) isolated adenylic acid (adenosine monophosphate, AMP) from muscle. ATP was first isolated by Fiske and Subbarow (1929). Lohmann (1929) demonstrated the metabolic importance of these compounds by showing an enzymic dephosphorylation and rephosphorylation of ATP in muscle. The structure of ATP (Fig. 1.8) was worked out over many years and confirmed when Todd and others (Baddiley *et al.*, 1949) synthesised the compound.

The role of ATP in metabolism may at first appear to be that of a phosphorylating or adenylating agent, e.g.:

$$ATP + RH \longrightarrow ADP + RP \qquad\qquad [1]$$

or $\quad ATP + RH \longrightarrow AMP + RPP \qquad\qquad [2]$

or $\quad ATP + RH \longrightarrow R\text{-}AMP + PP^2 \qquad\qquad [3]$

However, the significance of these and related reactions is understood by biochemists to lie more in the energetic aspects of the phosphates. In the decade following the isolation of the adenine nucleotides, Lipmann developed the concept of the 'high-energy phosphate bond' based on the thermodynamic studies of Meyerhof and others on the phosphorylation and dephosphorylation of creatine. The term is unfortunate, since to a chemist it suggests a bond of great strength which would require considerable energy for its disruption. However, the term 'high-energy bond' has been used (or misused!) by biochemists to refer to a structural feature that renders a molecule energy-rich with respect to its hydrolysis, i.e.

$$ATP + H_2O \longrightarrow ADP + Pi^2 + energy \qquad\qquad [4]$$

Lipmann was impressed by the high negative values for the standard free energies of hydrolysis $(\Delta G^\circ)^3$ for ATP and some other naturally occurring organic phosphates (see Table 1.2). In the hydrolysis of ATP, eqn [4], the energy of the reactants is substantially higher than that of the products. This in turn means that the equilibrium of the reaction is far to the right and the equilibrium constant, K, is high (and not easily measured).

$$\frac{[ADP][Pi]}{[ATP][H_2O]} = K \qquad\qquad [5]$$

---

[2] In this book P*i* and PP are used for inorganic phosphate and pyrophosphate respectively and (P) is used for the phosphate of organic phosphates.

[3] ΔG° or $\Delta G^{\circ\prime}$ is used for the standard free energy (Gibbs free energy) change at pH 7.0 or at a specified pH.

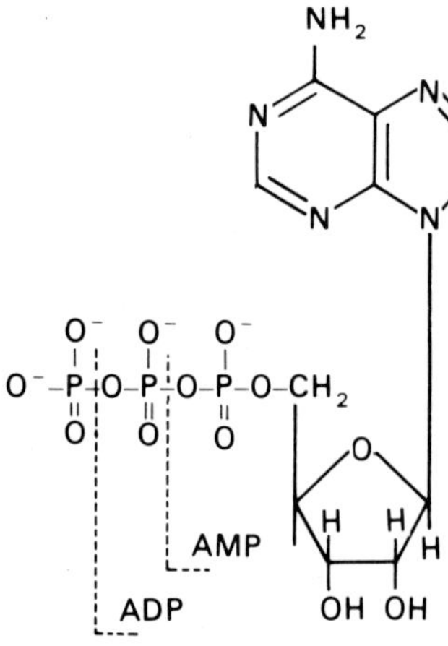

ATP

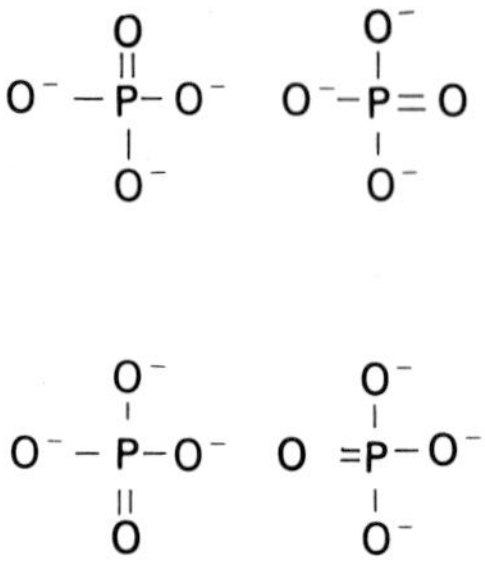

Inorganic phosphate: resonance forms

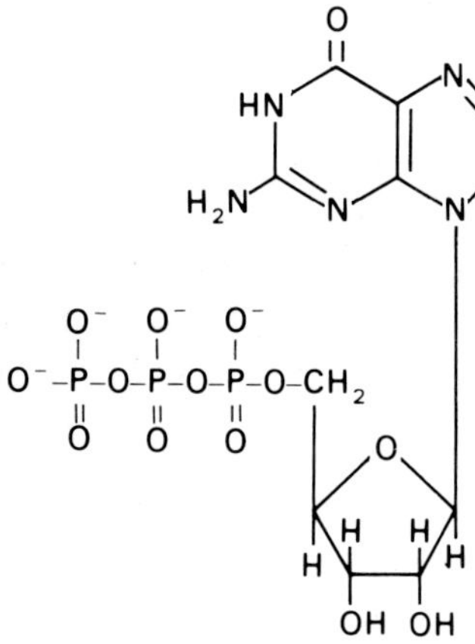

GTP

Fig. 1.8 The structures of the adenosine and guanosine triphosphates. The four resonance forms of inorganic phosphate. The reason for the relatively high negative value of $\Delta G^{\ominus}$ for these compounds lies in: (*a*) the repulsion of the negative charges on the phosphates; and (*b*) the greater resonance stabilisation (i.e. more possible resonance forms) which can be seen in inorganic phosphate as compared to the phosphate groups of the nucleotide phosphates.

The relationship between the actual free energy change in a reaction:

$$A + B = C + D \qquad [6]$$

and the standard free energy is given by

$$\Delta G = G^{\ominus} + RT \ln \frac{[C][D]}{[A][B]} \qquad [7]$$

where $\dfrac{[C][D]}{[A][B]}$ is the ratio of concentrations (or preferably activities)[4] of the products to the reactants. $\Delta G^{\ominus}$ is the standard free energy change at the relevant temperature, ΔG the actual free energy change at the temperature and activities of reactants and products specified, R is the gas constant and T is the absolute temperature. At equilibrium, eqn [7] becomes

$$0 = G^{\ominus} + RT \ln K, \qquad \Delta G^{\ominus} = -RT \ln K \qquad [8]$$

Thus the standard free energy of a reaction can be determined from a knowledge of the equilibrium constant. A value of zero for $\Delta G^{\ominus}$ is associated with a reaction in which the equilibrium constant is 1. High negative values for $\Delta G^{\ominus}$ are associated with reactions which tend to go virtually to completion and hence have a high equilibrium constant. Positive values of $\Delta G^{\ominus}$ denote reactions where at equilibrium, the reactants are favoured. If all the components of the system are present at unit activity (i.e. 1 mol l^{-1}), then eqn [7] becomes

$$\Delta G = \Delta G^{\ominus} \qquad [9]$$

$\Delta G^{\ominus}$ represents the free energy available from the reaction when all the reactants are present at some conventionally defined unit activity.

However, it should be noted that in a biological system, reactions cannot be considered in isolation but only as part of a complex metabolic process where they are seldom if ever at unit activity or where the ratio of products to reactants is rarely unity. As a consequence, the standard free energy for the reaction $\Delta G^{\ominus}$ is not directly applicable. Returning to eqn [7], we see that the actual free energy change is dependent on the activities of the components of the system. For the hydrolysis of ATP in the presence of excess Mg^{2+}, the free energy change will be:

$$\Delta G = \Delta G^{\ominus} + RT \ln \frac{[Mg \cdot ADP^-][HPO_4^{2-}][H^+]}{[Mg \cdot ATP^{2-}][H_2O]} \qquad [10]$$

Reliable estimates of the appropriate intracellular activities are not easy to obtain. $\Delta G^{\ominus}$ for the reaction at pH 7.0 and 37 °C in the presence of excess Mg^{2+} is 28.5 kJ/mol. However, ΔG for the intracellular reaction associated with mitochondria may rise to 50–60 kJ/mol after allowances have been made for the intracellular concentrations (or activities) of ATP, ADP and P*i*.

Further problems arise when the value of $\Delta G^{\ominus}$ itself is considered. The estimation of this value is based on eqn [8]. For this it is necessary to use a

---

[4] Although strictly the term $\dfrac{[C][D]}{[A][B]}$ should be expressed as a ratio of activities, in practice it is normally only concentrations of metabolites that can be reliably estimated in cells and subcellular particles.

reaction for which the equilibrium constant can be readily measured. One such reaction is that catalysed by the enzyme glutamine synthetase:

$$\text{glutamate} + NH_4^+ + ATP = \text{glutamine} + ADP + Pi \qquad [11]$$

The equilibrium for this reaction assuming excess magnesium is:

$$K = \frac{[\text{glutamine}][MgADP][Pi]}{[\text{glutamate}][NH_4^+][MgATP]} = 270 \text{ at } 37\,^\circ C \text{ and pH } 7.0 \qquad [12]$$

The reaction may be considered as the sum of two reactions:

$$MgATP + H_2O = MgADP + Pi \qquad [13]$$

and

$$\text{glutamate} + NH_4^+ = \text{glutamine} + H_2O \qquad [14]$$

The latter reaction is catalysed by glutaminase and the equilibrium constant can readily be measured: $K = 242$ at $37\,^\circ C$ and pH.7.0. The free energy changes for reactions [12] and [14] may be calculated from the equilibrium constants (eqn [8]). The $\Delta G^\oplus$ for ATP hydrolysis (eqn [13]) may then be calculated as the difference between the $\Delta G^\oplus$ values:

$$\Delta G^\oplus{}_{\text{eqn [13]}} = \Delta G^\oplus{}_{\text{eqn [12]}} - \Delta G^\oplus{}_{\text{eqn [14]}} = (-14.4) - (+14.1) =$$
$$-28.5 \text{ kJ/mol.}$$

This figure varies with pH, Mg^{2+} concentration and ionic strength (see Table 1.2).

   The discussion above has been concerned primarily with the standard free energy change for the hydrolysis of the terminal phosphate of ATP. The hydrolysis of the terminal pair of phosphates (eqns [2] and [3]) also involves a high negative free energy change (see Table 1.2). On the other hand, hydrolysis of the ribose–phosphate linkage is associated with a much smaller free energy change. Other nucleoside triphosphates (see Fig. 1.8) perform similar energetic functions in metabolism. They are metabolically related to ATP by the enzyme nucleoside diphosphate kinase (EC 2.7.4.6).

$$\text{ATP} + \text{nucleoside diphosphate} = \text{ADP} + \text{nucleoside triphosphate}$$

   We need to apply now the information on standard free energy changes to metabolic systems. We have already noted that the 'high-energy bond' of Lipmann is in reality a high free energy release on hydrolysis. Banks and Vernon (1970) have described this hydrolysis of ATP as given in eqn [4] as 'forbidden', that is it does not take place *in vivo* unless coupled to other metabolic events. A reaction sequence in which either ATP is used to promote synthesis or in which it is itself synthesised by oxidation can be found in the glycolytic pathway.

$$\text{3-phosphoglyceraldehyde} + NAD + Pi = \text{1,3-diphosphoglyceric}$$
$$\text{acid} + NADH_2 \quad [15]$$

$$\text{1,3-diphosphoglyceric acid} + ADP = \text{3-phosphoglyceric}$$
$$\text{acid} + ATP \quad [16]$$

Table 1.2 Free energies of hydrolysis of ATP and other phosphates

Phosphates at pH 7.0	$-\Delta G^{\circ}$ (kJ)*
Glucose-1-phosphate	20.9
Glucose-6-phosphate	13.8
Phosphoenolpyruvate	61.9
Acetyl phosphate	43.1
Pyrophosphate	33.4
Pyrophosphate in the presence of 5 mM Mg^{2+}	18.8
ATP† ($\rightarrow$ AMP + pyrophosphate) in the presence of excess Mg^{2+}	32.2

$-\Delta G^{\circ*}$ for ATP hydrolysis under various conditions‡ (ATP $\rightarrow$ ADP + Pi)

			Mg^{2+} concentration (mM)		
Temp.	*pH*	*I*§	*0*	*1*	*10*
			(kJ)	(kJ)	(kJ)
25	6.0	0.1	30.3	27.5	25.8
		0.2	30.8	28.0	26.3
	7.0	0.1	32.7	28.0	27.8
		0.2	33.1	28.4	28.4
	8.0	0.1	37.7	32.2	32.7
		0.2	38.1	32.7	33.2
37	6.0	0.1	30.3	27.1	25.7
		0.2	30.6	27.1	25.9
	7.0	0.1	33.1	28.0	28.5
		0.2	33.6	28.4	28.9
	8.0	0.1	38.5	32.8	33.8
		0.2	39.0	33.3	43.3

Notes

* The values for $-\Delta G^{\circ}$ are based on a standard state of 1 M concentration of reactants and products and an activity of water of 1.0 (Convention III).

† It is generally accepted that the value for the hydrolysis of ATP $\rightarrow$ AMP + PP will be the same as for ATP $\rightarrow$ ADP + Pi. This value is based on a different method to that used in the second part of this table.

‡ The values for ATP $\rightarrow$ ADP + Pi are based on the data of Rosing and Slater (1972) using the glutamine synthetase and glutaminase reactions. These values differ a little from those of Guynn and Veech (1973) who used a method based on acetyl phosphate.

§ The ionic strength $I = \frac{1}{2} \Sigma C_i z_i^2$ where C_i is the molar activity of any mobile ion, z_i its charge. The product $C_i z_i^2$ is summed for all mobile ions present.

The oxidation of 3-phosphoglyceraldehyde to 3-phosphoglyceric acid is coupled to the synthesis of ATP. This type of phosphorylation, in which ATP is synthesised from a phosphorylated metabolic intermediate, is described as substrate-level phosphorylation.

The synthesis of ATP can be considered the main function of catabolism. While other catabolic pathways may synthesise some ATP, quantitatively the most important source is oxidative phosphorylation which, unlike substrate-level phosphorylation, is ATP synthesis coupled to respiration. The mitochondrion utilises the energy of oxidation of pyruvate and fatty acids for the synthesis of ATP e.g.:

$$\text{pyruvate} + 2\tfrac{1}{2}O_2 + 15ADP + 15Pi \longrightarrow 3CO_2 + 17H_2O + 15ATP$$

By contrast, biosynthetic systems consume ATP which itself becomes a prime link between synthesis and degradation in cellular metabolism and between the mitochondrion and the metabolism of the cytosol. In the chloroplast, the energy for ATP synthesis is light energy. This drives an electron transport chain to which ATP synthesis is coupled. The ATP is used for the synthesis of carbohydrate.

In eqn [10], the value of ΔG will vary with variations of $[ATP]/[ADP][Pi]$, the phosphate potential. This is a simple and useful measure of the ease of formation of ATP from ADP and Pi. A measure of the energy available to drive metabolism has been proposed by Atkinson (1968). This takes account of the energy available from hydrolysis of ATP to ADP or AMP and inorganic phosphate and of ADP to AMP and inorganic phosphate. It is defined as $[ATP] + \tfrac{1}{2}[ADP]/[ATP] + [ADP] + [AMP]$ and termed the adenylate charge or energy charge.

1.3   Oxidation–reduction systems

Almost all the ATP synthesised in the mitochondrion is formed by a system in which the reaction

$$ADP + Pi \longrightarrow ATP + H_2O$$

is coupled to the oxidation–reduction reactions taking place in the inner mitochondrial membrane. A similar mechanism also occurs in the thylakoid membrane of the chloroplast. In the mitochondrial system the major substrate for the oxidation–reduction (or redox) reactions is reduced nicotinamide adenine dinucleotide, $NADH_2$ (see Fig. 1.9). This substance is oxidised by oxygen in a reaction catalysed by a multienzyme complex, the respiratory chain, composed of a series of components undergoing oxidation–reduction reactions some of which are coupled to ATP synthesis.

We now need to ask two general questions about such oxidation–reduction reactions. Firstly, how may we measure the ease with which one compound will oxidise another? Secondly, how can we determine the amount of free energy available during such oxidations? To do this, we shall consider a simple reaction in which A is oxidised by B:

$$A_{red} + B_{ox} = A_{ox} + B_{red} \tag{17}$$

At any time all four species may be present. We may consider the system as composed of the A_{red}/A_{ox} couple and the B_{red}/B_{ox} couple. The transfer of one or more electrons from A to B will depend on the potential of A to donate an

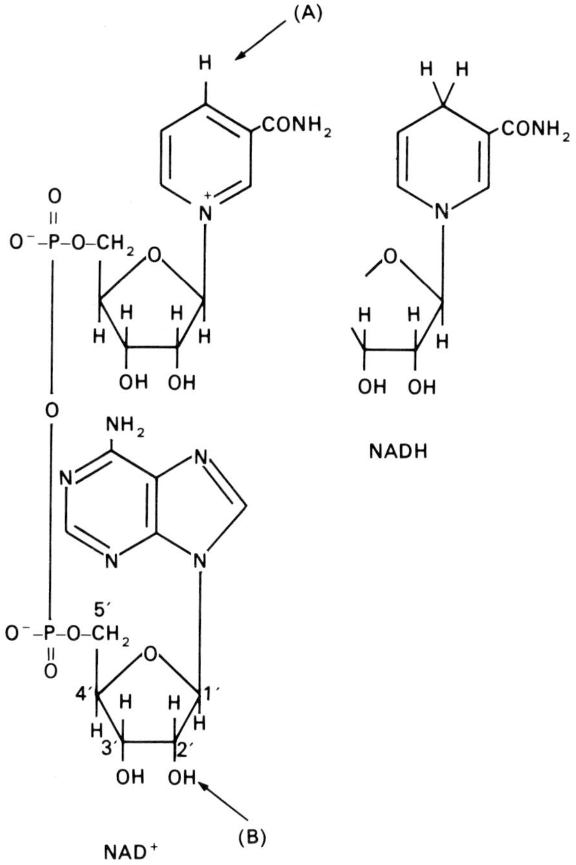

Fig. 1.9 The structures of the pyridine nucleotides. The full structure of NAD^+ (nicotinamide adenine dinucleotide) is shown. Reduction of the nicotinamide ring (arrowed A) gives rise to reduced NAD:

$$NAD^+ + SH_2 = NADH + H^+ + S$$

where SH_2 and S represent oxidised and reduced substrate. The reaction may alternatively be shown:

$$NAD + SH_2 = NADH_2 + S$$

Nicotinamide adenine dinucleotide phosphate (NADP) has the same structure as NAD, but with a phosphate on a ribose residue (arrowed B), and may be reduced in the same way.

electron and of B to receive one. This will be due in part to the nature of A and of B but also to the fraction of A in the reduced state and of B in the oxidised state. The ease with which A would donate an electron could be determined if it could be measured in relation to a suitable reference system. Formally this is the hydrogen half-cell which may be represented by:

$$\tfrac{1}{2}H_2 = H^+ + e \qquad [18]$$

The potential of this half-cell is arbitrarily taken as zero if the oxidised and reduced forms (H^+ and H_2) are both at unit activity, i.e. pH 0 and H_2 at one atmosphere. Thus the potential of an A_{red}/A_{ox} couple could be measured as shown in Fig. 1.10. The potential difference is a measure of the tendency of the two systems to transfer an electron.

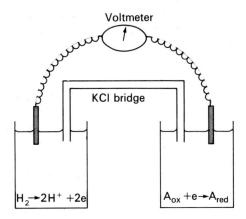

Fig. 1.10 Measurement of the potential of a redox couple. Theoretical approach using the hydrogen half-cell as a reference system.

The oxidation–reduction potential (E_h) of any electrode relative to a standard hydrogen electrode is given by:

$$E_h = E^\circ + \frac{RT}{nF} \ln \frac{[\text{oxidised}]}{[\text{reduced}]} \qquad [19]$$

where E° is the standard oxidation–reduction potential, n is the number of electrons transferred during the reaction and F is the Faraday (96 485 C/mol). When the activities of the oxidised and reduced forms are equal:

$$E_h = E^\circ \qquad [20]$$

Thus the standard oxidation–reduction potential is the potential of the redox couple when both forms are present at the same activity.

A number of redox reactions are particularly pH-dependent because directly or indirectly the reaction involves uptake or release of a proton. Since E° is defined as the standard potential at pH 0, it is necessary to define a biologically more useful potential $E^{\circ'}$ or $E_{m\,7.0}$[5] the standard potential at pH 7.0. At any pH the standard potential of the hydrogen electrode is given by

$$E = \frac{RT}{F} \ln \frac{(H^+)}{(P_{H_2})^{1/2}} \qquad [21]$$

---

[5] The term E_m, the mid-point potential, is now used more frequently in biochemical studies and will be used in this book. It is, however, equivalent to $E^{\circ'}$ or E_o', the latter being the standard potential at pH 7.0 or at any other specified pH.

where P_{H_2} is the partial pressure of hydrogen. Hence at 30 °C and at $P_{H_2} = 1$ atm (standard condition),

$$E = \frac{2.3\,RT}{F}\log_{10}(H^+) = -0.06\,\mathrm{pH}. \qquad [22]$$

Thus at pH 7.0, the potential of the hydrogen electrode is -0.42 V. At this pH, the ability of a redox couple to receive an electron, E_h, is dependent not only on the value of $E_{m\,7.0}$ but also on the activities of the oxidised and reduced forms, i.e.

$$E_h = E_{m\,7.0} + \frac{RT}{nF}\ln\frac{[\text{oxidised}]}{[\text{reduced}]} \qquad [23]$$

The more reducible a system is, the more positive E_h becomes, i.e. strong oxidising agents will have positive standard potentials; strong reducing agents will have negative standard potentials. Hence in the system shown in eqn [17], for electrons to be transferred from A to B, the value of E_h for B_{red}/B_{ox} must be more positive than the value of E_h for A_{red}/A_{ox}. It is not necessary that $E_{m\,7.0}$ for B should be more positive than $E_{m\,7.0}$ for A, because the actual potential E_h depends both on the $E_{m\,7.0}$ values and the activity values.

In biological systems, the standard potential of a redox couple (e.g. a membrane-bound cytochrome) can be determined by poising the system at a number of different potentials and measuring the ratio $\dfrac{[\text{oxidised}]}{[\text{reduced}]}$ which is usually achieved by spectroscopy. A plot similar to Fig. 1.11 can then be prepared in which the mid-point potential, E_m, is then evident. In poising the system at specific potentials, it is necessary to suspend the system in a medium of redox substances such as ferricyanide/ferrocyanide, which will react with the redox couple under investigation. The potential of the medium is measured in a manner

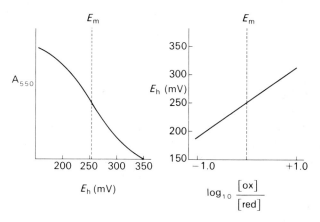

Fig. 1.11 Measurement of the mid-point potential of cytochrome *c*. The optical absorbance at 550 nm is measured in a system poised at a series of potentials. If the proportion of oxidised and reduced cytochrome is calculated from the absorbance data, a second plot may be prepared. The mid-point potential is the potential where the ratio of oxidised to reduced cytochrome is unity.

similar to pH. The electrode system consisting of a platinum electrode and a calomel reference electrode and the potential is measured direct (many pH meters are also designed for this function). The system may be calibrated with, for example, a saturated solution of quinhydrone which has an $E_{m\,7.0}$ of 293 mV.

In biological systems, oxidation–reduction reactions may be coupled to the synthesis of ATP. This implies a free energy change in the course of the oxidation–reduction reaction large enough to drive the phosphorylation of ADP with inorganic phosphate. We may relate oxidation–reduction potentials to free energy changes if we consider the potentials for the components A and B of eqn [17].

$$E_h^A = E_{m\,7.0}^A + \frac{RT}{nF} \cdot \ln \frac{[A_{ox}]}{[A_{red}]} \tag{24}$$

$$E_h^B = E_{m\,7.0}^B + \frac{RT}{nF} \cdot \ln \frac{[B_{ox}]}{[B_{red}]} \tag{25}$$

At equilibrium, $E_h^A = E_h^B$; subtracting eqn [24] from eqn [25], we obtain

$$E_{m\,7.0}^B - E_{m\,7.0}^A = \frac{RT}{nF} \left\{ \ln \frac{[A_{ox}]}{[A_{red}]} - \ln \frac{[B_{ox}]}{[B_{red}]} \right\}$$

$$\Delta E_{m\,7.0} = \frac{RT}{nF} \cdot \ln \frac{[B_{red}][A_{ox}]}{[B_{ox}][A_{red}]} \tag{26}$$

$$\Delta E_{m\,7.0} = \frac{RT}{nF} \cdot \ln K \tag{27}$$

In eqn [8], it was shown that the standard free energy change was related to the equilibrium constant: $\Delta G^\oplus = -RT \ln K$. Since the activities refer to the situation at equilibrium, the standard free energy change for the reaction shown in eqn [17] is given by

$$-\Delta G^\oplus = \Delta E_{m\,7.0} \cdot nF \tag{28}$$

This represents the free energy available during the passage of n equivalents of electrons from A to B when the oxidised and reduced forms of both A and B are present at unit activity.

In the mitochondrion the oxidation of $NADH_2$ by oxygen is coupled to the synthesis of ATP. We are now in a position to calculate the free energy available from this reaction if we *assume* unit activities of the reactants and products and a temperature of 25 °C.

$$NADH_2 + \tfrac{1}{2}O_2 = H_2O + NAD \tag{29}$$

$E_{m\,7.0}$ for $NAD/NADH_2 = -0.32$ V and for $O_2/H_2O = +0.82$ V

From eqn [28], $\Delta G^\oplus = -1.14 \times 2 \times 96\,485 = -220$ kJ/mol

This may be compared with the standard free energy change for the synthesis of ATP ($+28.5$ kJ/mol). It is clear that there could be an adequate free energy change in reaction [29] for the synthesis of several molecules of ATP if there were adequate coupling between oxidation and phosphorylation and if there were appropriate concentrations of reactants. In fact the large difference between the standard potential of $NADH_2$ and H_2O is bridged not by a single oxidation–reduction reaction but by a sequence involving a series of redox carriers. These carriers function essentially as catalysts for the transfer of electrons from donor to acceptor. It is desirable that the order of carriers is such that their standard oxidation–reduction potentials are increasingly positive from NAD (-0.32 V) to oxygen ($+0.82$ V). However, as we have seen above, there is no absolute necessity for this since all that is required is that the values of E_h should be increasingly positive in the direction of O_2. We have discussed these redox systems in terms of electron transfer. However, some redox processes may proceed primarily by transfer of one or two electrons, while in others hydrogen is transferred.

Further reading

(a) History of study of mitochondria and chloroplasts (see also Ch. 2)

Hill, R. (1965) The biochemists green mansions: the photosynthetic electron transport chain in plants, *Essays in Biochemistry*, Vol. 1, pp. 121–51. eds. P. N. Campbell and G. D. Greville, Academic Press, New York.

Lehninger, A. L. (1964) *The Mitochondrion: Molecular Basis of Structure and Function.* Benjamin, New York.

(b) Isolation of subcellular particles

Estabrook, R. W. and Pullman, M. E. (eds.) (1967) Oxidation and phosphorylation, *Methods in Enzymol.*, Vol. 10.

Walker, D. A. (1971) Chloroplasts (and grana) aqueous preparation (including high carbon fixation activity), *Methods in Enzymol.*, Vol. 23, pp. 211–20.

(c) Energetics

Clarke, W. M. (1960) *Oxidation–reduction Potentials of Organic Systems.* Bailliere, Tindall and Cox, London.

Dutton, P. L. (1971) Oxidation–reduction potential dependence of the interaction of cytochromes, bacteriochlorophyll and carotenoids at 77°K in chromatophores of *Chromatium* D and *Rhodopseudomonas gelatinosa, Biochim. Biophys. Acta*, **226**, 63–80.

Dutton, P. L., Wilson, D. F. and Lee, C. P. (1970) Oxidation–reduction potentials of cytochromes in mitochondria, *Biochemistry*, **9**, 5077–82.

Rosing, J. and Slater, E. C. (1972) The value of $\Delta G^{\ominus}$ for the hydrolysis of ATP, *Biochim. Biophys. Acta*, **267**, 275–90.

Chapter 2

Development of ideas on oxidation and phosphorylation

2.1 Early studies on cell oxidation

The fundamental question of how living cells and tissues utilise oxygen has attracted the attention of physiologists and chemists from the time of Lavoisier. Lavoisier himself thought of respiration as a slow combustion. However, by 1878 Claude Bernard wrote 'What is true, is that the exact role of oxygen which we believed we understood, is still unknown to us: we can hardly guess at it.' But Bernard did feel that oxygen was fixed in tissues as a constituent element. The view that cell oxidation proceeded by a combination of substrate with oxygen persisted into the twentieth century. The mechanisms suggested for the process often involved the activation of oxygen by formation of some intermediate such as a peroxide capable of oxidising substrate (S).

$$A + O_2 \longrightarrow A \diagdown \begin{matrix} O \\ | \\ O \end{matrix}$$

$$A \diagdown \begin{matrix} O \\ | \\ O \end{matrix} + S \longrightarrow SO + AO$$

Early this century, Heinrich Wieland began a series of publications which changed the whole approach to the biological oxidation of substrates. His early experiments showed that a number of organic substances such as lactate could be oxidised by colloidal palladium which became hydrogenated. The palladium could be oxidised by oxygen in the presence of an autoxidisable dye, methylene blue.

The substrate is therefore oxidised by removal of hydrogen and Wieland suggested that biological oxidations proceed in a similar way. Thunberg introduced an anaerobic method using an evacuated tube in which the oxidation of substrate by biological material was coupled to the reduction of methylene blue to its colourless form (leucomethylene blue). Using this technique, Thunberg was able to study a large number of dehydrogenases.

Subsequently in lectures *On the Mechanism of Oxidation* published in 1932, Wieland wrote 'A closer examination of those substances which are oxidised in the cell shows that, in general, they are substances which undergo the chemical changes involving the loss of hydrogen, that is by dehydrogenation. . . . There is no known example . . . of an unsaturated compound in the case of which it is necessary to assume direct addition of oxygen, that is additive oxidation.'

Concurrent with Wieland's studies were those of Otto Warburg who, in the years 1908–25, developed a view of respiration based on his discovery of the importance of iron in the process. In the nineteenth century interest had been shown in transition metals especially iron, in relation to oxidations. The central feature of Warburg's theory was the respiratory enzyme (*Atmungsferment*) which reacted with molecular oxygen and which contained iron as its most essential component. The iron combined with oxygen, became oxidised and subsequently was reduced by reacting with organic substrates. Much of Warburg's work concentrated on demonstrating the importance of iron in respiration. He drew attention to the level of iron in cells which was adequate for his theoretical predictions and also to the level of cyanide needed to inhibit iron oxidation *in vitro* which was of the same order as that required to inhibit cellular respiration. Of particular significance were observations with artificial systems in which iron-containing charcoals were shown to be capable of catalysing the oxidation of amino acids (e.g. leucine and cysteine). The oxidation was sensitive to cyanide and narcotics. Charcoal from sugar and silicate did not have the same properties. Addition of haemin to the sugar before charcoal formation gave an active charcoal, whereas addition of iron salts did not. This led Warburg to conclude that the iron must be bound to nitrogen for activity. Subsequently he wrote (see Warburg, 1949) of these experiments: 'The result of the charcoal model containing iron bound to nitrogen was that this combination produced a biological oxidation like that of the oxidative deamination of amino acids, an oxidation which is reversibly inhibited by narcotics and irreversibly and specifically inhibited by cyanide. Who could believe that this was only by chance in agreement with the behaviour of cell respiration?'

Warburg showed that carbon monoxide inhibited cell respiration and that this inhibition was lost in light. This he interpreted as indicative of the photolability of carbon monoxide–iron compounds, since similar results had previously been obtained for haemoglobin. By measuring the effectiveness of various wavelengths of light in reversing the inhibition by carbon monoxide, Warburg obtained a curve similar to that shown in Fig. 2.1. Since the effectiveness of any wavelength in reversing the inhibition also represents the degree to which that wavelength is absorbed by the carbon monoxide complex, the curve also represents the absorption spectrum of the complex. Thus by 1929, Warburg possessed an absorption spectrum of a derivative of the respiratory enzyme which, on the ground of its similarity to haem spectra, he supposed to be a haem compound.

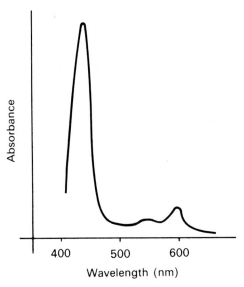

Fig. 2.1 Absorption spectrum of the carbon monoxide complex of the 'respiratory enzyme'.

2.2   The discovery of cytochromes

During the period of Wieland's and Warburg's pioneering investigations, another worker, David Keilin, commenced a study of respiration in Cambridge in 1919. In 1925, he published a paper entitled *On Cytochrome, a Respiratory Pigment Common to Animals, Yeasts and Higher Plants.* The discovery of cytochromes was, in fact, a rediscovery of pigments described as myohaematin in muscle and histohaematin in other tissues by a Scottish physician, C. A. MacMunn (1887). MacMunn observed the same optical absorbance bands as Keilin found subsequently and, from studies of their behaviour under oxidising and reducing conditions, concluded that they were due to substances with a respiratory function. The observations of MacMunn were attacked by Hoppe-Seyler and appeared at the time to be discredited. It was when Keilin was studying the fate of haemoglobin in the botfly (*Gasterophilus*), a parasitic dipteran, that he noted the four absorbance bands seen earlier by MacMunn. He then examined a number of species of animals and micro-organisms and found in each case the same four bands, although their precise position in the spectrum varied slightly. Hence the pigments are widely distributed in nature. The technique involved passing light through tissues (squashed, sliced or in divided suspension) and analysing the light by means of a spectroscope. The normal spectrum (violet to red) was seen but with dark bands indicating the regions of high absorbance (Fig. 2.2).

With a yeast suspension Keilin demonstrated the disappearance of the bands on shaking with air and their reappearance on standing, showing that it was reduced cytochrome which possessed the bands while the oxidised pigments did not. Thus the pigments were reduced endogenously and oxidised by air, indicating a respiratory function as proposed by MacMunn. Careful examination of muscle and yeast cytochrome bands indicated that there was not one component but three with different properties. The fourth band at about 520 nm, initially referred to as cytochrome *d*, was shown to be a composite of *β*-bands associated with the other pigments. The cytochromes are *a* with an *α*-band at 604, *b* at 564 nm and *c* at 550 nm. *γ*-Bands were later found at the blue end of the spectrum.

In 1930, Keilin isolated cytochrome *c* and described some of its properties. The spectrum of the reduced compound had maxima at 550 nm (*α*), 521 nm (*β*) and 415 nm (*γ*). It was shown to be a protein bearing an iron porphyrin. It was not oxidised in air at physiological pH but was very rapidly oxidised by a heart muscle preparation capable of oxidising succinate. Thus the muscle preparation possessed an enzyme capable of oxidising cytochrome *c*, which Keilin called cytochrome oxidase.

2.3 A simple respiratory chain

(a) Keilin's work on the cytochrome chain

During the 1930s, Keilin together with Hartree developed the studies on cytochrome. Figure 2.2 shows some of the results obtained and represents spectra resulting from passing light through a heart muscle preparation and then through a spectroscope (see Keilin and Hartree, 1939). The heart muscle preparation which oxidised succinate in the presence of oxygen, was prepared by homogenising and washing heart muscle.

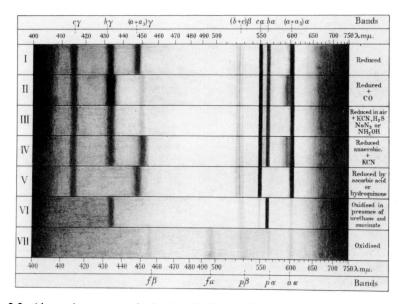

Fig. 2.2 Absorption spectra obtained by Keilin and Hartree for heart muscle preparations treated in various ways. Light is passed through the preparation and through a spectroscope; the resulting spectrum shows dark lines representing regions of high absorbance (from Keilin and Hartree, 1939, courtesy of Dr. J. Keilin–Whiteley).

In Fig. 2.2, comparison of I and VII shows that the reduced preparation possesses the α- , β- , and γ-bands of the cytochromes while the bands disappear almost completely in the oxidised state. Reduction in the presence of carbon monoxide (II), while not affecting bands of b and c results in the loss of the a γ-band at 444 nm and increased absorption in the region of the b γ-band as well as the appearance of a new band at 593 nm. The latter also appears when the system is inhibited by cyanide but only in anaerobic conditions (III and IV). However, in aerobic conditions, the main effect of cyanide, azide, H_2S or hydroxylamine is loss of the strong band at 448 nm leaving a weak band at 452 nm.

Keilin and Hartee's interpretation of these data was to propose a new component a_3 (bacterial cytochromes having been designated a_1 and a_2). It was assumed that the α- and γ-bands of a and a_3 were fused and that a predominated in the α-region and a_3 in the γ-region. Cytochrome a_3 is assumed to be autoxidisable and its oxidation decreases the absorption at 448 nm, the remaining band at 452 nm being attributed to cytochrome a. Carbon monoxide forms a complex with the autoxidisable cytochrome a_3 so that the α-band is shifted to about 590 nm while the γ-band shifts to about 432 nm to intensify the cytochrome b γ-band. The cyanide complex when fully reduced (anaerobic) also has its own characteristic spectrum which disappears on oxidation.

From the foregoing discussion, it is clear that Warburg's respiratory enzyme must be equated with the autoxidisable cytochrome a_3 of Keilin. However, the relationship between cytochromes a and a_3 has not been clear and it has become customary to refer to (a, a_3) together as the oxidase (see sect. 9.8).

Spectra V and VI give an indication of the sequence of the cytochromes. Reduction of the system with ascorbate reduces c and (a, a_3) while not affecting b. This indicates transfer of electrons in the following sequence:

$$\text{ascorbate} \longrightarrow \text{cytochrome } c \longrightarrow \text{cytochrome } a, a_3 \longrightarrow O_2$$

Finally succinate which normally reduces all components, reduces only cytochrome b in the presence of urethane suggesting a sequence

$$\text{succinate} \longrightarrow \longrightarrow \text{cytochrome } b \underset{\text{urethane}}{\longrightarrow\!\!\!/\!\!\!\longrightarrow} \text{cytochrome } c \longrightarrow \text{cytochrome } a, a_3$$

While the work of Keilin and Hartree allows a formulation of the simple cytochrome chain, confirmation remained a problem for some years. The work of Chance and Williams both provided this confirmation and gave a fuller understanding of the system. Before this is discussed we will consider some related matters.

(b) The discovery of cytochrome c_1

At least two major cytochromes were overlooked by Keilin and Hartree in their original studies. Two Japanese workers, Yakushiji and Okunuki (1940), observed in heart muscle treated with succinate and cyanide, an absorption band at 552 nm which they attributed to cytochrome c_1. This observation was largely ignored until its confirmation by Slater (1949), although it was not interpreted by him as being due to a new pigment but rather as a displacement of the α-band of cytochrome c. Using their newly introduced low-temperature spectroscopy, which sharpens the absorption bands, Keilin and Hartree (1955) observed a sharp band at 552 nm between those of c and b. This band was initially designated cytochrome e. Further studies by several groups of workers confirmed the existence of this cytochrome with an α-band at 553 nm which, in deference to the original observations, become known as cytochrome c_1. It functions between cytochromes b and c in the respiratory chain.

(c) b-Cytochromes

The existence of more than one b-cytochrome was not fully realised until the late 1960s, when it was shown that at least two b-type pigments functioned between the flavoprotein dehydrogenases and cytochrome c_1 (see sect. 9.7).

(d) Isolation of flavoprotein dehydrogenases

Succinate dehydrogenase activity in biological systems was first observed by Thunberg. The heart muscle preparation of Keilin and Hartree oxidised succinate with oxygen via the cytochrome chain. The enzyme for succinate oxidation, succinate dehydrogenase, was firmly bound to the particulate fraction and not isolated until Singer and Kearney (1954) solubilised and purified it. The enzyme which is part of the citric acid cycle, is an iron-containing flavoprotein with covalently-bound flavin adenine dinucleotide as its prosthetic group (Fig. 2.3).

A more general problem arises when the link between substrates other than succinate and the respiratory chain is considered. The isolation of the pyridine nucleotides (NAD, NADP) provided an indication of the first step in substrate oxidation. How are these reduced pyridine nucleotides oxidised by the respiratory system?

During a study of hexose monophosphate metabolism by erythrocyte preparations, Warburg and Christian obtained evidence for the role of a coenzyme. Their system may be represented:

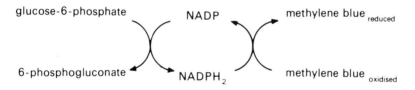

Their investigations led to the discovery of NADP and the glucose-6-phosphate dehydrogenase and to a search for enzymes catalysing the oxidation of pyridine nucleotides. Warburg and Christian (1932) isolated a flavoprotein from yeast which was subsequently shown to oxidise $NADPH_2$ using oxygen as an electron acceptor. This enzyme possessed flavin mononucleotide as prosthetic group (Fig. 2.3) and was only isolated from yeast. This was clearly not the pyridine nucleotide oxidising enzyme of the respiratory chain, but Warburg and Christian's work pointed to flavoproteins as the enzymes responsible for pyridine nucleotide oxidation.

Numerous attempts were made to isolate pyridine nucleotide oxidising flavoproteins. One example is the $NADH_2$–cytochrome c reductase of the Wisconsin group (Mahler *et al.*, 1952). This enzyme, solubilised from pig heart muscle, contained a prosthetic group similar to FAD and four atoms of non-

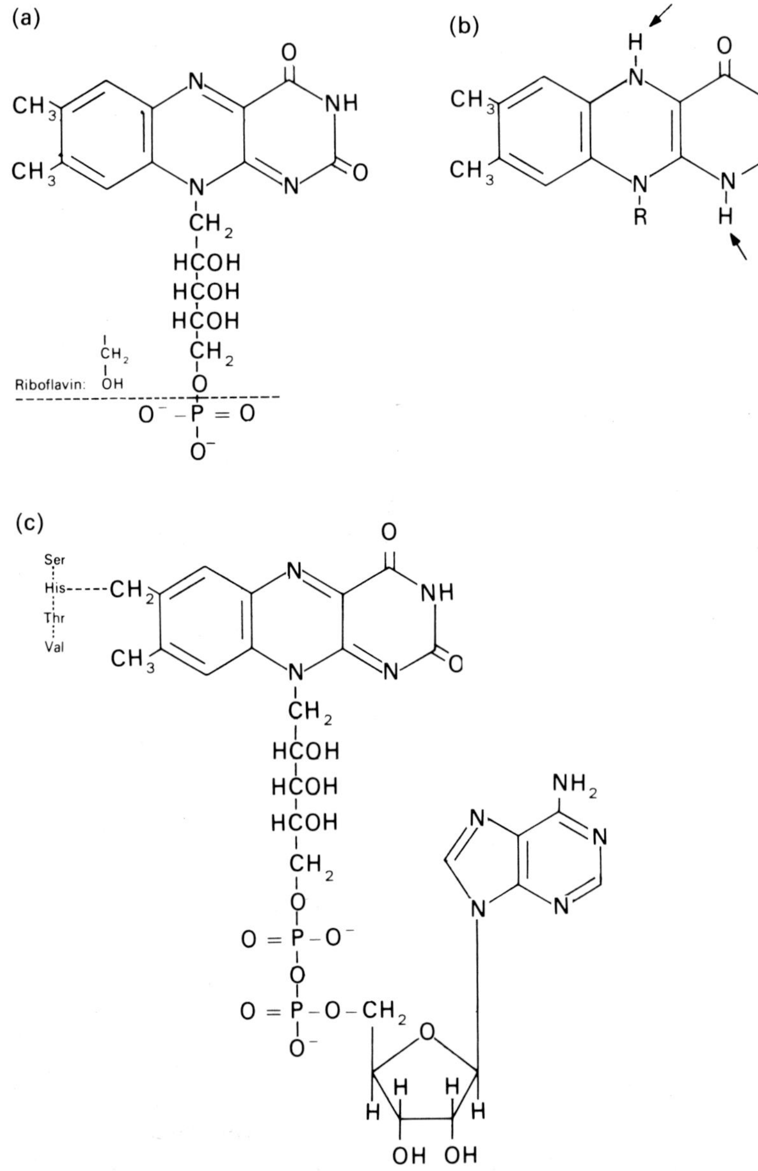

Fig. 2.3 Structure of flavin mononucleotide (FMN) and flavin adenine dinucleotide (FAD). (a) FMN or riboflavin phosphate. Riboflavin (vitamin B_2) lacks the terminal phosphate. (b) The reduced flavin. (c) FAD which has the same structure as FMN but with ADP added. The covalent binding of FAD to protein in succinate dehydrogenase is shown.

haem iron. It oxidised $NADH_2$ and reduced cytochrome c, but unlike the native mitochondrial system was insensitive to inhibition by amytal or antimycin (see Fig. 2.15).

A flavoprotein having appropriate properies was finally isolated from heart muscle mitochondrial particles by Singer's group (Ringler *et al.*, 1963). This enzyme did not reduce cytochrome *c* significantly except when allowed to decay, but like the membrane-bound enzyme, reduced ferricyanide

$$NADH_2 + 2Fe(CN)_6^{3-} \longrightarrow NAD + 2H^+ + 2Fe(CN)_6^{4-}$$

It possessed FMN as a prosthetic group together with 16–18 atoms of non-haem iron per flavin. Thus the first step in oxidation of $NADH_2$ by the respiratory system is the NADH dehydrogenase, a flavoprotein.

(e) Ubiquinone

A further component of the respiratory system is the quinone, ubiquinone or coenzyme Q (Fig. 2.4). In about 1955, Morton's group in Liverpool isolated a lipid-soluble substance with an absorption maximum at 272 nm (see Morton, 1958). At the same time Crane and coworkers (see Crane *et al.*, 1959) in Wisconsin isolated a quinone which proved identical with Morton's compound. Subsequently the structure was determined. The quinone was found to be widely distributed in nature and present in reasonably large amounts in aerobic tissues.

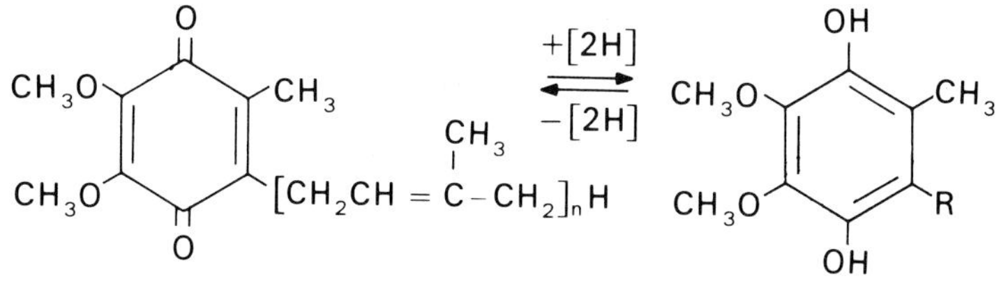

Fig. 2.4 Ubiquinone (coenzyme Q).

It was early realised, particularly by Crane's group, that the quinone underwent oxidation and reduction and its possible role in respiration was examined. In the presence of substrate and cyanide, conditions which reduce respiratory components, the endogenous mitochondrial quinone became reduced. Reduced quinone could be oxidised but the oxidation was sensitive to respiratory inhibitors such as antimycin and cyanide. The extraction of quinone from mitochondria with lipid solvents (iso-octane, pentane or acetone) resulted in a loss of succinate oxidase activity, but the activity could be restored by addition of quinone (and cytochrome *c*) to the system. After restoration succinate oxidation was still sensitive to antimycin. In molar terms, the quinone is present in mitochondria in greater amounts than the succinate dehydrogenase.

From submitochondrial particles oxidising $NADH_2$ and reducing cytochrome *c*, Hatefi and others (1962b) isolated complexes oxidising $NADH_2$ and reducing ubiquinone ($NADH_2$–ubiquinone reductase) and oxidising ubiquinone and reducing cytochrome *c* (ubiquinone–cytochrome *c* reductase). These and other experiments led to a formulation of the respiratory chain shown in Fig. 2.10. The role of quinone can be seen as oxidising flavoproteins of the membrane and reducing the *b* cytochrome complex.

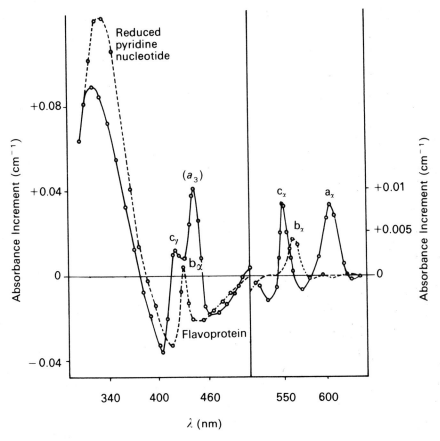

Fig. 2.5 Difference spectra of rat liver mitochondria. *Solid line*: reduced (substrate added in anaerobic conditions) minus oxidised mitochondria. *Dashed line*: reduced minus oxidised in the presence of antimycin. (From Chance and Williams, 1956).

(f) The use of difference spectra

From 1953 onwards, Chance (see Chance and Williams, 1956) developed the spectrophotometric approach using the difference spectrum, for example the difference in absorption of reduced and oxidised preparations of liver mitochondria shown in Fig. 2.5 (solid line). This shows essentially the same bands as seen by Keilin and Hartree but these may now be examined quantitatively. Reduction gives an increase in absorption of the cytochrome bands as seen earlier, but a decrease in absorption at around 450 nm due to flavoproteins which absorb more strongly in the oxidised state, resulting in a negative region at the blue end of the spectrum. The pyridine nucleotides absorb more strongly in the reduced state at 340 nm.

Not only is it possible to confirm Keilin's work by this technique, but new approaches now become possible. For example, the addition of oxygen to a reduced preparation results in the oxidation of the carriers in sequence, a_3 being oxidised more rapidly than a, the overall sequence being: a_3, a, c, b, flavoproteins.

The oxidation of cytochrome c_1 can be followed in mitochondria partially depleted of cytochrome c which enables the c_1 absorption to be measured. Cytochrome c is now seen to be oxidised more rapidly than cytochrome c_1. This sequence of carriers is also supported by their oxidation–reduction state. In the succinate oxidase system of a normally respiring preparation at steady state, the percentage reduction of the carriers was found to be about 13, 31 and 36 for a_3, a and c respectively and to be even greater at the substrate end of the respiratory chain.

The quantitative nature of the difference spectrum enabled Chance to estimate the ratios of the various components of the chain. In liver mitochondria the concentration of $NADH_2$ especially, and also flavoprotein and ubiquinone was much higher than that of any of the cytochromes. The ratios of cytochromes have been determined several times with varying results. There appears to be a simple stoicheiometric relationship, for example $2a_3 : 2a : 1c_1 : 2b$. The amount of cytochrome c is less reliably estimated since it is partially lost during preparation of the mitochondria.

(g) Inhibitors and cross-over points

The type of spectroscopic study just described may also be used to demonstrate cross-over points in the respiratory chain consequent on inhibition at a specific site. This may be illustrated with reference to the inhibitor antimycin, which inhibits the chain between cytochromes b and c_1 (see Fig. 2.15). What happens to the reduction state of the carriers when respiration is inhibited by addition of antimycin? Firstly, the carriers on the substrate side of the point of inhibition must become more reduced since their reoxidation can no longer occur. Secondly, the carriers on the oxidising side of the site of inhibition become more oxidised since their reduction by substrate is blocked. Thus the site of action of the inhibitor is defined as the point where carriers on the oxidising side become more oxidised and on the substrate side become more reduced. Such a point is the cross-over point. It is illustrated in Fig. 2.5 where it can be seen that cytochromes a and c have become more oxidised while b, the flavoproteins and pyridine nucleotides have become more reduced. Using cytochrome-c-depleted mitochondria, the cross-over point can be seen to lie between cytochromes b and c_1. Thus antimycin blocks the oxidation of the b-cytochromes by cytochrome c_1. The sites of other inhibitors are shown in Fig. 2.15.

2.4 The cytochromes

Cytochromes are proteins which have as their prosthetic group an iron-containing porphyrin, iron-protoporphyrin IX or a closely related haem (Table 2.1). The original classification of cytochromes by Keilin was based on the position of the α-band. Cytochromes are now distinguished more on the basis of their prosthetic groups (Fig. 2.6). Thus cytochromes a have haem a as a prosthetic group while b has protohaem. Unlike the other cytochromes, those of the c type have a covalently bound prosthetic group (Fig. 2.7). In all cases it is the iron which may exist in a reduced (Fe^{2+}) or an oxidised (Fe^{3+}) valency state. Thus cytochromes require the addition of, or loss of, a single electron to reduce

Table 2.1 Cytochromes of the respiratory chain

Cytochrome	Prosthetic group	$E_{m\,7.0}$ (mV)	Molecular weight	Main absorption maxima of reduced form in visible spectrum		
				α	β	γ
a, a_3	Haem a	+290	—	604	(519)	444
$b*$	Protohaem	+70	18–30 000	563	531	429
c	Protohaem (covalently bound)	+255	12–13 000	550	521	415
c_1	Protohaem (covalently bound)	+226	37 000	553	524	418

* Although the *b* complex appears to consist of two cytochromes, only one has been isolated.

(a)

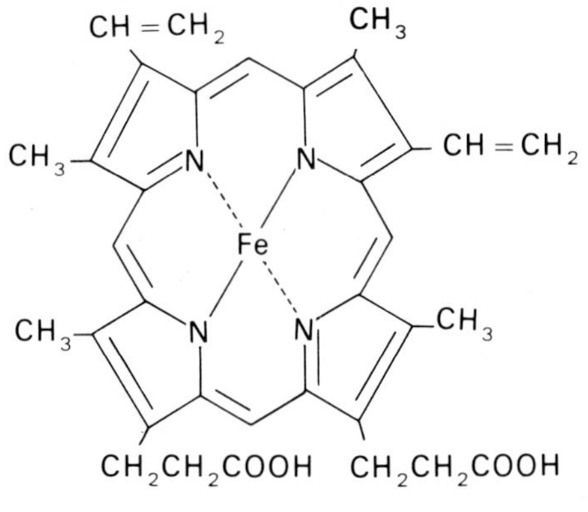

(b)

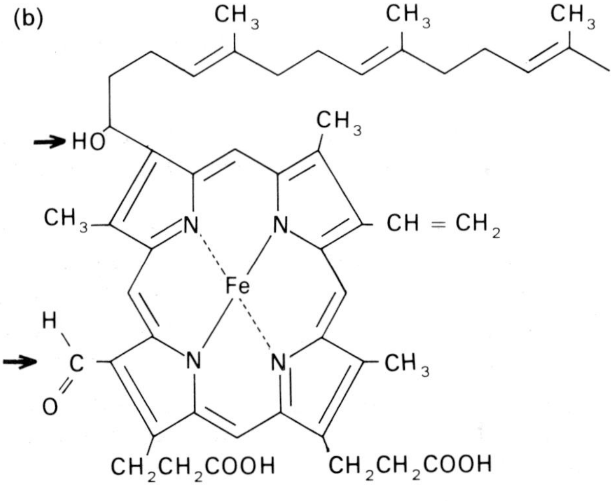

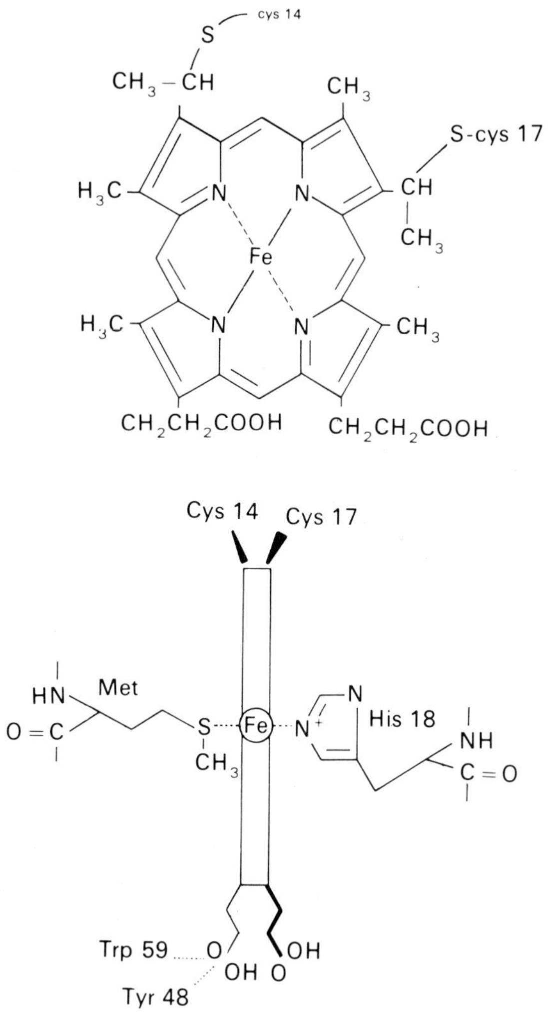

Fig. 2.7 Prosthetic group of cytochrome *c*. This is a covalently bound form of protohaem as shown.

or oxidise them. The oxidation–reduction potential depends on the structures surrounding the central iron which not only includes the porphyrin but also the protein.

   Cytochrome *c* has been more fully studied than the other pigments. It is less firmly bound to the mitochondrial membrane than components *b*, c_1 or *a*, a_3 and, as a result, may be partially lost from mitochondria during preparation; this ease of solubilisation has promoted its isolation and study. Both Keilin and

Fig. 2.6 Prosthetic groups of cytochromes *a* and *b*. (a) Protohaem (iron-proto-porphyrin IX) of cytochrome *b*. (b) Haem *a* of cytochrome *a*. Note that this structure involves only two modifications (arrowed) of protohaem.

Theorell with their coworkers were able to prepare c in almost pure state in the years 1935 to 1945 although crystallisation was not achieved until later. The cytochrome is stable, non-autoxidisable ($E_{m\,7.0} = 255$ mV) and of relatively low molecular weight (13 370 isolated from ox heart); its absorption spectrum is shown in Fig. 2.8. The protein from mammalian systems possesses 104 amino acid residues although the number may vary as in the case of some fishes, which possess a protein with 103 residues and higher plants with 111. Although there are a number of variations in amino acid sequence between species, in mammals 84 of the 104 are invariable and in only 11 positions are major changes in amino acid residue found. Outside mammals there is a greater variation in the number and sequence of the amino acids.

The prosthetic group is bound covalently by attachment to cysteine residues 14 and 17 and to tyrosine (48) and tryptophan (59). In addition, the iron atom in the haem is linked to residues on either side of the plane of the haem, to residue 80 (methionine) and to 18 (histidine) (see Fig. 2.7). The three-dimensional structure of the cytochrome is shown in Fig. 2.9 where the porphyrin can be seen largely confined to a single plane and inserted in a groove in the protein.

Mitochondrial cytochrome c is one of a closely linked group of 'high-potential' c cytochromes ($E_{m\,7.0} = 150\text{–}380$ mV) which include cytochrome c_2 of photosynthetic bacteria (sect. 15.5). They are characterised by small size (a single polypeptide of about 85 to 135 residues) and with methionine and histidine as the amino acid residues which act as axial ligands for the iron.

Several attempts have been made to understand the process of oxidation and reduction. The iron itself is buried deep within the molecule and would not be accessible for direct reduction by, say, cytochrome c_1 or direct oxidation by cytochrome oxidase. Alterations to a number of amino acid residues appear to hinder reduction of the cytochrome by a succinate–cytochrome c reductase complex, but do not affect oxidation by the oxidase. This has led to the conclusion that electrons are transferred to the iron through the protein, possibly via a route involving aromatic residues. Further, alteration of other amino acid residues hinders oxidation and not reduction, suggesting that different pathways are followed by electrons in oxidation. However, direct reduction of the haem now seems more probable on thermodynamic grounds. The haem itself is partly accessible at the surface of the molecule and a direct interaction between haem and the prosthetic groups of the adjacent respiratory electron carriers is proposed. This would necessitate rotation of the cytochrome in the membrane to enable the haem to interact first with an adjacent donor and then with an electron acceptor. It is assumed that an electron will readily migrate from the periphery of the haem to the iron. From X-ray analysis of cytochrome crystals, it has been found that the conformation of the oxidised and reduced states is the same. (Dickerson *et al.*, 1971, Swanson *et al.*, 1977, Takano *et al.*, 1977.) In solution the conformation of the two states is different (Bosshard and Zurrer 1980).

Cytochromes (a, a_3) are firmly bound to the membrane but may be solubilised with detergents such as cholate or deoxycholate. After this treatment, however, the haems are no longer attached to their proteins. There are two views on the a cytochromes. Either they are different chemical entities or, more

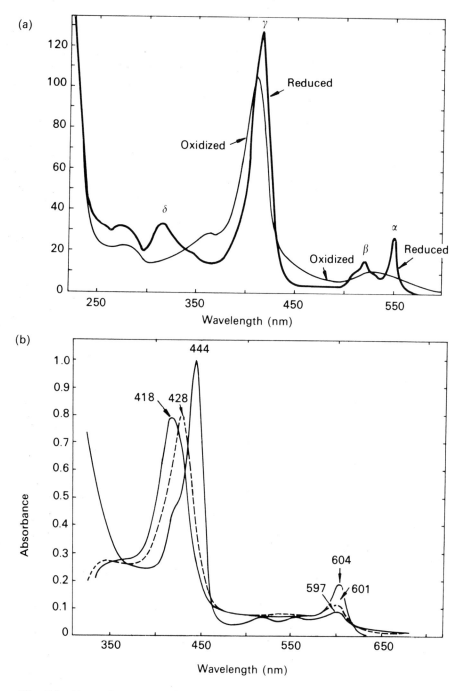

Fig. 2.8 Absorption spectra of: (a) cytochrome *c* (reduced and oxidised forms). From Keilin (1966). (b) cytochrome oxidase, reduced (ferrous), maxima at 604 and 444 nm; oxidised (ferric), maxima at 597 and 418 nm; oxygenated, maxima at 601 and 428 nm. With permission from Lemberg and Barrett (1973).

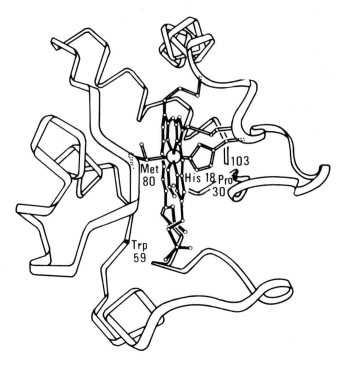

Fig. 2.9 Schematic representation of cytochrome *c* from tuna heart mitochondria. From Salemme (1977). Reproduced with permission from the *Annual Review of Biochemistry* volume 46 © 1977 by Annual Reviews Inc.

probably, they are chemically identical but functionally distinct, one of them acting as cytochrome *a*, the other acting as a_3 and reacting with oxygen and carbon monoxide. Also associated with this cytochrome are two atoms of copper which are tightly bound to protein and probably also play a role in the overall reaction. (The absorption spectrum is shown in Fig. 2.8.) Cytochromes (a, a_3) are the major terminal oxidase for respiratory systems in invertebrates, vertebrates, plants, yeasts and algae. In bacteria, the oxidase systems are more complex and will be discussed separately.

As noted earlier, there are two *b* cytochromes present in mammalian mitochondria firmly bound to the membrane.

2.5   The respiratory chain and mitochondrial oxidation

Figure 2.10 summarises the respiratory chain as outlined in this chapter. $NADH_2$ and succinate are the major substrates for this multienzyme complex bound to the inner mitochondrial membrane and catalysing the reactions:

$$NADH_2 + \tfrac{1}{2}O_2 \longrightarrow NAD + H_2O$$
$$succinate + \tfrac{1}{2}O_2 \longrightarrow fumarate + H_2O$$

The mechanism of the reaction involves a series of steps in which a component oxidises its neighbour on the substrate side, becoming itself reduced. Reoxidation occurs by reducing the other neighbour. Such a simple view presupposes some linear spatial relationship in the membrane and also a simple stoicheiometry of components of the chain. Evidence for the latter was seen in sect. 2.3(f) above.

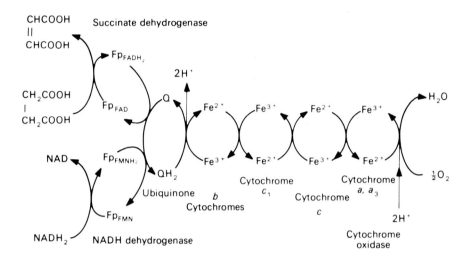

Fig. 2.10 Simple scheme for the respiratory chain.

A moment's consideration of the terminal reaction, where one molecule of oxygen is reduced to two molecules of water by the cytochrome oxidase, will indicate the need for a more complex arrangement involving the transfer of not one but four electrons to oxygen.

$$4\,\text{cytochrome}\,c\,(\text{Fe}^{2+}) + 4\text{H}^+ + \text{O}_2 \longrightarrow 4\,\text{cytochrome}\,c\,(\text{Fe}^{3+})$$
$$+\,2\text{H}_2\text{O}$$

In fact cytochrome oxidase is itself a multimolecular complex consisting of several proteins, two haems and two copper atoms. As will be discussed in Chapter 8, the respiratory chain is composed of four principal complexes.

The close relationship of the respiratory chain to the oxidation of substrates is shown in Fig. 2.11. A number of substrates are shown to undergo oxidation coupled to NAD reduction. In addition there are several flavoproteins which oxidise substrates (succinate dehydrogenase, acyl coenzyme A dehydrogenase) and which transfer their reducing equivalents to ubiquinone. Thus the quinone may be seen as a mobile carrier (dissolved in the lipid phase of the membrane) shuttling reducing equivalents between flavoproteins and cytochrome *b* complexes.

The initial steps of the respiratory chain associated with the flavoprotein dehydrogenases involve the transfer of pairs of hydrogens, whereas the subsequent (cytochrome) steps are single electron transfers. The early reactions

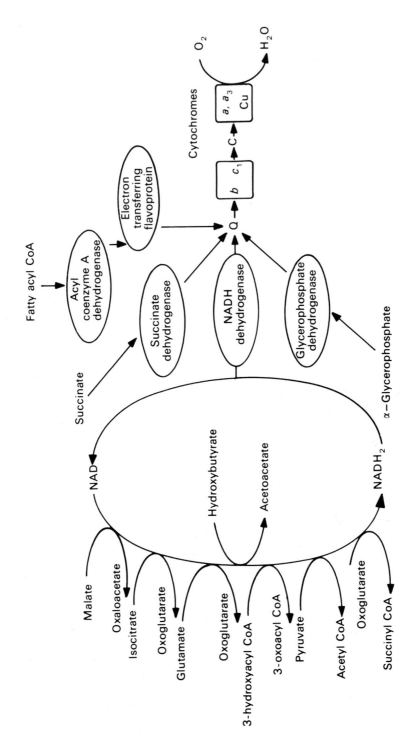

Fig. 2.11 Oxidation of a variety of substrates by the respiratory chain.

must therefore generate protons, while the cytochrome oxidase reaction (see above) will require protons. Thus there must be a proton flow associated with the electron transport.

The approximate mid-point potentials, $E_{m\,7.0}$ (standard potentials), for the various components of the chain are shown in Fig. 2.12. It will be seen that there is a difference of 1.14 V between $NADH_2$ and O_2 (see sect.1.3). Furthermore, with the exception of the values for ubiquinone and cytochrome *b*, the potentials become progressively more positive from reduced pyridine nucleotide to oxygen. As will be shown later, the energy made available by this potential gradient can be used for ATP synthesis.

2.6 Phosphorylation coupled to respiration

(a) Types of phosphorylation

It is conventional to distinguish between substrate-level phosphorylation and oxidative phosphorylation. The former term is applied to the synthesis of ATP intimately associated with metabolic pathways, for example in the glycolytic pathway:

$$\text{glyceraldehyde-3-phosphate} + Pi + NAD = \text{1,3-diphosphoglycerate} + NADH_2$$

$$\text{1,3-disphosphoglycerate} + ADP = \text{3-phosphoglyceric acid} + ATP$$

also: $$\text{phosphoenolpyruvate} + ADP = \text{pyruvate} + ATP$$

or in the citric acid cycle:

$$\text{2-oxoglutarate} + NAD + CoA = \text{succinyl CoA} + NADH_2 + CO_2$$

$$\text{succinyl CoA} + GDP + Pi = \text{succinate} + CoA + GTP$$

Although in two of the above cases phosphorylation is obviously linked to a prior oxidation, this type is known as substrate-level phosphorylation. Oxidative phosphorylation is normally restricted to a synthesis of ATP closely linked to respiration itself.

(b) Discovery of oxidative phosphorylation

During the 1930s, several workers studied changes in levels of phosphates, particularly hexose phosphates and creatine phosphate, in tissue preparations. Kalckar noted that 'phosphorylation (measured as ester accumulation) in kidney cortex is proportional to the oxygen consumption'. Stimulation of respiration using glutamate stimulated phosphorylation. However, these data were frequently interpreted as evidence of glycolytic phosphorylation which is also associated with a reduction of NAD for subsequent oxidation. The oxidation of succinate to fumarate, already known to cause cytochrome reduction, was found to promote phosphorylation, but this puzzled workers at the time.

The classic paper on the subject, *The Mechanism of Phosphorylation Associated with Respiration*, was published by the Russian workers V. A. Belitzer and E. T. Tsybakova in 1939. They concluded from earlier work, including

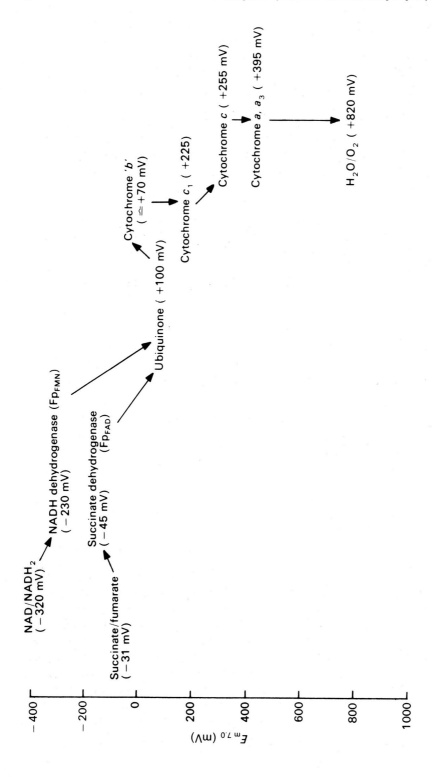

Fig. 2.12 Mid-point potentials ($E_{m\ 7.0}$) of the main components of the respiratory chain (see sect. 9.7).

Kalckar's, that 'a "strictly respiratory" synthesis of these esters might be inferred'. The paper demonstrated that in minced muscle, for each molecule of oxygen consumed, between four and seven phosphorylations occurred. The high values were later criticised and rejected, but they demonstrated that the observed phosphorylations were quantitatively in considerable excess of those expected from glycolysis. For each molecule of glyceraldehyde phosphate converted to pyruvate in the glycolytic pathway, one NAD is reduced and this is associated with two phosphorylations corresponding to a P/O (P/2H) ratio of 2. The P/O ratio has been widely used in studies of oxidative phosphorylation and is defined:

$$P/O = \frac{\text{number of phosphorylations}}{\text{number of } \textit{atoms} \text{ of oxygen consumed}}$$

Belitzer and Tsybakova's view that phosphorylation was coupled to respiration was investigated by several workers, among them Severo Ochoa (1943), who obtained a more precise estimate for the P/O ratio. Ochoa was aware that mince preparations used in the measurement of P/O ratios contained ATPases catalysing ATP hydrolysis and reducing the yield of phosphate esters. He therefore measured the ratio in a system where he was also able to estimate losses due to hydrolysis of phosphate esters. Using a cat heart mince and with pyruvate as the main substrate, he measured the P/O (or P/2H) ratio in anaerobic conditions where the oxidation and phosphorylation of glyceraldehyde was coupled to the reduction of pyruvate through NAD/NADH$_2$ (formation of pyruvate from the glyceraldehyde phosphate was blocked by the presence of fluoride).

3-phosphoglyceraldehyde + NAD + Pi = 1,3-diphosphoglyceric
$$\text{acid} + \text{NADH}_2$$

1,3-disphosphoglyceric acid + ADP = 3-phosphoglyceric acid + ATP

pyruvic acid + NADH$_2$ = lactic acid + NAD

The equivalent of the P/O ratio, the P/2H ratio (measured as P/lactic acid) is theoretically 1. The P/O ratio was also measured for the total oxidation of pyruvate under aerobic conditions.

pyruvic acid + 2.5O$_2$ + xPi + xADP = 3CO$_2$ + (x+2)H$_2$O + xATP

The results showed that the observed ratio for the anaerobic experiment was 0.62 indicating a loss of 38 per cent of the ATP formed. The P/O ratio for the total oxidation of pyruvate was 1.9 which, on the basis of a similar loss gives a true value of 3.1. Ochoa concluded that 3 ATP per atom of oxygen were synthesised during oxidative phosphorylation. Since the oxidation of pyruvate involved five dehydrogenations, Ochoa concluded that 15 ATP would be synthesised for each pyruvate oxidised. The source of these ATPs will be considered further.

A later method of measuring oxidative phosphorylation includes glucose and the enzyme hexokinase (EC 2.7.1.1.) in the incubation mixture. The formation of ATP is measured as formation of glucose-6-phosphate:

glucose + ATP $\longrightarrow$ glucose-6-phosphate + ADP

By this method ATP is removed as soon as it is formed.

(c) Measurement of P/O ratios

In the late 1940s, it was shown in Green's laboratory that phosphoryl-ation was coupled to the mitochondrial oxidation of oxoglutarate, malate, pyruvate and citrate with a P/O ratio greater than 2. Succinate oxidation had a P/O ratio less than 2. A ratio approaching 4 for oxidation of oxoglutarate was subsequently obtained in Lardy's laboratory. From these data it could be deduced that the P/O ratio for all oxidative steps in the citric acid cycle was 3 except for the oxidation of succinate, 2, and for oxoglutarate, 4 (see Slater, 1953b). In regard to oxoglutarate oxidation, it was shown that under anaerobic conditions, a phosphorylation still occurred and that dinitrophenol, which uncoupled oxidative phosphorylation, inhibited only three of the phosphoryl-ations associated with oxoglutarate oxidation (Judah, 1951). Thus there is a substrate-level phosphorylation as well as three oxidative phosphorylations associated with oxoglutarate metabolism.

All these experiments led to the conclusion that the oxidation of $NADH_2$ was coupled to phosphorylation, although this remained to be demonstrated. Lehninger (1951) succeeded in showing that the oxidation of $NADH_2$ by oxygen in specially prepared mitochondria was coupled to phosphorylation with an uncorrected P/O ratio of 1.89. Lehninger found that the hypotonic pretreatment of mitochondria was necessary, as the particles were otherwise impermeable to pyridine nucleotides. Several workers used 3-hydroxybutyrate as a substrate to generate $NADH_2$ internally in mitochondria by the 3-hydroxybutyrate dehy-drogenase and obtained P/O ratios in excess of 2.

(d) Phosphorylation associated with regions of the respiratory chain

ATP synthesis is coupled not to one but to several discrete regions of the chain. Succinate oxidation has a P/O ratio of 2, while NADH oxidation has a ratio of 3. Since most of the respiratory chain is common to both these substrates, some ATP synthesis must be associated with the initial dehydrogenase complex of the chain.

The use of added oxidised cytochrome c as an electron acceptor in place of oxygen gives a P/O ratio approaching 2 for $NADH_2$ oxidation (using 3-hydroxybutyrate as substrate) and approaching 1 for succinate oxidation. This could imply a site common to both succinate and $NADH_2$ oxidation on the substrate side of cytochrome c as well as one in the cytochrome oxidase region. The oxidation of reduced cytochrome c (kept reduced by ascorbate) has been used to assay the cytochrome oxidase and this also gives a P/O ratio less than 1, suggesting a single phosphorylation associated with the oxidase.

These data lead to the conclusion that three segments of the respiratory chain are coupled to phosphorylation (see Fig. 2.13). Each of these segments may be assayed independently. Segment I may be assayed using ubiquinone as a terminal electron acceptor, $NADH_2$ as substrate and antimycin to inhibit other parts of the chain. Segment II may be assayed with succinate as substrate and cytochrome c as electron acceptor and with azide to inhibit the cytochrome oxidase. More specifically, reduced ubiquinone may be used as substrate in place

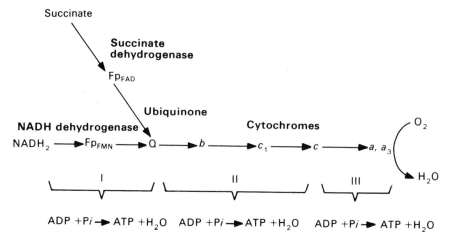

Fig. 2.13 Respiratory chain showing the three segments whose oxidation-reduction reactions are coupled to ATP synthesis.

of succinate. Segment III is assayed with ascorbate–cytochrome c as electron donor and oxygen as electron acceptor with antimycin to prevent electron transport in the earlier part of the chain.

2.7 Coupled mitochondria, respiratory control

Freshly prepared mitochondria in good condition show a phosphorylation of ADP coupled to electron transport:

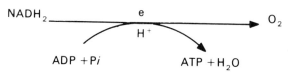

When mitochondria age or become damaged, the coupled synthesis of ATP disappears and respiration proceeds, usually at a faster rate. Keilin's heart-muscle preparations described earlier were uncoupled mitochondrial fragments and did not phosphorylate. Thus respiration and phosphorylation are separate processes which in the mitochondrion are linked together, the former driving the latter. Respiration is said to be *coupled* or *uncoupled* to phosphorylation and mitochondria are described as being coupled or uncoupled for the same reason.

Most mitochondria *in vivo* are regarded as coupled. Uncoupling is brought about by ageing, by an uncoupling agent or by structural changes caused by preparative or manipulative techniques. Dinitrophenol was early discovered to be an uncoupler which inhibits phosphorylation and promotes respiration. The labile coupling of the phosphorylation process suggests the presence of a chemical or physical intermediate state in which the energy of the redox reactions is conserved and used for ATP synthesis. The nature of the coupling will be discussed in detail later, but for the moment it will be convenient to refer to an intermediate high-energy state.

Implicit in the concept of coupling is that of respiratory control. If, in a coupled system, the synthesis of ATP requires that oxidation of substrate should take place concurrently, it is also true that respiration requires a supply of ADP and inorganic phosphate. In coupled mitochondria the rate of respiration can be controlled by the supply of ADP. In the absence of ADP, the respiratory rate is low. Thus the oxidation of $NADH_2$ is correctly shown as:

$$NADH_2 + \tfrac{1}{2}O_2 + 3ADP + 3Pi = NAD + 3ATP + 4H_2O$$

The term respiratory control is used to describe the control of respiration by the phosphorylation process. Chance and Williams (1956) defined the metabolic states of the mitochondrion. Table 2.2. shows the dependence of respiratory process on substrate, ADP and oxygen. In the presence of all three, state 3, respiration is fast, limited only by the maximum rate of the system as a whole. If the ADP is allowed to fall to a low level (state 4), this deficiency limits respiration.

Table 2.2 Metabolic states of mitochondria (Chance and Williams, 1956)

State	Oxygen concn.	ADP level	Substrate level	Respiration rate	Rate-limiting factor
1.	normal	low	low	slow	ADP
2.	normal	high	zero	slow	substrate
3.	normal	high	high	fast	respiratory chain
4.	normal	low	high	slow	ADP
5.	zero	high	high	zero	oxygen

2.8   Sites of phosphorylation

The experiments described earlier led to the conclusion that phosphorylation was distributed along the length of the respiratory chain and gave an indication of the regions of the chain which are coupled to ATP synthesis. Those conclusions were confirmed and extended by an ingenious experiment of Chance and Williams (1956). We have already noted that in coupled mitochondria, respiratory activity and oxidative phosphorylation are obligatorily linked: full oxidative activity requires an adequate level of ADP and Pi. In effect the absence of ADP inhibits respiration. This inhibitory effect might be expected to act at the coupling site, specifically blocking the redox reactions linked to phosphorylation. As described in sect. 2.3(g), the site of action of an inhibitor can be determined by locating the cross-over point. Thus by allowing the ADP level of a respiring coupled mitochondrial preparation to fall (transition from state 3 to state 4), it should be possible to determine a cross-over point indicating a phosphorylation site. Chance and Williams found that on passing from state 3 to state 4, cytochrome c became more reduced and cytochrome a more oxidised. This suggested a phosphorylation site at the cytochromes $c-a$ point in the respiratory chain. When low levels of azide were used in this experiment, the cross-over point was found to shift to the $b-c_1$ position on transition from state 3 to state 4. Higher levels of azide gave a cross-over point in the region of the $NADH_2$ dehydrogenase.

Thus the three sites of phosphorylation are associated with (1) the $NADH_2$ dehydrogenase, (2) the cytochrome $b-c_1$ oxidation–reduction, and (3) the cytochrome-c–cytochrome oxidase oxidation–reduction (Fig. 2.15).

A more theoretical approach to this problem can be made by a study of the redox potentials. Chance found that the ratio of oxidised/reduced states of the redox carriers was, in most cases, not far from unity in mitochondria in state 3. Thus the actual potential of the carriers is close to their standard potential. Conservation of chemical energy in the form of ATP synthesis might be expected in those redox reactions where there is a substantial difference in redox potential. Experimentally, the ATP/ADP ratio varies widely, but using a ΔG of 50 kJ for ATP synthesis, this corresponds to a difference in potential $\Delta E_{m\,7.0}$ of 260 mV. From $NADH_2$ to ubiquinone, $\Delta E_{m\,7.0} = 420$ mV (at pH 7.0 and 30 °C), while from cytochrome b to cytochrome c, $\Delta E_{m\,7.0} = 185$ mV and from cytochrome c to oxygen $\Delta E_{m\,7.0} = 565$ mV. On very simple thermodynamic grounds the synthesis of ATP coupled to respiration at three sites is feasible. We shall see later that there is a precise relationship between the potential span of the respiratory chain and the synthesis of ATP (sect. 9.11).

2.9   The link between oxidation and phosphorylation

(a) The mitochondrial ATPase

Mitochondria possess ATPase activity. In freshly prepared coupled mitochondria, this activity is low under most conditions. Ageing or mechanical damage of the mitochondrial structure results in a significant increase of ATPase activity, associated with loss of ability to carry out oxidative phosphorylation. This latent ATPase (catalysing $ATP + H_2O \longrightarrow ADP + Pi$) is generally regarded as a modified form of the enzyme responsible for ATP synthesis in oxidative phosphorylation ($ADP + Pi \longrightarrow ATP + H_2O$). Treatment of coupled mitochondria with an uncoupling agent, dinitrophenol, also stimulates this ATPase activity as well as inhibiting oxidative phosphorylation. Oxidative phosphorylation itself is strongly inhibited by the antibiotic, oligomycin which also inhibits the dinitrophenol-stimulated ATPase.

The ATPase can be isolated and purified as a high molecular weight complex, the properties of which will be discussed later.

(b) Reverse electron flow driven by ATP hydrolysis

ATP hydrolysis may be shown to drive a reversal of normal electron transport. The mid-point potential of the succinate–fumarate couple is considerably more positive than that for $NADH_2/NAD$. Thus, on thermodynamic grounds, significant reduction of NAD by succinate appears unlikely. Nevertheless, Chance and Hollunger (1960) reported a rapid reduction of NAD by succinate in coupled guinea pig liver mitochondria.

Later several groups described a reduction of NAD by succinate which was ATP-dependent in the presence of antimycin (Low and Vallin, 1963). This reaction was shown to be sensitive to oligomycin, amytal and uncoupling agents (see Fig. 2.14). The experiment demonstrates the reversible link between the ATPase and electron transport, i.e. ATP hydrolysis will drive electron transport and electron transport will drive ATP synthesis.

In the studies of reverse electron flow, it was found that NAD reduction could be achieved by succinate oxidation in the absence of added adenine nucleotide, as long as succinate was also oxidised through the normal respiratory route. Under these conditions NAD reduction was not sensitive to the ATPase inhibitor, oligomycin, but was still sensitive to uncouplers. The intermediate high-energy state created by the passage of electrons through sites II and III could be used to drive the reverse flow of electrons through site I. For this process, the ATPase is unnecessary. Sites I, II and III are equivalent as far as generation and dissipation of the intermediate high-energy state are concerned (Fig. 2.15). Oxidation of either succinate or cytochrome c may be coupled to an energy-dependent reduction of NAD. Hence all three sites differ only in respect of the redox reactions to which they are coupled, but apparently not in respect of their coupling or phosphorylation mechanisms.

(c) The inorganic phosphate–ATP exchange reaction

The ATPase in coupled mitochondria can be assayed as a separate reaction by measurement of exchange reactions. One of these only will be mentioned here. Boyer and others (1954) demonstrated a rapid exchange between the γ-phosphate of ATP and labelled inorganic phosphate (^{32}Pi) in washed coupled liver mitochondria:

$$\text{adenosine}-\overset{\alpha}{(P)}-\overset{\beta}{(P)}-\overset{\gamma}{(P)} + H_2O \;\rightleftharpoons\; ADP + {}^{32}Pi + \text{energy}$$

This exchange is inhibited by oligomycin and the uncoupler dinitrophenol, but not by inhibitors of the respiratory chain. It implies the storage within the overall system of the necessary energy for resynthesising ATP from the inorganic phosphate pool after hydrolysis. This does not involve the respiratory chain itself and therefore suggests the existence of some intermediate high-energy state.

(d) The intermediate high-energy state

The nature of this state has been a matter for speculation but its existence is amply confirmed by a wide variety of experimental approaches including those described in sect. 2.9(b) and (c) above. The reactions we have discussed enable us to formulate a generalised tentative scheme for energy transformations in oxidative phosphorylation (Fig. 2.15). In this scheme, oxidation–reduction reactions are reversibly related to the phosphorylation system by the intermediate high-energy state.

2.10 Inhibitors of oxidative phosphorylation

Oxidative phosphorylation may be inhibited at the three levels shown in Fig. 2.15. Firstly, phosphorylation is inhibited by an inhibitor of the respiratory chain which stops the oxidation–reduction reactions coupled to the generation of the intermediate high-energy state. Thus amytal, rotenone, antimycin, azide,

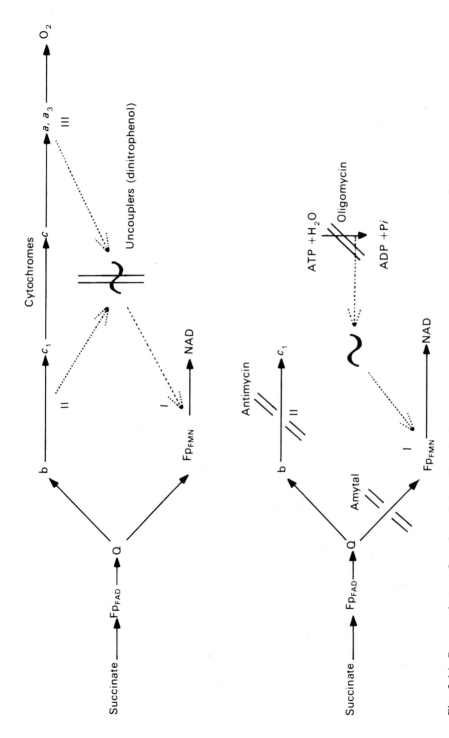

Fig. 2.14 Reverse electron flow through coupling site I.

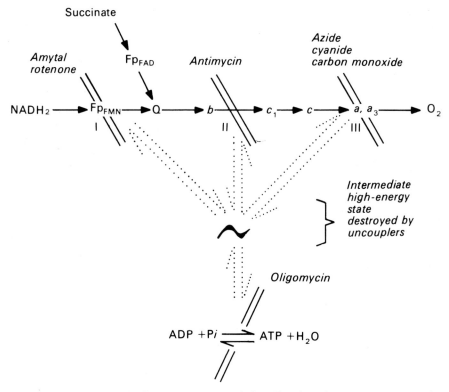

Fig. 2.15 The coupled electron transport chain. The dotted arrows represent the reversible exchange of energy between the ATPase, the intermediate high-energy state ($\sim$) and the respiratory chain.

cyanide and carbon monoxide are all inhibitors of oxidative phosphorylation in this sense. Secondly, dissipation of the intermediate high-energy state by uncouplers such as dinitrophenol, also inhibits phosphorylation. Under most conditions, the addition of uncouplers to coupled mitochondria produces some stimulation of respiration. A pronounced increase of respiratory rate is observed when uncouplers are added to mitochondria in state 4 (ADP limited), owing to a removal of respiratory control. The third class of inhibitors are concerned with the phosphorylation process itself. Oligomycin is one of a group of inhibitors which inhibit phosphorylation and consequently respiration in coupled mitochondria; addition of an uncoupler releases respiration from inhibition by oligomycin.

2.11   Phosphorylation and control of metabolism

As we have noted, the supply of ADP controls the rate of respiration and hence will regulate the oxidation of substrates. Since in the mitochondrion the size of the adenine nucleotide pool (ADP + ATP) is approximately constant, we can consider the ATP/ADP ratio as a potential regulator of respiration.

Conversely, respiration in coupled systems will increase the ATP/ADP ratio. In other words there is a self-correcting balance between the ATP/ADP ratio and the rate of respiration. The evidence for such a homeostatic system will be considered later (sect. 9.11). However, the influence of the ATP/ADP ratio extends much further than the respiratory chain itself; as we shall see in the next chapter, the metabolic pathway supplying reducing equivalents to the respiratory chain, the citric acid cycle, is also subject to regulation by the ATP/ADP ratio. The formation of pyruvate, a major substrate for mitochondrial metabolism which is formed extramitochondrially from many sources, but particularly by glycolytic breakdown of carbohydrates, is similarly regulated.

In the nineteenth century, the French scientist, Louis Pasteur, observed that the rate of glycolysis (fermentation) in yeast was much greater in anaerobic conditions (in which there is only limited ATP synthesis during substrate-level phosphorylation, 1 ATP per pyruvate produced from glucose) than in aerobic conditions where substantial ATP synthesis occurs (15 ATP per pyruvate oxidised). Although when Pasteur observed this phenomenon, known as the Pasteur effect, neither glycolysis nor phosphorylation was understood, it now seems that the ATP/ADP ratio plays a major role in regulating breakdown of sugars by the glycolytic pathway and in particular in controlling the activity of a key glycolytic enzyme, phosphofructokinase (EC 2.7.1.11). Thus, in addition to regulating the citric acid cycle, the ATP/ADP ratio also controls the formation of the substrate for mitochondrial metabolism, pyruvate, high ratios inhibiting pyruvate formation.

Further reading

Dickerson, R. E. (1972) The structure and history of an ancient protein, *Scientific American*, **226**, 58–70.

Keilin, D. (1966) *The History of Cell Respiration and Cytochrome.* Cambridge University Press.

Lemberg, R. and Barrett, J. (1973) *Cytochromes.* Academic Press, New York.

Salemme, F. R. (1977) Structure and function of cytochromes C, *Annu. Rev. Biochem.*, **46**, 299–330.

Chapter 3
Mitochondrial oxidative metabolism

3.1   The formulation of the citric acid cycle

There are two major pathways of oxidative metabolism in the mitochondrion, the oxidation of pyruvate through the citric acid or tricarboxylic acid cycle and the oxidation of fatty acids initially by β-oxidation and finally also through the citric acid cycle. The pyruvate pathway will be considered first.

The elucidation of the metabolism of pyruvate, citrate and related compounds stemmed from the work on the biological oxidation of carbohydrates. Krebs in a review (1943) wrote 'The chief aim of the theory of carbohydrate oxidation is to describe, step by step, the chemical changes of the carbohydrate molecule leading to the formation of carbon dioxide and water.' When Krebs wrote this, it was already clear that the first stage in the process was the production of pyruvate by the glycolytic pathway and that the final stages involved the citric acid cycle formulated by Krebs and Johnson (1937). What became clear only subsequently was the means whereby pyruvate was converted to citrate.

The background to the formulation of the citric acid cycle was in experiments of Thunberg, Szent-Györgyi, Krebs and others. Both Thunberg and Knoop had proposed a cycle for the oxidation of carbohydrate involving the C_4 dicarboxylic acids (succinate, fumarate, malate and oxaloacetate). This had been revised by Toeniessen and Brinkmann (1930). Szent-Györgyi in 1936 (see Annau *et al.*, 1936) introduced a novel approach to the function of the C_4 dicarboxylic acids by proposing that their role was to serve as a catalytic hydrogen carrier between foodstuff and cytochrome.

These studies made use of slices or minces of muscle tissue from mammalian heart and pigeon breast where a high respiratory rate could be observed *in vitro*. Although in this early work the significance of the mitochondrion in respiratory studies was not realised, it was noted that 'extracts of muscle and other tissues contain the enzyme systems responsible for the anaerobic conversion of carbohydrate into lactic acid' and that oxidations by molecular oxygen were 'associated with the water-insoluble constituents of the tissue' (Krebs, 1943).

Krebs and Johnson (1937) formulated the citric acid cycle (Fig. 3.1) on the basis of four sets of experimental observations (also see Krebs, 1943). We will briefly examine each of these.

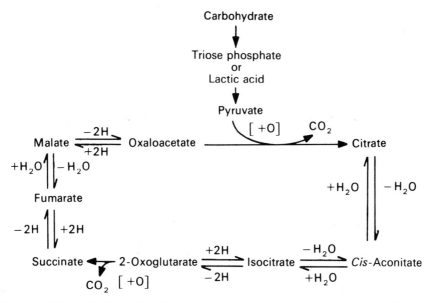

Fig. 3.1 The oxidation of carbohydrate through the citric acid cycle as originally conceived.

(a) The rapid oxidation of citrate, isocitrate, cis-aconitate and 2-oxoglutarate in pigeon breast muscle. The importance of these observations lay in linking these carboxylic acids with the C_4 dicarboxylic acids (succinate, fumarate, malate and oxaloacetate) which featured in earlier schemes for carbohydrate oxidation. In the period between 1909 and 1920, Thunberg had examined the oxidation of over 60 organic substances by animal tissues using the methylene blue reduction technique. Most were only metabolised slowly, but succinate, fumarate, malate and citrate were rapidly oxidised. These observations were extended by later workers who added *cis*-aconitate, isocitrate, 2-oxoglutarate, oxaloacetate, pyruvate, glutamate, aspartate and alanine to the list of rapidly metabolised substrates. The significance of the amino acids lies in the fact that they are interconvertible with oxo acids (i.e. alanine $\rightleftharpoons$ pyruvate; glutamate $\rightleftharpoons$ oxoglutarate; and aspartate $\rightleftharpoons$ oxaloacetate).

(b) The catalytic effect of citrate on the respiration of pigeon breast muscle. This was shown to be of the same order of magnitude as that of succinate and its oxidation products. In 1936 Szent-Györgyi had found that the addition of succinate, fumarate, malate or oxaloacetate catalytically increased the oxygen uptake and carbon dioxide evolution of respiring muscle tissues. The addition of fumarate to muscle suspensions resulted in an oxygen uptake which could approach five times that required for the complete oxidation of the fumarate itself. Krebs and Johnson found that citrate had a similar effect. Furthermore, these five substances all catalysed the removal of pyruvate from muscle preparations.

(c) Synthesis of citrate from oxaloacetate. Krebs and Johnson demonstrated the synthesis of citrate from added oxaloacetate in muscle tissue. Later

it was shown that 2-oxoglutarate, a product of citrate metabolism, also arose from oxaloacetate.

(d) Formation of succinate from fumarate. The most important observation supporting the formulation of the cycle was the *oxidative* formation of succinate from fumarate or oxaloacetate. Succinate could be formed from fumarate by a simple reduction and this reaction is specifically inhibited by low concentrations of malonate (originally demonstrated by Quastel and Wooldridge in 1928). Krebs showed that succinate could still be formed from fumarate in the presence of malonate and that the reactions required oxygen. Thus there are two pathways from fumarate and oxaloacetate to succinate; one reductive and malonate-sensitive (succinate dehydrogenase) and the other oxidative and malonate-insensitive.

When considering the evidence for the original formulation of the cycle, it should be noted that a number of the individual reactions had already been demonstrated. For example, Einbeck (1919) had shown succinate to be converted to fumarate and malate while Hahn and Haarmann (1928) demonstrated the formation of oxaloacetate from malate. Krebs and Johnson showed the formation of succinate from oxoglutarate thus relating a number of organic acids metabolically:

$$\text{oxoglutarate} \rightleftharpoons \text{succinate} \rightleftharpoons \text{fumarate} \rightleftharpoons \text{malate} \rightleftharpoons \text{oxaloacetate}$$

Formation of oxoglutarate by oxidative decarboxylation of citrate (catalysed by a 'citrate dehydrogenase') was proposed by Martius and Knoop. Later, Martius (1938) showed that the substrate for the oxidative decarboxylation was not citrate but isocitrate and that *cis*-aconitate was apparently an intermediate in the conversion of citrate to oxoglutarate:

$$\text{citrate} \rightleftharpoons \text{\emph{cis}-aconitate} \rightleftharpoons \text{isocitrate} \rightleftharpoons \text{oxoglutarate}$$

Thus, by the late 1930s, the citric acid cycle or Krebs cycle was clearly formulated as a system for the oxidation of carbohydrate. After initial uncertainty, it was eventually agreed that pyruvate (rather than, say, lactate) was the substrate for the cycle. Further confirmation came from experiments using radioisotopes. For several years there was doubt about which of the tricarboxylic acids was initially formed from oxaloacetate and pyruvate. A later scheme (Krebs, 1943) showed *cis*-aconitate as the first tricarboxylic acid in the cycle and citrate involved only in a side reaction of little quantitative importance:

$$\text{oxaloacetate} + \text{pyruvate} \longrightarrow \text{\emph{cis}-aconitate} \longrightarrow \text{isocitrate} \longrightarrow \text{oxoglutarate}$$
$$\searrow \qquad\qquad \updownarrow$$
$$CO_2 \qquad \text{citrate}$$

The purpose of the scheme was to remove the one symmetrical member of the tricarboxylates, citrate. Evidence from experiments in which labelled CO_2 and pyruvate were incubated with liver preparations showed label in only one of the carboxyl carbons of oxoglutarate. It was argued that if citrate was metabolised as a symmetrical intermediate both carboxyl carbons would be labelled (Fig. 3.2). The inevitable conclusion seemed to be that citrate was not an intermediate.

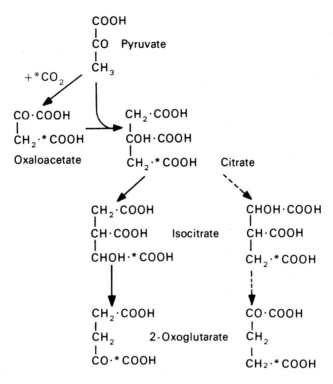

Fig. 3.2 Citrate metabolism. Citrate was labelled with $^{13}CO_2$ in the presence of pyruvate. The oxoglutarate formed from citrate was found to be labelled in C-1 but not in C-5. Hence citrate is not metabolised symmetrically.

However, Ogston (1948) showed the fallacy in the argument. Citrate could be metabolised symmetrically if attached to the enzyme at one or two sites but it would be metabolised asymmetrically if attached at three sites (Fig. 3.3). The matter was finally settled when a preparation from liver virtually free from aconitase (see Fig. 3.9) was shown to synthesise citrate, and not isocitrate, from oxaloacetate and pyruvate, thus confirming the original scheme.

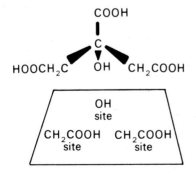

Fig. 3.3 Diagram of hypothetical attachment of citrate to the enzyme surface to give asymmetric metabolism.

THAMES POLYTECHNIC LIBRARY

3.2   The oxidation of pyruvate

(a) The role of acetyl coenzyme A

The steps by which pyruvate is converted to citrate remained a mystery for some time after Krebs formulated the citric acid cycle. Pyruvate oxidation was shown by Krebs and Eggleston (1940) to be inhibited by malonate and the inhibition was relieved by addition of fumarate. The metabolism of pyruvate requires oxaloacetate normally produced from fumarate in the cycle:

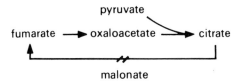

An intermediate between pyruvate and citrate was sought. A hypothetical C_7 compound could not be found, but when acetate, like pyruvate, was shown to be converted to citrate in the presence of oxaloacetate, the intermediate in pyruvate oxidation was seen as a form of acetate, 'active acetate'.

Interest in 'active acetate' grew in the 1940s as biochemists realised that such a compound was involved not only in citrate synthesis but also in acetoacetate, fatty acid and steroid synthesis and in acetylation reactions, particularly the acetylation of choline to acetyl choline. Lipmann (1945) observed that for acetylation in pigeon liver, a coenzyme was needed in addition to ATP and acetate. Further, except in bacteria, acetyl phosphate would not act as an acetylating agent. Lipmann *et al.* (1947) showed that the new coenzyme, coenzyme A, was a pantothenic acid derivative. Later he suggested that 'active acetate' was acetyl coenzyme A (Fig. 3.4) and subsequently this was shown to be converted to citrate in the presence of oxaloacetate and also to be the product of the pyruvate dehydrogenase reaction.

As indicated above, acetate can be converted to citrate in the presence of ATP, coenzyme A and oxaloacetate. In mammalian systems, plants and some micro-organisms, acetate is metabolised by acetyl coenzyme A synthetase (EC 6.2.1.1)

$$\text{acetate} + \text{ATP} + \text{CoA} \longrightarrow \text{acetyl CoA} + \text{AMP} + \text{PP}$$

In bacteria, acetyl coenzyme A is formed by means of the acetate kinase (EC 2.7.2.1) and the phosphate acetyltransferase (phosphotransacetylase, EC 2.3.1.8) reactions.

$$\text{acetate} + \text{ATP} \longrightarrow \text{acetylphosphate} + \text{ADP}$$
$$\text{acetylphosphate} + \text{CoA} \longrightarrow \text{acetyl CoA} + \text{P}i$$

(b) Cofactors and prosthetic groups

Six cofactors are involved in the conversion of pyruvate to acetyl coenzyme A: thiamin pyrophosphate, lipoic acid, coenzyme A, FAD, NAD and Mg^{2+}. Following early demonstrations of pyruvate metabolism in crude particulate preparations (containing mitochondria), a soluble pyruvate dehy-

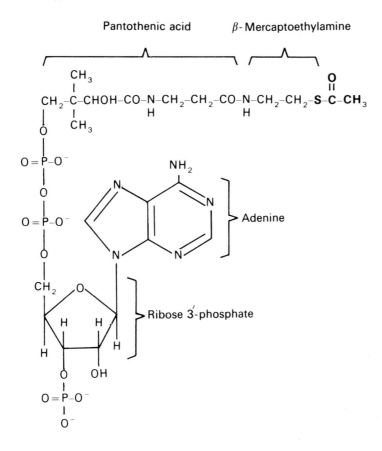

Fig. 3.4 Acetyl coenzyme A.

drogenase was prepared in Ochoa's laboratory from bacteria (Korkes *et al.*, 1951) and from muscle by Jagannathan and Schweet (1952). Coenzyme A and NAD were shown to be essential for the reaction and acetyl coenzyme A to be the product. Later lipoic acid, discovered originally as a growth factor in bacteria and isolated by Reed *et al.* (1951), was shown to be a component of the enzyme complex. It is covalently bound to a lysine residue of the enzyme (Fig. 3.5) and undergoes reduction to dihydrolipoate during the oxidation of pyruvate. Reoxidation is achieved by a flavoprotein dehydrogenase which is part of the complex.

   A role for thiamin (vitamin B_1) was demonstrated by R. A. Peters between 1929 and 1939 in experiments on pyruvate metabolism by pigeon brain from vitamin-B_1-deficient birds. These studies led to the discovery of thiamin pyrophosphate (TPP) as the active form of the cofactor (see Banga *et al.*, 1939). The hydrogen on C-2 of TPP readily exchanges with deuterium in the medium and C-2 of TPP makes a strongly nucleophilic attack on C-2 of pyruvate as the first step in the enzyme reaction. Two derivatives of TPP, lactyl-TPP and

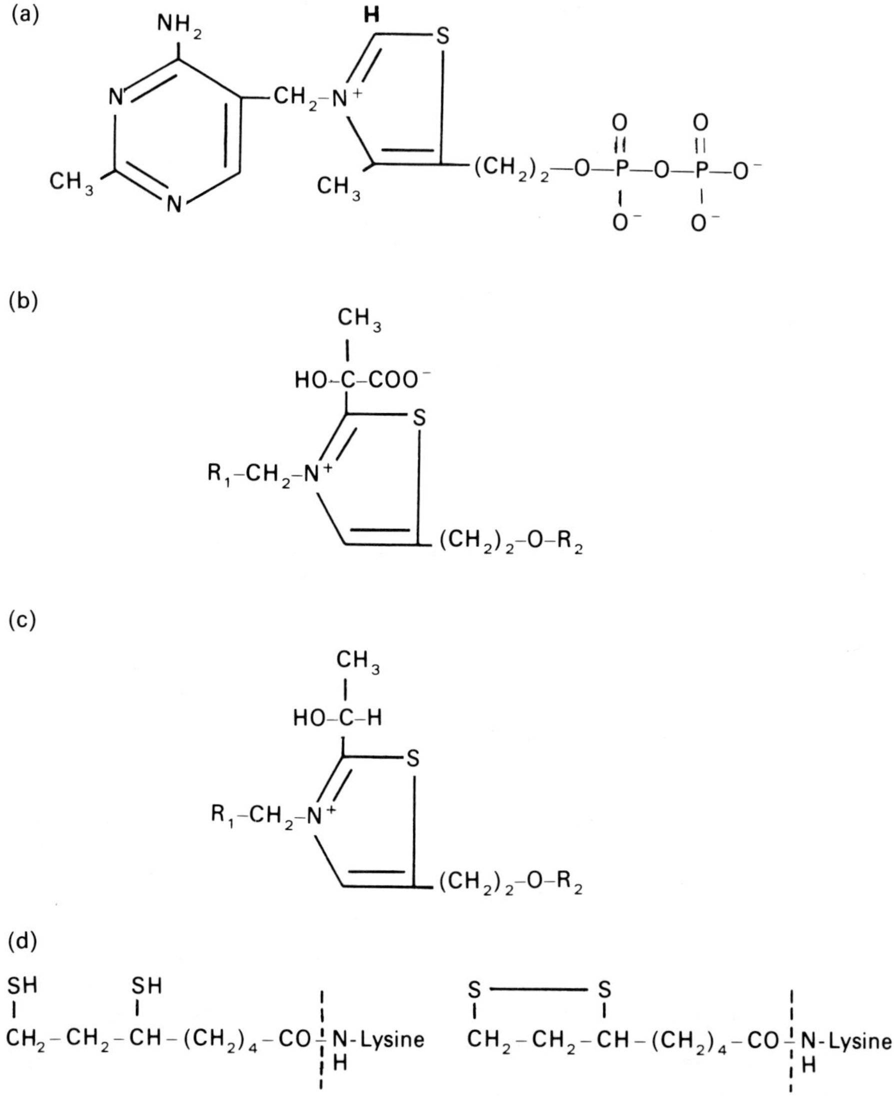

Fig. 3.5 Thiamin pyrophosphate and lipoic acid: (a) thiamin pyrophosphate; (b) lactylthiamin pyrophosphate; (c) ethylthiamin pyrophosphate; and (d) reduced and oxidised lipoic acid linked to lysine.

hydroxyethyl-TPP have been isolated by chromatography of incubation mixtures containing pyruvate and enzyme (Fig. 3.5.).

(c) Pyruvate dehydrogenase

$$\text{pyruvate} + \text{NAD} + \text{CoA} \longrightarrow \text{acetyl CoA} + \text{NADH}_2 + \text{CO}_2$$

Pyruvate dehydrogenase has attracted the attention of investigators firstly because of its intricate structure and secondly because it is a major point of

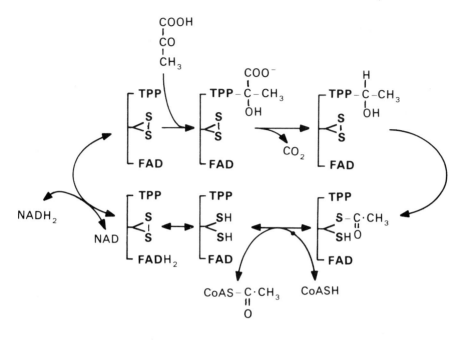

Fig. 3.6 The pyruvate dehydrogenase reaction. The enzyme is represented by the prosthetic groups which take part in the reaction, thiamin pyrophosphate (TPP), lipoic acid ($\langle^S_S$) and flavin adenine dinucloetide (FAD).

regulation in mitochondrial metabolism. The enzyme is a multienzyme complex composed of three separate enzymes. The overall reaction is shown in Fig. 3.6. Decarboxylation of pyruvate occurs when the substrate is covalently bound to the TPP. The ethyl group is transferred to the lipoic acid which reacts with coenzyme A to form the acetyl coenzyme A, leaving the lipoic acid residue reduced. The initial form of the enzyme is regenerated by oxidising the lipoic acid and reducing NAD. Thus pyruvate dehydrogenase is a complex of three enzymes catalysing the following reactions:

Pyruvate decarboxylase[1] (EC 1.2.4.1)

$$CH_3COCOOH + E\text{-}TPPH \longrightarrow E\text{-}TPP\text{-}CHOHCH_3 + CO_2$$

Lipoate acetyltransferase (EC 2.3.1.12)

$$E\text{-}TPP\text{-}CHOH \cdot CH_3 + E\text{-}Lip\langle^S_S \rightleftharpoons E\text{-}TPPH + E\text{-}Lip\langle^{S\text{-}CO\cdot CH_3}_{SH}$$

$$E\text{-}Lip\langle^{S\text{-}CO\cdot CH_3}_{SH} + HS\cdot CoA \rightleftharpoons E\text{-}Lip\langle^{SH}_{SH} + CH_3\cdot CO\text{-}S\cdot CoA$$

---

[1] Although pyruvate dehydrogenase is the recommended trivial name for this enzyme, 'pyruvate decarboxylase' is used to avoid confusion with the complex of the three enzymes referred to as the pyruvate dehydrogenase complex.

Dihydrolipoate dehydrogenase (EC 1.6.4.3)

$$\text{E-Lip}\!\!<\!\!{}^{\text{SH}}_{\text{SH}} + \text{NAD} \longrightarrow \text{E-Lip}\!\!<\!\!{}^{\text{S}}_{\text{S}} + \text{NADH}_2$$

The decarboxylase reaction is irreversible while the second and third reactions are reversible. The complex can be dissociated under suitable conditions; for example the bacterial enzyme treated at pH 7.5 will release the pyruvate decarboxylase and treatment with urea will dissociate the lipoate acetyltransferase from the dihydrolipoate dehydrogenase. The detailed structure of the complex in bacteria differs from that in mitochondria of higher organisms. In either form there is uncertainty about the number of subunits making up the complex.

The bacterial enzyme complex has a molecular weight of about 4×10^6. The core of the complex is composed of 24 (or 16) lipoate acetyltransferase polypeptides (MW $= 65\,000$–$70\,000$ or $80\,000$) each possessing two or three lipoic acid residues. To this core are attached the decarboxylase polypeptides (MW $= 90\,000$) as dimers and the dehydrogenase subunits (12 or 16 polypeptides, MW $= 56\,000$). The last two enzymes probably exist as dimers of identical subunits. The complex can be seen under the electron microscope (Fig. 3.7). Three-dimensional models of the enzyme complex can be built mainly on the basis of the electron micrographs of the complex and the self-aggregating lipoate acetyltransferase subunits, together with analysis of the subunit composition (see Fig. 3.7). However, agreement on the detailed form of the model has not been achieved. (Hale and Perham, 1979; Reed *et al.*, 1975; Vogel, 1977).

The variation in the experimental values for ratios of subunits has suggested the possibility that the number of subunits in the complex may vary in the native enzyme, particularly since complexes with reduced numbers of subunits can retain enzyme activity.

The mammalian mitochondrial enzyme is much larger (MW $= 10^7$) and may have a structure composed of a core of 60 lipoate acetyltransferase subunits (MW $= 76\,100$) to which 30 subunits of the decarboxylase and perhaps 20–24 subunits of the dehydrogenase (MW $= 58\,000$) are attached. The decarboxylase has a structure composed of four polypeptides of two types, i.e. $\alpha_2\beta_2$ (MW $\alpha = 40\,000$, $\beta = 35\,00$) (Sugden and Randle, 1978).

Fig. 3.7 The pyruvate dehydrogenase complex.
(a) An electron micrograph of the complex isolated from *Escherichia coli*, $\times 254\,000$.
(b) A model of the complex showing the central cubical core of lipoate acetyltransferase subunits (white spheres) to which are attached the decarboxylase (black spheres) and the dihydrolipoate dehydrogenase (grey spheres). The model is based on a composition of 24 lipoate acetyltransferase polypeptides (MW $= 65\,000$), 12 pyruvate decarboxylase dimers (monomer MW $= 96\,000$) and 6 dihydrolipoate dehydrogenase dimers (monomer MW $= 56\,000$).
(c) An electron micrograph of a preparation of lipoate acetyltransferase showing the formation of cubical aggregates, $\times 254\,000$.
(Courtesy of Dr Lester Reed, University of Texas, U.S.A.). Reprinted in part with permission from *Accounts of Chemical Research* **7**, 40–46 copyright American Chemical Society.

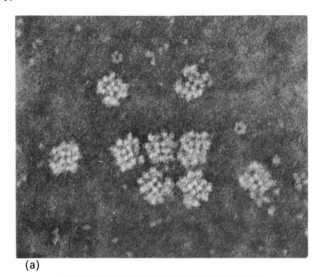

(a)

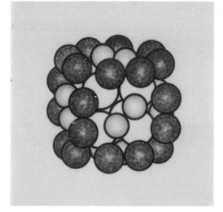

(b)

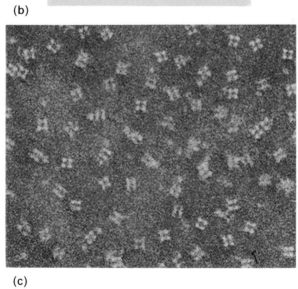

(c)

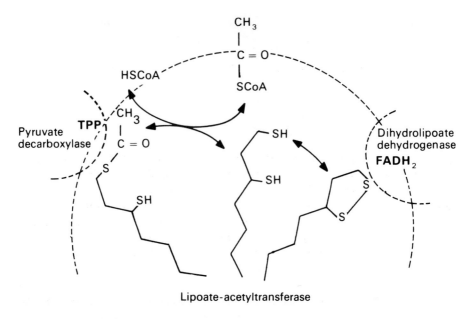

Fig. 3.8 Diagrammatic representation of proposed mechanism for lipoic acid, which behaves as a flexible arm interacting with both the decarboxylase and the dehydrogenase.

The reaction scheme (Fig. 3.6) shows that the lipoic acid residue has to react in turn with the decarboxylase, acetyl coenzyme A and the dehydrogenase. This has led to the proposal of the 'swinging arm' model for the lipoyl residue (Fig. 3.8).

(d) Regulation

The pyruvate dehydrogenase reaction irreversibly commits the products of carbohydrate metabolism to acetyl coenzyme A. In mammalian systems pyruvate cannot be synthesised from acetyl coenzyme A, although in plants and bacteria such pathways exist. It is not surprising to find that pyruvate metabolism by the dehydrogenase is subject to extensive regulation, particularly in mammalian systems.

The activity of the bacterial enzyme complex is regulated by a number of effectors. Reduced NAD strongly inhibits the dihydrolipoate dehydrogenase and acetyl coenzyme A weakly inhibits the decarboxylase; these inhibitory effects are reversed by oxidised NAD and coenzyme A respectively. Thus, under conditions where the products of the reaction are not being adequately metabolised, the activity of the enzyme is decreased. The degree of phosphorylation of the adenine nucleotides (the energy charge) also affects activity, high ATP concentrations being inhibitory.

The primary means of regulation of the enzyme complex in mammalian mitochondria is by a phosphorylation–dephosphorylation system first demonstrated by Linn *et al.* (1969)

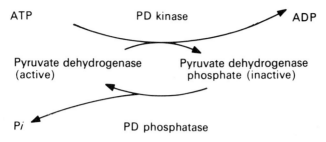

The pyruvate dehydrogenase kinase (PD kinase) is bound to the complex while the phosphatase appears to be free in the mitochondrial matrix. The α-subunits of the decarboxylase are phosphorylated by the kinase, one phosphorylation per tetramer being adequate to inactivate the enzyme. Further phosphorylations can occur up to three and these appear to inhibit reactivation by the phosphatase (Sugden and Randle, 1978). Regulation of the activation/inactivation system is mainly through the kinase, the inactivating enzyme, which is activated by high ratios of acetyl CoA/CoA, $NADH_2/NAD$ and ATP/ADP but inhibited by pyruvate, Ca^{2+}, Mg^{2+} or thiamin pyrophosphate. The phosphatase is activated by Ca^{2+}. 3,5-Cyclic AMP, which is a powerful regulator of enzyme phosphorylations in the cytosol, is apparently without effect on the mitochondrial enzyme system. The mammalian dehydrogenase is also regulated by the same effectors as the bacterial enzyme.

A number of factors are known to lead to changes in the degree of activation of the dehydrogenase complex. In particular insulin appears to activate and the oxidation of fatty acids and ketone bodies to inactivate, but these effects are probably indirect. For example, fatty acid oxidation in a tissue such as heart muscle will tend to increase the ratios of acetyl CoA/CoA, $NADH_2/NAD$ and ATP/ADP. This will activate the kinase and inactivate the dehydrogenase complex, thus reducing the oxidation of pyruvate in the presence of an alternative energy source, fatty acids.

The regulation of the pyruvate dehydrogenase in plants is probably broadly similar to that in animals (Randall *et al.*, 1977).

3.3 Citric acid cycle reactions (Fig. 3.9, see also Table 4.2)

(a) Tricarboxylate metabolism

Acetyl coenzyme A is incorporated into citrate by the citrate condensing enzyme, citrate synthase, an enzyme found exclusively in the mitochondria in animals although it may also be found in glyoxysomes or peroxisomes in plants.

$$\text{acetyl CoA} + \text{oxaloacetate} + H_2O \longrightarrow \text{citrate} + \text{CoA}$$

The equilibrium is strongly in favour of citrate formation; although reversible *in vitro*, the reaction is essentially irreversible *in vivo*. Activity is inhibited by ATP except in Gram-negative bacteria, where NADH rather than ATP inhibits. In micro-organisms capable of anaerobic growth, oxoglutarate is also inhibitory. An extramitochondrial enzyme, ATP-citrate lyase (EC 4.1.3.8), provides a source of acetyl CoA in the cytoplasm for lipid and other syntheses, since citrate

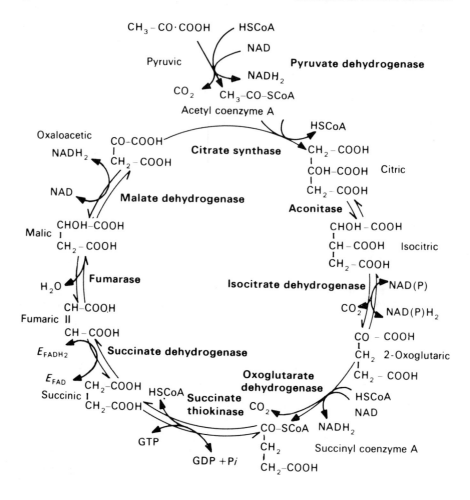

Fig. 3.9 The citric acid cycle (the intermediates are shown as free acids)

but not acetyl coenzyme A can permeate the mitochondrial inner membrane.

$$\text{citrate} + \text{CoA} + \text{ATP} \longrightarrow \text{acetyl CoA} + \text{oxaloacetate} + \text{ADP} + \text{P}i$$

Some bacteria possess a further enzyme which is important for growth on citrate, citrate lyase (EC 4.1.3.6)

$$\text{citrate} \longrightarrow \text{acetate} + \text{oxaloacetate}$$

Citrate is converted to isocitrate by aconitase, a ferrous iron-containing protein found both inside and outside the mitochondrion. The reaction is fully reversible although, at equilibrium, citrate is favoured. *cis*-Aconitate, which may also act as a substrate, is not normally a free intermediate in the conversion of citrate to isocitrate. Thus the enzyme is both an isomerase interconverting citrate and isocitrate and a hydratase in its ability to hydrate *cis*-aconitate to citrate and isocitrate. The asymmetric metabolism of citrate by aconitase has already been noted.

The interconversion of isocitrate to 2-oxoglutarate is catalysed by isocitrate dehydrogenases specific for either NAD or NADP as cofactor.

$$\text{isocitrate} + \text{NAD(P)} = \text{2-oxoglutarate} + CO_2 + \text{NAD(P)H}_2$$

The NAD-specific enzyme appears to be intramitochondrial while the NADP-specific is mainly extramitochondrial although significant amounts are found in some mitochondria. The reaction itself involves an oxidation and a decarboxylation. The NADP-dependent enzyme has been shown to catalyse the reaction in two stages:

$$\text{isocitrate} + \text{NADP} = \text{oxalosuccinate} + \text{NADPH}_2$$

$$\text{oxalosuccinate} \xrightarrow{\text{Mn}^{2+} \text{ or Mg}^{2+}} \text{2-oxoglutarate} + CO_2$$

Under normal conditions in the presence of NADP and cations, oxalosuccinate does not dissociate from the enzyme and thus is not a free intermediate. However, omission of the divalent cation from the reaction mixture results in accumulation of oxalosuccinate which may be converted to oxoglutarate on addition of the cation. The NAD-dependent mitochondrial enzyme from animal tissues requires ADP as an activator and is relatively inactive in its absence; other nucleotides are without effect. An NAD-linked enzyme from bacteria is insensitive to all adenine nucleotides, while a yeast enzyme is activated by AMP.

(b) Conversion of oxoglutarate to succinate

Several enzymic steps are involved in the formation of succinate from oxoglutarate. The enzyme oxoglutarate dehydrogenase is a complex (MW about 2.4 to 2.7×10^6) similar to the pyruvic dehydrogenase complex although less well studied: the product of these reactions is succinyl coenzyme A. (see Fig. 3.10 and compare with Fig. 3.6).

$$\text{2-oxoglutarate} + \text{CoA} + \text{NAD} \longrightarrow \text{succinyl CoA} + \text{NADH}_2 + CO_2$$

The overall reaction is irreversible. The complex is composed of three separate enzymes, the lipoate-succinyltransferase, 24 subunits of which make up the core of the complex and to which are attached the oxoglutarate decarboxylase and the dihydrolipoate dehydrogenase. The last enzyme is similar to, or identical with, the enzyme found in the pyruvate dehydrogenase complex. So far evidence is lacking for a phosphorylation of the enzyme resembling that found with pyruvate dehydrogenase. However, enzyme activity is inhibited by high ratios of succinyl CoA/CoA and NADH/NAD; AMP activates.

Two enzymes catalyse the conversion of succinyl coenzyme A to succinate: *succinate thiokinase*

$$\text{succinyl CoA} + \text{GDP} + \text{P}i \rightleftharpoons \text{succinate} + \text{GTP} + \text{CoA}$$

and *succinyl CoA–acetoacetate CoA transferase*

$$\text{succinyl CoA} + \text{acetoacetate} \rightleftharpoons \text{succinate} + \text{acetoacetyl CoA}$$

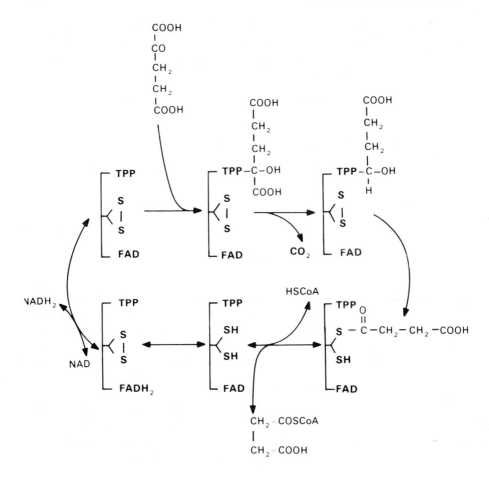

Fig. 3.10 The oxoglutarate dehydrogenase reaction (cf. Fig. 3.6)

In liver most of the succinyl coenzyme A is metabolised by the thiokinase and only very little by the transferase. In skeletal muscle to a limited extent, but particularly in kidney and heart, the transferase activity which is involved in ketone body metabolism (see sect. 3.12(b)), is of considerable importance. In most eukaryotes both enzymes are concentrated in the mitochondria, although the transferase may occur in the cytoplasm.

Succinate thiokinase catalyses the only phosphorylation step in the citric acid cycle itself. Acyl coenzyme A compounds have a high standard free energy of hydrolysis. In mitochondrial acetyl coenzyme A metabolism, this is used to favour citrate synthesis, while in succinyl coenzyme A metabolism it is used for a substrate level phosphorylation to form GTP. While animal cells possess the GDP-dependent enzyme, plant mitochondria and *Escherichia coli* possess an ATP-dependent enzyme. Photosynthetic bacteria possess a succinate thiokinase accepting ADP, GDP or inosine diphosphate.

The reaction involves a phosphorylated histidine residue on the enzyme protein as an intermediate in the reaction. Incubation with either ^{32}P-labelled GTP and Mg^{2+} or ^{32}Pi, succinyl coenzyme A and Mg^{2+} results in phosphorylation of the enzyme. The reaction may occur as follows:

$$succinyl\ CoA + Enz\text{-}his \rightleftharpoons Enz\text{-}(his, CoA) + succinate$$

$$Enz\text{-}(his, CoA) + Pi \rightleftharpoons Enz\text{-}his\text{-}(P) + CoA$$

$$Enz\text{-}his(P) + GDP \rightleftharpoons Enz\text{-}his + GTP$$

The GTP can react with ADP in the mitochondrial nucleoside diphosphokinase reaction.

$$GTP + ADP \rightleftharpoons GDP + ATP$$

Mammalian mitochondria also contain a GTP–AMP phosphotransferase

$$GTP + AMP \rightleftharpoons ADP + GDP$$

(c) Dicarboxylate metabolism

The succinate dehydrogenase which oxidises succinate to fumarate is found exclusively in the mitochondrion in eukaryotes where it is firmly membrane-bound. In bacteria, the enzyme is usually bound to the plasma membrane. The reversible oxidation proceeds by transfer of two hydrogens to the covalently bound flavin (FAD), which is the prosthetic group of the enzyme.

$$succinate + E_{FAD} = fumarate + E_{FADH_2}$$

Reoxidation is through the respiratory chain and involves iron–sulphur centres which are also a part of the enzyme (see Ch. 9). The enzyme is slowly activated by substrate (succinate, fumarate) or the competitive inhibitor malonate. Phosphate, ATP and reduced ubiquinone-10 also activate the enzyme, while low concentrations of oxaloacetate are strongly inhibitory.

The reversible asymmetric hydration of fumarate to form L-malate (Fig. 3.11) is catalysed by fumarase, an enzyme found both inside and outside the mitochondrion.

$$fumarate + H_2O = \text{L-malate}$$

The affinity of the enzyme for fumarate is decreased by ATP. Thus if fumarate is present at less than saturating level, ATP acts as an inhibitor of fumarase.

Oxaloacetate is formed from malate by an NAD-dependent enzyme, malate dehydrogenase

$$malate + NAD = oxaloacetate + NADH_2$$

The reaction is reversible, although the equilibrium is strongly in favour of malate. The rapid removal of oxaloacetate drives the reaction in the forward direction. The enzyme is found both inside and outside the mitochondrion, but the properties of the intramitochondrial enzyme are different from those of the extramitochondrial one.

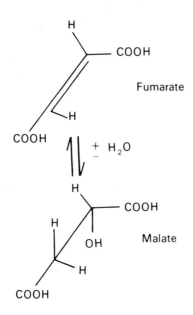

Fig. 3.11 The fumarase reaction.

3.4   The glyoxylate cycle

In mammalian systems, both carbons of the acetyl group are oxidised to CO_2 by the citric acid cycle; hence acetate cannot give rise to a net synthesis of oxaloacetate or products of oxaloacetate metabolism such as phosphoenol-pyruvate and carbohydrate. However, many bacteria will grow on acetate as a sole source of organic carbon. Plant tissues, particularly developing seedlings such as those of the castor oil plant, can use acetyl coenzyme A for biosynthesis of carbohydrate. Both plants and micro-organisms possess a modification of the citric acid cycle, the glyoxylate cycle (Fig. 3.12; Kornberg and Krebs, 1957). The utilisation of ^{14}C-labelled acetate was examined in aerobic and facultatively anaerobic bacteria. After a short incubation, two metabolic intermediates were significantly labelled, citrate (as expected) and malate (Kornberg and Elsden, 1961) although, succinate, a precursor of malate was not appreciably labelled.

Two enzymes previously described in bacteria are responsible for the synthesis of malate from labelled acetate:

Isocitrate lyase (EC 4.1.3.1)

isocitrate $\longrightarrow$ succinate + glyoxylate

Malate synthase (EC 4.1.3.2)

acetyl CoA + glyoxylate + H_2O $\longrightarrow$ malate + CoA

Cell extracts incubated with isocitrate form either glyoxylate and succinate or, if acetate, ATP and coenzyme A are added, malate and succinate. These and other lines of evidence led to the formulation of the glyoxylate cycle as the means of obtaining carbohydrate from acetate.

The main features of this modification of the citric acid cycle are firstly that the decarboxylation steps are bypassed (isocitrate and oxoglutarate dehydrogenases), secondly that two molecules of acetyl coenzyme A are fed into the cycle in each revolution and thirdly that two of the four oxidative steps (isocitrate and oxoglutarate dehydrogenases) are bypassed and one (succinate dehydrogenase) partially so. The cycle may be summarised:

$$2 \text{ acetyl CoA} + 2\text{NAD} + E_{FAD} + 3H_2O \longrightarrow$$
$$\text{oxaloacctate} + 2\text{NADH}_2 + E_{FADH_2} + 2\text{CoA}$$

In plants the key enzymes of the glyoxylate cycle, malate synthase and isocitrate

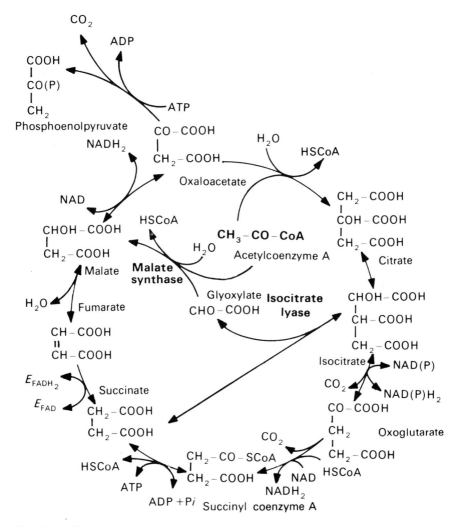

Fig. 3.12 The glyoxylate cycle

lyase, are found not in the mitochondria but in glyoxysomes or peroxisomes. These particles may also contain a malate dehydrogenase but do not possess complete citric acid cycle activity, nor do they have a respiratory chain.

3.5   Carboxylation reactions

In mammalian systems the reactions discussed so far envisage a cycle which requires one molecule of oxaloacetate for each acetyl coenzyme A oxidised and which regenerates that oxaloacetate. Clearly reactions are needed which regulate the level of the C_4 dicarboxylic acids by removal of excess intermediates when they are derived from another source (i.e. amino acid catabolism) or by addition of intermediates when they have been removed from the cycle for biosynthetic pathways (e.g. porphyrin synthesis). The concentrations of the cycle intermediates are maintained by carboxylation and decarboxylation reactions.

The importance of CO_2 for cellular metabolism and for biosynthesis in heterotrophic organisms first became apparent from a study of fermentations in propionic acid bacteria. These organisms were able to form succinate and propionate from carbonate and glycerol (Fig. 3.13). If labelled carbonate was used, the label could be found in the carboxyl carbons of the succinate. Wood and Werkman (1940) suggested that pyruvate was formed from glycerol and that this was carboxylated to oxaloacetate and reduced to succinate. In the same year evidence was obtained for CO_2 fixation in pigeon liver and the use of isotopes confirmed that pyruvate and CO_2 could give rise to oxaloacetate by a 'Wood–Werkman reaction'.

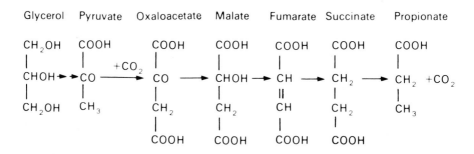

Fig. 3.13 Outline pathway for propionate synthesis in bacteria.

Several enzymes have now been described which interconvert either pyruvate or phosphoenolpyruvate with oxaloacetate or malate (Fig. 3.14).

The malic enzyme or NADP-dependent malic dehydrogenase (decarboxylating) was described by Ochoa *et al.* (1947)

$$malate + NADP = pyruvate + NADPH_2 + CO_2$$

This type of enzyme is found extramitochondrially in most animal tissues although some malic enzyme activity is found in brain mitochondria. In plants the enzyme is mitochondrial where malate is oxidised by two complementary

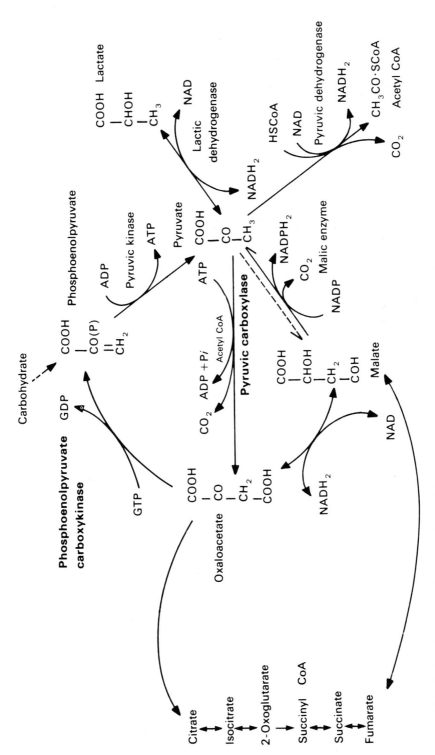

Fig. 3.14 Carboxylation reactions associated with pyruvate and the citric acid cycle.

routes; one forms pyruvate and then acetyl coenzyme A while the other forms oxaloacetate, which together with the acetyl coenzyme A, gives rise to citrate (see Palmer, 1979). The malic enzyme also occurs in micro-organisms. The K_m for CO_2 is relatively high and the enzyme probably functions primarily as a source of reduced NADP for lipid syntheses. The enzyme is also able to decarboxylate oxaloacetate. Two other malic enzymes are known, both NAD-dependent but differing in their ability to decarboxylate oxaloacetate. The enzyme unable to decarboxylate oxaloacetate is found in mammalian mitochondria.

The phosphoenolpyruvate carboxykinase has a variable location in mammalian tissues, being found wholly or partly present or absent in mitochondria

$$phosphoenolpyruvate + CO_2 + GDP = oxaloacetate + GTP$$

In guinea pig, rabbit, sheep, beef and human liver, the enzyme is found both in mitochondria and in the cytosol, in rats and mice the enzyme is in the cytosol while in pigeon it is mitochondrial. The K_m for HCO_3^- is high and probably above the concentration found intracellularly. The bacterial enzyme is ATP-dependent. The enzyme functions as a means of decarboxylating oxaloacetate.

In contrast, the pyruvate carboxylase has a low K_m for CO_2 and is responsible for the synthesis of oxaloacetate in mitochondria from pyruvate.

$$H_2O + ATP + CO_2 + pyruvate \xrightarrow{\text{acetyl coenzyme A}} oxaloacetate$$
$$+ ADP + Pi$$

The enzyme is probably exclusively mitochondrial. It contains biotin as a prosthetic group which takes part in the carboxylation by forming an enzyme—biotin–CO_2 complex at the expense of ATP hydrolysis. The reaction is magnesium-dependent and the enzyme requires acetyl coenzyme A for activity (see Utter *et al.*, 1975).

In plants and also in autotrophic bacteria, oxaloacetate is formed not from pyruvate but from phosphoenolpyruvate by the phosphoenolpyruvate carboxylase (EC 4.1.1.31).

$$phosphoenolpyruvate + CO_2 + H_2O \longrightarrow oxaloacetate + Pi$$

The reaction is irreversible.

In mitochondria there are two main functions for these enzymes. Firstly, they serve to increase the level of oxaloacetate when this is required to act as an acetyl coenzyme A acceptor. The pyruvate carboxylase, with a low K_m for CO_2 and a requirement for acetyl coenzyme A, carries out this function in animal mitochondria, while the phosphoenolpyruvate carboxylase has a similar role in plants. In animals, an increase in mitochondrial acetyl coenzyme A will activate the carboxylase, thus promoting the oxidation of the acetyl coenzyme A. Secondly, they serve to lower the level of cycle intermediates by formation of phosphoenolpyruvate catalysed by the carboxykinase.

3.6   Regulation of the citric acid cycle

The oxidation of acetyl coenzyme A by the citric acid cycle may be represented as

1. $CH_3-CO\cdot CoA + 3H_2O + 3NAD + E_{FAD} + GDP + Pi$
$$\longrightarrow 2CO_2 + E_{FADH_2} + 3NADH_2 + CoA + GTP$$

2a. $NADH_2 + \frac{1}{2}O_2 + 3ADP + 3P_i \longrightarrow NAD + 4H_2O + 3ATP$

2b. $E_{FADH_2} + \frac{1}{2}O_2 + 2ADP + 2P_i \longrightarrow E_{FAD} + 3H_2O + 2ATP$

The metabolic function of the cycle can be seen ultimately as the synthesis of ATP which can then be used for other cellular purposes. If one asks how the rate of cycling is regulated, there are clearly several possibilities which include the following.

(a) The supply of oxidised NAD, i.e. the $NADH_2/NAD$ ratio. Failure to reoxidise NAD at an adequate rate will result in a limitation of the rate of the dehydrogenase reactions.

(b) The supply of ADP necessary for reactions 2a and 2b above which could regulate the rate of $NADH_2$ and succinate oxidation.

(c) The effect of adenine nucleotides on some cycle reactions, for example the requirement of ADP by the isocitrate dehydrogenase or the inhibition of citrate synthase by ATP. In most of these systems, the effect of one adenine nucleotide is reduced in the presence of another so that it is the ATP/ADP ratio or the energy charge which is important (see sect. 1.3).

(d) The regulatory effect of the $NADH_2/NAD$ ratio and the succinyl CoA/CoA ratio on the oxoglutarate dehydrogenase.

(e) The inhibitory effect of oxaloacetate on the succinate dehydrogenase.

The problem is to decide which of these regulatory systems operate *in vivo*. All of them can be demonstrated *in vitro*. Although the consumption of oxygen is variable, it can be shown in mammalian cells that reduction of the oxygen supply has to be extreme before the $NADH_2/NAD$ ratio is significantly affected. It therefore seems unlikely that this ratio is important in most higher animal tissues, although it could be significant in micro-organisms.

In mammalian cells it has been shown that an equilibrium exists between the redox state of the carriers of the respiratory chain and the ATP/ADP ratio (see sect. 9.11). However, when rates of oxidation in tissues are varied, it seems that the enzyme reactions which may be expected to reflect the $NADH_2/NAD$ ratio, such as the malate/oxaloacetate ratio or the hydroxybutyrate/acetoacetate ratio, do not change significantly. Thus, although the rate of cycle activity must increase when the rate of oxygen utilisation rises, this increase does not appear to be mediated through an effect of the pyridine nucleotide ratio on the rates of dehydrogenase reactions.

The argument for adenine nucleotides playing a key regulatory role is much stronger. In coupled mammalian mitochondria, the ATP/ADP ratio is in equilibrium with the $NADH_2/NAD$ ratio and transition from state 4 (ADP-limited, high ATP/ADP ratio) to state 3 (low ATP/ADP ratio) results in NADH oxidation. Uncoupling mitochondria substantially lowers the ATP/ADP ratio and also rapidly increases the rate of acetyl coenzyme A oxidation, i.e. citric acid cycle activity. The ratios of adenine nucleotides are also known to have a strong regulatory effect on cycle enzymes, high ATP/ADP ratios being inhibitory. In insect flight muscle mitochondria, transition from state 4 to state 3 results in NAD reduction which indicates increased cycle activity. In this case it seems that the ATP/ADP ratio must be acting primarily in the citric acid cycle (Johnson and Hansford, 1977).

The possibility that citrate synthase plays a key role has been explored. Enzyme activity is regulated by several factors, especially by adenine nucleotides or by pyridine nucleotides in some micro-organisms (Weitzman and Danson, 1970). The supply of acetyl coenzyme A is regulated by the effects of the acetyl CoA/CoA ratio on fatty acid oxidation. Several factors also control the conversion of pyruvate to acetyl coenzyme A. The levels of oxaloacetate are low (see Table 3.1) and hence the rate of reaction will be sensitive to the oxaloacetate concentration. A body of experimental evidence supports the view that oxalo-acetate may regulate its own formation from succinate by feedback inhibition, while its formation from pyruvate is controlled by acetyl coenzyme A levels.

Table 3.1 Approximate concentrations of citric acid cycle intermediates in mammalian tissues

	nMol/g fresh weight
Citrate	400
Isocitrate	20
Oxoglutarate	130
Succinate	700
Fumarate	70
Malate	300
Oxaloacetate	1

In discussing the regulation of the citric acid cycle, a wide variety of conditions and cells need to be taken into account. It is probable that all of the factors discussed above could play a part in the regulatory process and that in any given cell under its normal environmental conditions, more than one of the factors will be involved.

Regulation of the glyoxylate pathway in bacteria is primarily at the branchpoint, i.e. isocitrate metabolism. When acetate is the sole carbon source, the glyoxylate/citric acid cycle functions both to provide net synthesis of oxaloacetate and to oxidise substrate resulting in oxidative phosphorylation. The former role is promoted by isocitrate lyase activity, the latter by isocitrate dehydrogenase. A coarse form of regulation is provided by induction of the lyase. When bacteria are grown on media, such as glucose and salts or nutrient broth, enzyme synthesis is repressed; when they are grown on acetate, the enzyme is synthesised. Both succinate and oxaloacetate are inhibitors of lyase activity, so that the C_4 dicarboxylic acids will regulate their own formation, thus ensuring a balance between the oxidative and biosynthetic activities of the cycle.

3.7   Gluconeogenesis

Under conditions where lactate has accumulated in mammalian tissues, it is metabolised primarily in the liver. Part of this lactate is oxidised through the citric acid cycle providing the necessary cofactors for glycogen synthesis. Most of the lactate is synthesised to carbohydrate by the glycolytic pathway. Since pyruvate kinase has an equilibrium strongly in favour of pyruvate, the pyruvate carboxylase and phosphoenolpyruvate carboxykinase reactions serve as an alternative route from pyruvate to phosphoenolpyruvate (Fig. 3.14). This pathway utilises both an ATP-dependent and a GTP-dependent reaction, thus rendering the formation of the phosphoenolpyruvate energetically favourable.

The carboxykinase reaction also serves as a means of removing intermediates from the citric acid cycle for glycogen synthesis.

3.8   Mitochondrial nitrogen metabolism

In animals, unwanted nitrogen is excreted usually as ammonia or urea or uric acid, the animals being referred to as ammoniotelic, ureotelic or uricotelic respectively. In general, birds and some reptiles are uricotelic, mammals and amphibians are ureotelic while a number of the lower groups are ammoniotelic.

Most amino acids are transaminated in the cytosol with oxoglutarate to form glutamate which enters the mitochondrion. The carbon skeletons of the amino acids give rise to compounds which are intermediates in the citric acid cycle or related to cycle intermediates (Fig. 3.15). While most of the pathways for amino acid metabolism are extramitochondrial, proline can be metabolised to glutamate by mitochondria.

Glutamate is deaminated in the mitochondria by glutamate dehydrogenase.

$$\text{glutamate} + NAD(P) + H_2O = \text{oxoglutarate} + NH_3 + NAD(P)H_2$$

The enzyme accepts NAD or NADP, is inhibited by GTP and activated by ADP and leucine. In ureotelic species, ammonia is converted in the liver to urea by the urea cycle which is partly mitochondrial and partly cytoplasmic (Fig. 3.16). In uricotelic species, the ammonia is converted to glutamine by a mitochondrial glutamine synthetase.

$$\text{glutamate} + NH_3 + ATP = \text{glutamine} + ADP + Pi$$

The glutamine is used for uric acid synthesis in the cytosol.

The established view of glutamate dehydrogenase described above, has been challenged by McGivan and Chappell (1975) who observed a decrease in urea synthesis on treatment of liver preparations with leucine, an activator of the dehydrogenase. They have suggested that the enzyme functions primarily in nitrogen storage.

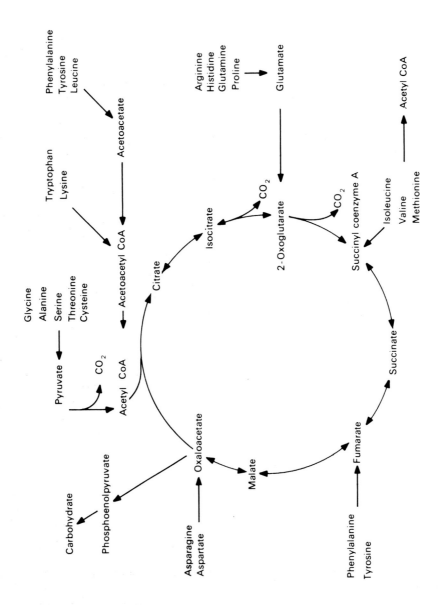

Fig. 3.15 Metabolism of the carbon skeletons of the amino acids through the citric acid cycle.

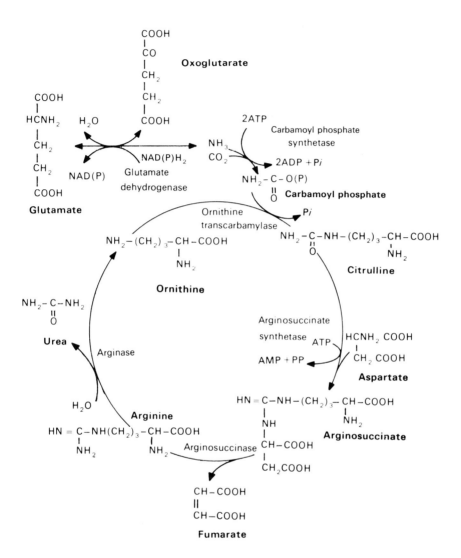

Fig. 3.16 The urea cycle.

Hence ammonia for urea synthesis might come from another source, possibly an extramitochondrial one.

In kidney cortex, glutamine or glutamate is used for ammonia formation particularly under conditions of starvation, which lead to substantial ketone body formation (see sect. 3.12(b)) and to excretion of acetoacetate and hydroxybutyrate. The excretion of ammonium ions serves to balance these anions in the urine. Glutamate and some of the glutamine metabolised within the kidney cortex mitochondria leads to gluconeogenesis (Fig. 3.17).

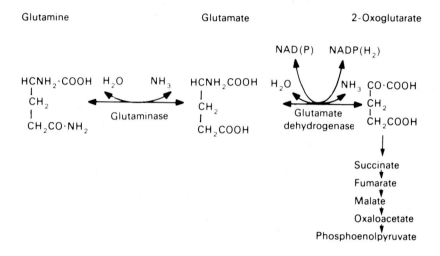

Fig. 3.17 Glutamine metabolism in kidney.

The synthesis of glutamate as a means of converting ammonia into amino nitrogen for amino acid synthesis, appears to be the major function of the enzyme in plants, where it occurs mostly in the mitochondria, sometimes in the cytosol and also in the chloroplasts.

The fungal enzyme is specific for NAD and is extramitochondrial. Transaminases are found both in mitochondria and in the cytosol. The main mitochondrial transaminase is the aspartate transaminase.

3.9   Fatty acid oxidation

Fatty acids are also major precursors of acetyl coenzyme A by β-oxidation and are preferred substrates in some tissues. The β-oxidation pathway for degradation of fatty acids was first proposed by Franz Knoop in a classic paper published in 1905. He suggested that the β-carbon (C-3) was more readily oxidised than the α-carbon contrary to previous views. Knoop's theory was not readily acceptable until Dakin (1909) showed that the β-carbon of fatty acids was more readily oxidised by purely chemical means using hydrogen peroxide.

Knoop's theory of β-oxidation was based on experiments in which phenyl derivatives of fatty acids were fed to dogs and the products of metabolism were isolated as the glycyl derivatives in the urine (Fig. 3.18). The phenyl group itself was not attacked and therefore served as a means of labelling the fatty acid. These experiments were interpreted as demonstrating the oxidation of two C_2 fragments from the fatty acid. Longer chain fatty acids were later shown to be oxidised similarly.

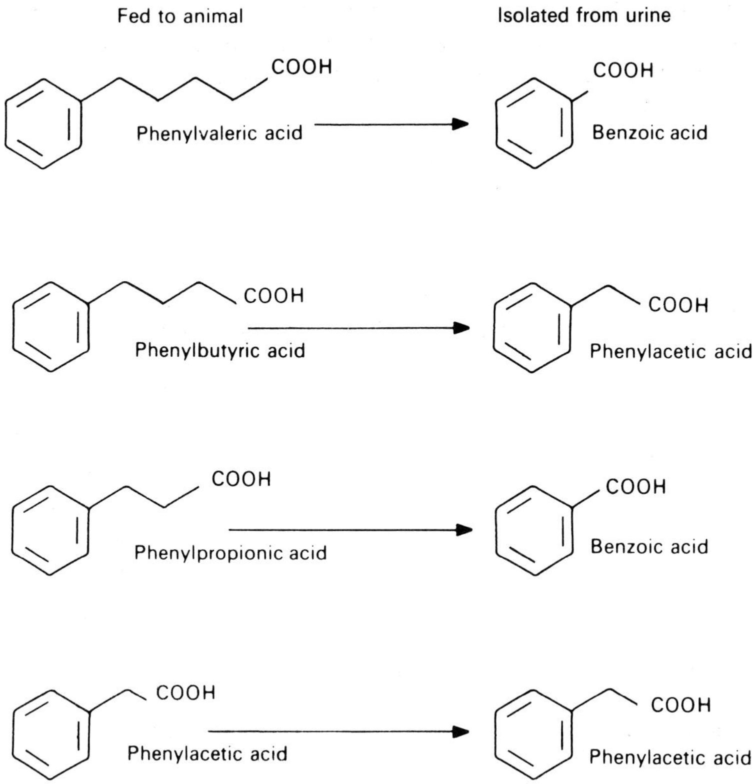

Fed to animal   Isolated from urine

Fig. 3.18 Metabolism of phenyl derivatives of fatty acids by dogs.

Knoop's β-oxidation pathway, as developed by Dakin (Fig. 3.19), was supported by the isolation of some of the expected intermediates such as phenyl-3-oxopropionic acid. Further significant progress was not made until 1939, when Leloir and Munoz were able to demonstrate fatty acid oxidation in a cell-free preparation from rat liver. Their experiments showed that fatty acid oxidation was stimulated by addition of fumarate which also reduced the formation of ketone bodies. Ketone bodies (acetoacetate and hydroxybutyrate) are products of pyruvate and fatty acid metabolism, and are formed in the livers of many vertebrates. They are transferred mainly to the peripheral tissues for oxidation. The role of fumarate here is to supply oxaloacetate to promote the oxidation of the main product of fatty acid metabolism, acetyl coenzyme A (see sect. 3.1 above and Fig. 3.20).

The relationship between ketone body formation and β-oxidation remained unclear. However, in 1944 Weinhouse and coworkers showed that if octanoic acid labelled in C-1 with ^{13}C was metabolised by liver slices, the

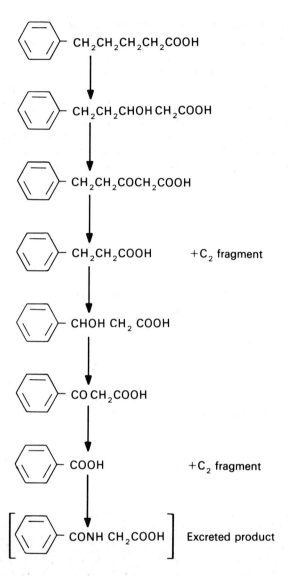

Fig. 3.19 The β-oxidation theory of Knoop and Dakin for the oxidation of phenyl-valeric acid.

resultant acetoacetate was labelled in both the carboxyl and carbonyl carbons. Contrary to other views then held, this established a sequence:

$$CH_3(CH_2)_6C^* OOH \longrightarrow \underset{(C_2 \text{ fragment})}{C-C^*} \longrightarrow CH_3C^*OCH_2C^*OOH$$

Such labelling could only have arisen by β-oxidation to a C_2 fragment and subsequent condensation. Further, acetoacetate itself cannot be an intermediate in fatty acid breakdown.

Other labelling experiments showed that the carbon atoms in fatty acids were metabolised to citric acid cycle intermediates. For example, with a particulate liver preparation oxidising octanoate, the addition of fumarate and malonate decreased acetoacetate formation but increased the formation of citrate, oxoglutarate and succinate. These and other experiments confirmed the notion that the oxidation of fatty acids involved the citric acid cycle. Until the discovery of coenzyme A by Lipmann, the nature of the C_2 fragment was obscure. In a lecture to the Harvey Society in 1948, Lipmann suggested that acetyl coenzyme A might be the elusive C_2 fragment and this was subsequently confirmed, thus demonstrating the relationship of β-oxidation with the citric acid cycle (Fig. 3.20).

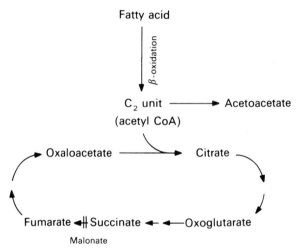

Fig. 3.20 Relationship of fatty acid oxidation to ketone body formation (acetoacetate) and the citric acid cycle in liver.

β-Oxidation was shown to occur in mitochondria by Kennedy and Lehninger (1949). More recently substantial fatty acid oxidative activity was found in peroxisomes of rat liver (Krahling *et al.*, 1978). Drysdale and Lardy (1953) obtained extracts from rat liver mitochondria which proved a rich source of enzymes for β-oxidation and provided a means of identifying the cofactor requirements. These experiments stimulated the isolation and study of the individual enzymes involved in fatty acid oxidation (see Lynen and Ochoa, 1953; Green, 1954). Fatty acid oxidation of long-chain fatty acids proceeds firstly by activation and transfer of the fatty acyl group to the mitochondrial matrix, secondly by the cycle of reactions known as β-oxidation and thirdly by the metabolism of the products, mostly acetyl coenzyme A.

3.10 Fatty acid activation

Knoop and Dakin based their theory on the oxidation of free fatty acids. However, it is the coenzyme A ester which is the substrate for fatty acid oxidation. The initial step in fatty acid catabolism is activation to the acyl coenzyme A ester.

$$R \cdot COOH + HSCoA + ATP \longrightarrow R \cdot CO \cdot SCoA + AMP + PP$$

This general reaction is catalysed by at least five enzymes found inside and outside the mithochondrion in a wide range of animals, plants and micro-organisms. The major enzyme is a long-chain acyl coenzyme A synthetase (EC 6.2.1.3) which acts on substrates of chain length about C_6 to C_{22}. In most tissues this enzyme is found attached to the endoplasmic reticulum and also to the outside of the mithochondrion. In rat and human skeletal muscle it is mitochondrial, while in liver it is part microsomal and part mitochondrial.

A medium chain acyl coenzyme A synthetase (EC 6.2.1.2) has been demonstrated in beef liver. This enzyme is widely distributed and will act on fatty acids of chain length $C_4 - C_{11}$; it occurs in the mitochondrial matrix.

At least three short-chain acyl coenzyme A synthetases are now known. The acetyl coenzyme A synthetase (EC 6.2.1.1) utilises acetate, propionate and acrylate as substrates in decreasing order of activity. The enzyme is widely distributed and occurs in the mitochondrial matrix in animal tissues, but in liver it is also found in the cytosol. A butyryl coenzyme A synthetase which also acts on acetoacetate is found in the mitochondrial matrix in mammalian tissues, particularly heart. More recent studies have shown in some plant and mammalian tissues, particularly guinea pig liver, a propionyl coenzyme A synthetase.

All of these enzymes catalyse the reaction above and are ATP-dependent. The mitochondrial matrix in mammals also possesses an acyl coenzyme A synthetase which is GTP-dependent (EC 6.2.1.10):

$$R \cdot COOH + GTP + HSCoA \longrightarrow R \cdot CO \cdot SCoA + GDP + Pi$$

Quantitatively, this enzyme appears to be of minor importance.

The membrane surrounding the inner compartment of the mito-chondrion, the matrix, is relatively impermeable to acyl coenzyme A compounds. However, long-chain fatty acids are esterified outside this membrane and oxidised inside. The transfer of acyl groups across this membrane requires carnitine (Fig. 3.21), which is found both in the mitochondrial matrix and in the cytosol. Carnitine stimulates fatty acid oxidation in liver homogenates and is essential for the oxidation of acyl coenzyme A esters by intact mitochondria. Addition of acyl carnitine to mitochondria results in acylation of intramitochon-drial coenzyme A. In 1955, Friedman and Fraenkel found an enzyme in pigeon liver which utilises carnitine as substrate, carnitine acyltransferase.

$$acyl\ CoA + carnitine = acyl\ carnitine + CoA$$

Two carnitine acyltransferases are bound to the inner mitochondrial membrane, one being tightly bound on the inside and the other loosely bound on the outside. The enzymes are not identical functionally, since the outside transferase has a low K_m for carnitine and the interior enzyme has a high K_m. A transport system carries carnitine across the membrane. External carnitine will exchange with internal carnitine in a $1:1$ manner.

The system for transfer of acyl groups from the cytosol to the mitochondrial matrix shown in Fig. 3.21 will tend to be unidirectional, since the difference in K_m of the acyltransferases will result in formation of acyl carnitine on the outside of the membrane and formation of acyl coenzyme A on the inside.

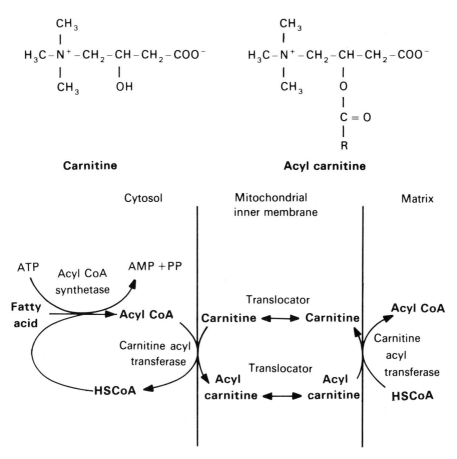

Fig. 3.21 The role of carnitine in fatty acid metabolism by mitochondria.

3.11 The β-oxidation pathway

The overall pathway is shown in Fig. 3.22 (see also Table 4.2). The first step in the oxidation of fatty acyl coenzyme A compounds is 2,3-dehydrogenation catalysed by acyl-coenzyme A dehydrogenases, which are flavoproteins with FAD as prosthetic group:

$$R \cdot CH_2CH_2CO \cdot CoA + E_{FAD} \longrightarrow \textit{trans-}R \cdot CH = CHCO-CoA + E_{FADH_2}$$

In mammalian mitochondria, three enzymes of this type have been isolated differing in their chain-length specificities, a butyryl coenzyme A dehydrogenase and two long chain acyl coenzyme A dehydrogenases with similar but not identical specificities. Although the enzyme resembles the succinate dehydrogenase which catalyses an analogous reaction, there are no iron–sulphur centres or other metals present. An electron-transferring flavo-protein (ETF) also possessing FAD (but no iron–sulphur centres) as a prosthetic group is able to accept reducing equivalents from the acyl coenzyme A dehydrogenase (as well as from other dehydrogenases). Finally the ETF dehydrogenase, an (Fe–S)-containing flavoprotein, oxidises ETF and probably reduces ubiquinone. Thus a chain of three flavoproteins is involved in the oxidation of acyl coenzyme A.

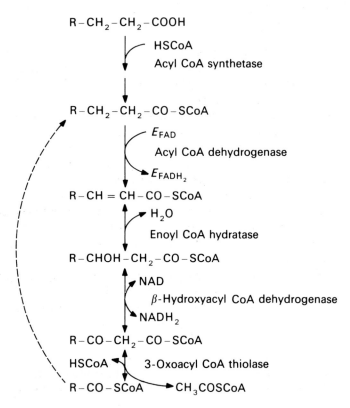

Fig. 3.22 β-Oxidation. The sequence of four reactions is repeated, the product of the thiolase reaction being the substrate for the acyl coenzyme A dehydrogenase.

The 2,3-unsaturated acyl coenzyme A is hydrated by enoyl coenzyme A hydratase also known as crotonase forming a 3-hydroxyacyl coenzyme A.

$$\textit{trans-}RCH = CHCO-SCoA + H_2O = L-RCHOH \cdot CH_2CO-SCoA$$

This is an enzyme with a wide substrate specificity, capable of hydrating *cis-* and *trans-*2,3-unsaturated fatty acid esters. The hydration is stereospecific giving the L-isomer from the *trans-*ester and the D-isomer from the *cis-*ester.

A second oxidation converts the 2-hydroxyacyl coenzyme A to a 3-oxoacyl ester. The enzyme, L-3-hydroxyacyl coenzyme A dehydrogenase is NAD-dependent and stereospecific for the L-isomer. Mammalian mitochondria may contain enzymes for both long- and short-chain substrates.

$$\text{L-R·CHOH·CH}_2\text{CO–CoA} + \text{NAD} = \text{R·CO·CH}_2\text{CO–CoA} + \text{NADH}_2$$

Acetyl coenzyme A is formed from the 3-oxoacyl coenzyme A through a thiolytic cleavage by 3-oxoacyl coenzyme A thiolase. The reaction proceeds by two steps in which an intermediate acyl-S-enzyme complex is formed with a thiol group on the enzyme.

$$\text{R·COCH}_2\text{CO–SCoA} + \text{HS-E} = \text{R·CO–S–E} + \text{CH}_3\text{CO–SCoA}$$
$$\text{R·CO–S–E} + \text{HSCoA} = \text{R·CO–SCoA} + \text{HS–E}$$

Two types of mitochondrial 3-oxoacyl coenzyme A thiolase are found in mammalian systems, one having a wide range chain-length specificity, although its activity towards acetoacetyl coenzyme A is relatively weak, the other being highly specific to acetoacetyl coenzyme A.

The oxidation of naturally occurring unsaturated fatty acids, which normally possess *cis*-double bonds several carbon atoms removed from the carboxyl end, requires additional enzymes. In fatty acids where there are two or more double bonds, these are not usually conjugated but separated by an intervening methylene group. In the case of the common unsaturated fatty acid, oleic acid, with a *cis*-double bond in the 9–10 position, β-oxidation removes pairs of carbons until the double bond is in the 3–4 position. Mitochondria possess an isomerase (*cis*-3, *trans*-2 enoyl coenzyme A isomerase), which converts the 3,4-*cis*-double bond to 2,3-*trans* (Fig. 3.23). Hydration by the hydratase then occurs in the normal way.

A similar sequence applies to linoleic acid oxidation, except that here the additional double bond becomes *cis*-2,3 (see Fig. 3.23) which, as noted above, can be hydrated to the D-hydroxyacyl ester.[2] This is not however, a substrate for the dehydrogenase. The D-3-hydroxyacyl coenzyme A epimerase converts the D- to the L-form:

$$\text{D-3-hydroxyacyl CoA} = \text{L-3-hydroxyacyl CoA}$$

The oxidation of the C_{16} saturated fatty acid, palmitate, can be summarised:

$$\text{C}_{15}\text{H}_{31}\text{COOH} + 8\text{HCoA} + \text{ATP} + 7\text{Fp} + 7\text{NAD} + 7\text{H}_2\text{O} \longrightarrow$$
$$8\text{CH}_3\text{CO–CoA} + \text{AMP} + \text{PP} + 7\text{FpH}_2 + 7\text{NADH}_2$$

The FpH_2 and $NADH_2$ are substrates for reoxidation by the respiratory chain, giving a net gain of 129 molecules ATP/palmitate oxidised (2 ATP equivalents being required for activation).

The presence of intermediates in this pathway has proved difficult to demonstrate in intact mitochondria. Those that have been found are probably

---

[2] Alternatively, the double bond might be reduced directly (Kunau and Dommes, 1978).

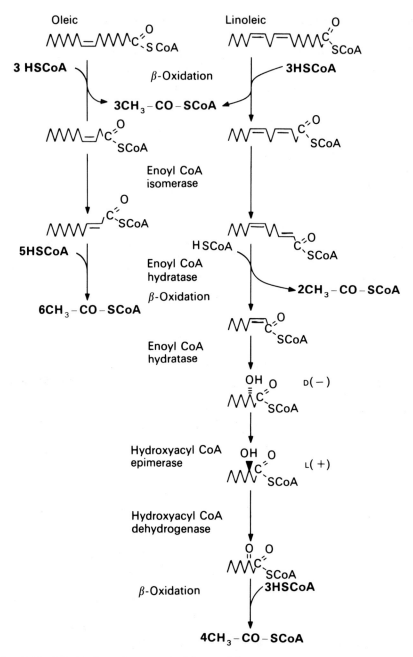

Fig. 3.23 β-Oxidation of unsaturated fatty acids.

not part of the main flow of metabolites in β-oxidation but by-products (Stanley and Tubbs, 1975). This has promoted the suggestion that, like the fatty acid synthesising system, the enzymes of β-oxidation may form a complex from which intermediates are not readily released. However, such a complex has not so far been isolated.

The regulation of the pathway is poorly understood. The rate of β-oxidation is stimulated by glucagon and influenced by the supply of fatty acids to cells. The supply of coenzyme A and hence the intramitochondrial acetyl CoA/CoA ratio may also be a significant regulatory factor.

3.12 Metabolism of the products of β-oxidation

(a) Oxidation of acetyl coenzyme A

Acetyl coenzyme A from fatty acid breakdown may be oxidised in the citric acid cycle, provided the cycle is primed with adequate levels of the dicarboxylic acids.

(b) Synthesis and metabolism of ketone bodies

Under diabetic conditions or in cases of starvation when oxaloacetate is used for gluconeogenesis or where there is a high fat diet, acetyl coenzyme A is converted to ketone bodies in liver mitochondria (Fig. 3.24). Ketone bodies

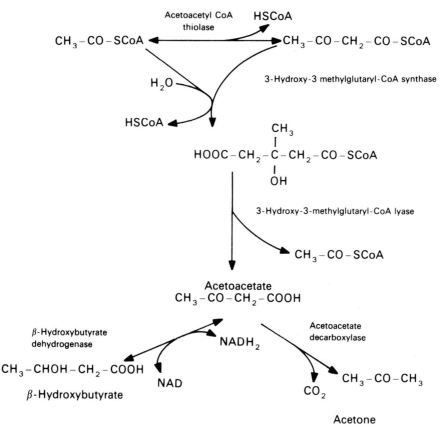

Fig. 3.24 Ketone body formation in liver.

formed in the liver are transferred to the peripheral tissues and metabolised in the mitochondria or in the cytoplasm, mainly the former where they act as a source of energy (Fig. 3.25). The acetoacetyl coenzyme A is formed from aceto-acetate by the succinyl coenzyme A transferase, which is of negligible importance in liver mitochondria but found in significant amounts in peripheral tissues.

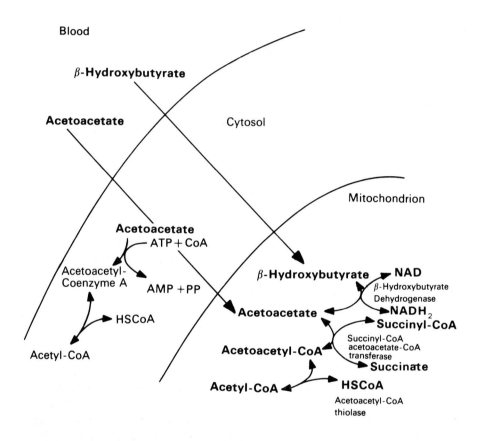

Fig. 3.25 Ketone body metabolism.

(c) Propionyl coenzyme A metabolism

The end product of β-oxidation of fatty acids with even numbers of carbon atoms will be solely acetyl coenzyme A. With fatty acids with odd numbers of carbon atoms (C_{15}, C_{17}) found in many tissues, β-oxidation will generate one molecule of propionyl coenzyme A for every fatty acid oxidised. A common fate of the propionyl coenzyme A is carboxylation to methylmalonyl coenzyme A (a biotin-containing enzyme) and isomerisation to succinyl coenzyme A (Fig. 3.26).

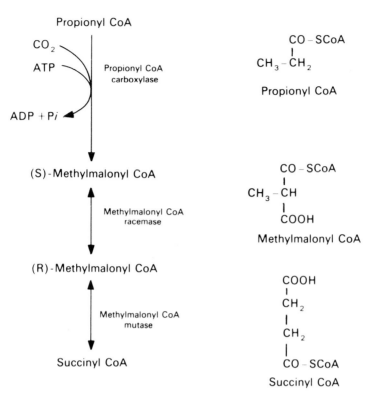

Fig. 3.26 Metabolism of propionyl coenzyme A.

(d) Gluconeogenesis in plants

The enzymes for the β-oxidation pathway are found in mitochondria and in glyoxysomes in plants. In the germinating castor oil seed which uses stored lipid, the β-oxidation enzymes are in the glyoxysomes. The products of fatty acid oxidation can, however, be used for carbohydrate synthesis through the glyoxylate pathway. Some protozoa have also been shown to possess β-oxidation systems in peroxisomes.

3.13 Ethanol metabolism

Mammalian liver and yeast metabolise ethanol. In liver, the alcohol dehydrogenase is found in the cytosol and oxidises ethanol to acetaldehyde. Liver mitochondria possess a wide-specificity aldehyde dehydrogenase which oxidises acetaldehyde to acetate:

$$acetaldehyde + NAD + H_2O \ \rightleftharpoons \ acetate + NADH_2$$

The acetate can be activated to acetyl coenzyme A as described in sect. 3.10 and metabolised through the citric acid cycle.

3.14   The relationship between carbohydrate and lipid metabolism

This relationship is summarised in Fig. 3.27 which also shows the flow of coenzyme A, frequently regarded as a controlling factor in fatty acid oxidation and ketogenesis. In liver, acetyl coenzyme A is used for ketone body synthesis where the oxaloacetate supply is inadequate for oxidation through the citric acid cycle. Thus ketone bodies will be formed under conditions where oxaloacetate is used for gluconeogenesis. The key enzyme in gluconeogenesis from citric acid cycle intermediates is the phosphoenolpyruvate carboxykinase which is re-

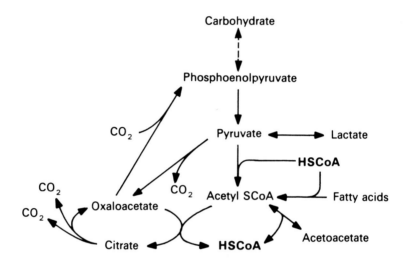

Fig. 3.27 Relationships between lipid and carbohydrate metabolism in vertebrate tissues. It should be noted that the relative importance of the pathways will differ from tissue to tissue and will vary with conditions.

gulated by several factors and thus serves to control the level of dicarboxylic acids. A further factor influencing the formation of ketone bodies, which is normally low, is the inhibition of the citrate synthase by ATP. As noted earlier (sect. 3.2(d)) β-oxidation of fatty acids inhibits oxidation of pyruvate to acetyl coenzyme A and therefore carbohydrate oxidation.

3.15   Appendix: fatty acid elongation

Fatty acid elongation is a function of the mitochondrion and of the endoplasmic reticulum. Extramitochondrial fatty acid synthesis produces palmitate as a major product. Intramitochondrial systems can extend chain lengths, C_{10} to C_{22}, by a pathway which is essentially the reverse of β-oxidation. Three of the enzyme reactions of β-oxidation are reversible (the hydratase, dehydrogenase and the thiolase) while the fourth, the acyl coenzyme A dehydrogenase, is irreversible and in fatty acid elongation is replaced by the NADP-dependent enoyl coenzyme A reductase (Fig. 3.28). However, in liver mitochondria it is claimed that both the inner and the outer membranes possess an NAD-dependent elongation system. Further, the evidence for fatty acid elongation involving enzymes of β-oxidation is not strong.

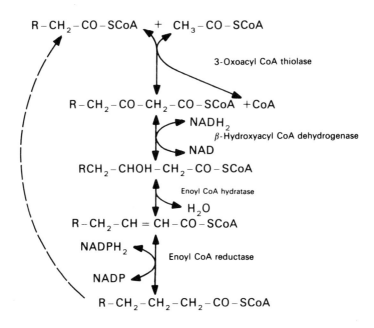

Fig. 3.28 Synthesis of long-chain fatty acids by elongation.

Further Reading

Atkinson, D. E. (1977) *Cellular Energy Metabolism and its Regulation.* Academic Press, New York.

Biochemical Society Colloquium (1978) Recent progress in the β-oxidation of fatty acids. *Biochem. Soc. Trans.*, **6**, 68–88.

Florkin, M. (1975) History of the identification of sources of free energy in organisms. *Comprehensive Biochemistry*, Vol. 31, Florkin, M. & Statz, E. H., eds. Elsevier, Amsterdam.

Groot, P. H. E., Scholten, H. R. and Hulsmann, W. C. (1976) Fatty acid activation: specificity, localisation and function. *Adv. Lipid. Res.*, **14**, 75–126.

Gurr, M. I. and James, A. T. (1975) *Lipid Biochemistry*. Chapman & Hall, London.

Randle, P. J. (1978) Pyruvate dehydrogenase complex – meticulous regulator of glucose disposal in animals. *TIBS*, **3**, 217–9.

Williamson, D. H. (1979) Recent developments in ketone-body metabolism. *Biochem. Soc. Trans.*, **7**, 1313–21.

Also:

McGarry, I. D. and Foster, D. W. (1980) Regulation of hepatic fatty acid oxidation and ketone body production. *Ann. Rev. Biochem.*, **49**, 395–420.

Chapter 4

The structure of the mitochondrion

4.1   The mitochondrion: development of ideas

Towards the end of the nineteenth century, improvements in the quality of the light microscope enabled cytologists to recognise the presence of granules in cells. In 1898, the cytologist, C. Benda,[1] applied the name 'mitochondrion' (from Greek, $\mu\iota\tau\sigma\sigma$ (thread) and $\chi o\nu\delta\rho o\sigma$ (grain) to describe the 'thread granules' which he observed in sperm and egg cells. He later concluded that mitochondria were permanent cellular structures and were involved in the hereditary mechanism – a view which persisted for many years. Cytological study of mitochondria was further promoted through the introduction by Michaelis[1] in 1900 of the dye, Janus green, for staining mitochondria.

The first step in the direction of a chemical understanding of mitochondria came from Regaud (1908)[1] whose paper had some influence on the early studies of isolation of these particles. Regaud had suggested on cytological evidence that mitochondria were composed primarily of proteins and phospholipids.

Probably the earliest isolation and biochemical investigation of mitochondria by a cellular fractionation technique is due to Otto Warburg (1913) in a paper entitled *Oxygen-Respiring Granules in Liver Cells*. Warburg crushed liver cells in water or saline and, after centrifugation, obtained a supernatant containing a suspension of small granules. He demonstrated that the granules themselves possessed respiratory activity and concluded 'From mammalian liver, suspensions of small granules may be obtained which show Brownian movement and which consume oxygen and evolve CO_2.' Warburg did not, however, relate these particles to mitochondria and his findings appear to have been largely ignored until Stern (1939) drew attention to them.

The modern study of isolated mitochondria derives from experiments of Bensley and Hoer (1934) who isolated these particles in saline by a differential centrifugal technique applied to liver which had been washed free of blood and ground in a mortar. Bensley's main interest was in the chemical composition of the isolated particles which he claimed to contain two proteins and fatty

---

[1] For references and a fuller account of the development of ideas on the mitochondrion, see A. B. Novokoff (1961) *Mitochondria (Chondriosomes)* in 'The Cell'. Vol. 3, pp. 391–421. Brachett and Mirsky (eds), Academic Press, New York.

substances although little phospholipid was found. Bensley also noted that mitochondria swelled in isotonic saline.

Bensley's work was influential on Albert Claude working at the Rocke-feller Institute in New York. Claude's initial interest in the 1930s was in tumour-producing agents which he isolated from chicken tumours using a differential centrifugal technique. He obtained an extract which not only caused tumours but also contained granules. Identical granules could also be obtained from chicken embryo. Partly on the basis of Bensley's studies, he concluded that these granules were mitochondria (Claude, 1940). Concurrently with this work on the isolation of granules, other investigators were examining the particles associated with respiratory enzymes such as cytochrome oxidase and succinate dehydrogenase.

From about 1940 Claude turned his attention to analysing the granular components obtained in cell homogenates, using liver tissue as his experimental material. Initially he described large and small granules obtained by differential centrifugation. Early studies concentrated on the identification, appearance, chemical composition and general properties of these large granules (mitochondria) but by 1946 a group at the Rockefeller Institute were examining the subcellular distribution of enzymes in various fractions. Hogeboom *et al.* (1946) showed the presence of succinoxidase and cytochrome oxidase in the large granule fraction. Simultaneously, other workers, inspired by the techniques developed by Bensley and Claude, also commenced studies of subcellular distribution of metabolic activities.

The isolation of mitochondria up to 1947 was generally carried out in water or in isotonic saline. However, these isolated particles did not possess two important characteristics of mitochondria in liver cells; the rod-like shape and the ability to stain vitally with Janus green. In 1947–48, the group from the Rockefeller Institute (Hogeboom *et al.*, 1948) examined the effect of the isolating medium on the properties of mitochondria. Whereas mitochondria isolated in water or saline were spherical, those isolated in hypertonic sucrose (0.88 M) were elongated structures and were stained with Janus green. This group proposed a revised method for the isolation of mitochondria by differential centrifugation of a liver homogenate prepared in 0.88 M sucrose. Particles isolated by the new technique appeared to be whole mitochondria and to contain all of the cell's succinoxidase activity. They could be demonstrated to possess a limiting membrane, apparently semipermeable, and to be osmotically sensitive.

Kennedy and Lehninger (1949), using a new method for isolation, showed that mitochondria oxidise citric acid cycle intermediates and possess 'all the demonstrable fatty acid oxidase activity of whole rat liver'. Simultaneously with these observations Green and his collaborators (1948) launched an investigation of the 'cyclophorase system', a mitochondrial preparation containing the enzymes of the citric acid cycle and other oxidative activities.

By 1952, the main metabolic activities of the mitochondrion had been demonstrated with particle preparations prepared by the method of Hogeboom *et al.* (1948), i.e. fatty acid oxidation, pyruvate oxidation and the citric acid cycle, the respiratory chain and oxidative phosphorylation. Later the use of 0.88 M sucrose was replaced by lower concentrations (e.g. 0.25 M).

4.2 Mitochondrial structure

(a) General structure

The complexity of the metabolic processes in the mitochondrion called for a knowledge of the detailed structure of the particle. In 1952 Palade published the first high resolution electron micrographs of mitochondria using thin sections of a variety of tissues. A few months later, January 1953, Sjöstrand independently published his electron micrographs. This double event initiated the study of the detailed structure of the mitochondrion. Curiously, Palade saw 'a membrane 7–8 nm thick that separates them [mitochondria] from the rest of the cytoplasm' while Sjöstrand recognised the double nature of the membrane.

The mitochondrion may be described as rod-shaped or sausage-shaped (Fig. 4.1). Mitochondria vary somewhat in shape even in the same tissue; in some the length of the rod in comparison with its diameter is relatively short. While rod-shaped mitochondria are found in many mammalian cells, more exotic forms are known such as cup-shaped mitochondria in hamster adrenocortex and dormant tubers of Jerusalem artichoke. Branched shapes forming a reticulate structure are found in rat diaphragm and a number of micro-organisms. In rat liver both simple and branched particles are found. (See Bakeeva *et al.*, 1978, for summary.)

The mitochondrion consists basically of two membranes and the enclosed spaces[2] (Fig. 4.1.). An outer membrane without folds surrounds the particle. An inner membrane surrounds the inner space (matrix) but is invaginated to form double membranes which cross the mitochondrion. These transverse membranes (called cristae mitochondriales by Palade) are known as cristae (crista sing.). Thus the two membranes enclose the two major spaces within the mitochondrion, the matrix or inner space and the intermembrane space between the inner and outer membranes. This latter space appears to be continuous with the space in the middle of the cristae. The result of this architecture is to provide a set of closely related regions which include aqueous spaces, membrane surfaces and mainly hydrophobic regions within the membranes themselves.

(b) Membrane surfaces

A major development in the study of mitochondrial membrane structure came in 1962 with the observations of Fernandez-Moran (1962) on negatively stained preparations of the inner membrane (Fig. 4.2). The technique of negative staining involves treatment of a preparation with an electron-dense material such as phosphotungstic acid which gives the preparation a dark background when viewed in the electron microscope. Some phosphotungstate also penetrates the hydrophobic regions of the membrane. The remainder of the membrane material

---

[2] A radically different interpretation is proposed by Sjostrand (1978). In the native mitochondrion, the membranes of the cristae are closely apposed as are the peripheral inner membrane and the outer membrane so that there is no intermembrane space and no cristal space. The outer membrane is of the fluid mosaic type but the inner membrane has a protein structure with areas of lipid on the inner (matrix) surface reducing water permeability. The membrane interior has polar and non-polar regions but a very low water content.

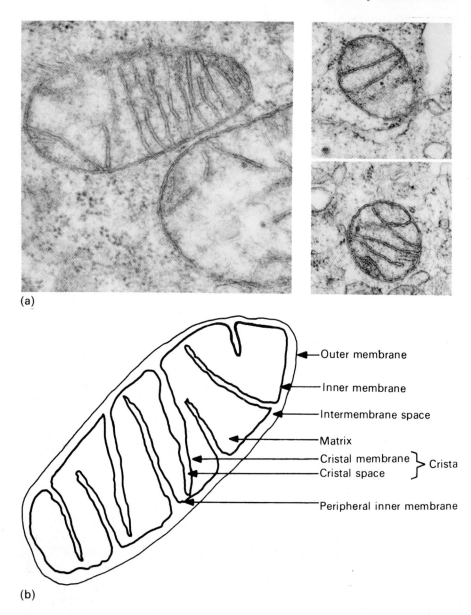

(a)

(b)

Outer membrane

Inner membrane

Intermembrane space

Matrix

Cristal membrane ⎫
                 ⎬ Crista
Cristal space   ⎭

Peripheral inner membrane

Fig. 4.1 Mitochondrial structure. (a) Electron micrographs of thin sections through human bladder cancer cells; left-hand × 52 500, right-hand × 35 000. (Courtesy of Dr. D. Hockley, National Institute of Biological Standards and Control, London). (b) Diagram of mitochondrial structure. (c) Electron micrograph of a thin-section of the digestive diverticulum of *Amphioxus* showing the invagination of the inner mitochondrial membrane to form the cristae, × 80 500. (Courtesy of Prof. S. Bullivant, University of Auckland, N.Z). (d) A negatively stained whole mitochondrion from bovine adrenal medulla, × 143 000. (Courtesy of Dr. R. Lang, National Institute for Medical Research, Mill Hill, London).

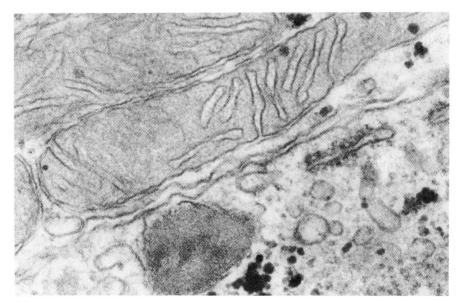

(c)

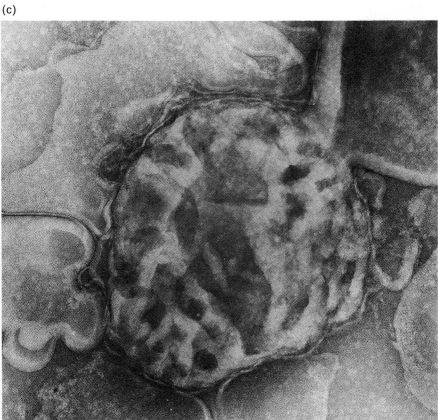

(d)

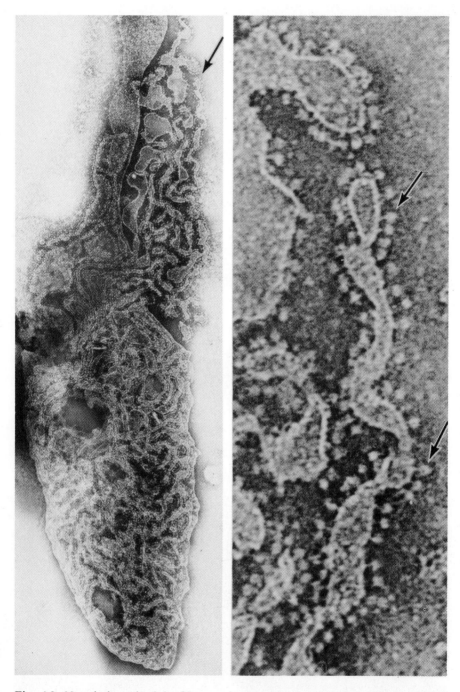

Fig. 4.2 Negatively stained beef-heart mitochondria showing the stalked particles (arrowed) identified as the F_1 ATPase, $\times 50\,000$ (left hand), $\times 340\,000$ (right hand). (Courtesy of Prof. Fernandez-Moran, University of Chicago, U.S.A., from Fernandez-Moran *et al.*, 1964).

then shows up as pale structures against a dark background. Preparations of the inner membrane, negatively stained, show rows of small particles apparently on short stalks distributed along the inner membrane surface. The projections have been found on negatively stained inner membranes of mitochondria from a variety of animal, plant and microbial sources. However, they are not seen in sections of whole mitochondria. In liver mitochondria there are about 2000–4000 particles per μm^2 of inner membrane inner surface.

Initially, it was suggested that the particles represented the complexes of the electron transport chain and that electrons were passed from complex to complex by the particles swinging on their stalks so as to interact with each other. However, this theory had to be abandoned when the particles were removed from the membrane by ultrasonic treatment and shown to possess ATPase activity, the electron-transporting activity remaining in the membrane. Subsequently, this ATPase particle proved to be identical with the F_1 ATPase isolated by Racker from mitochondria and shown to be involved in oxidative phosphorylation.

The stalked particles found on the inner membrane appear to be much more frequent on the cristae than on the peripheral parts of the inner membrane. This has led to a view of mitochondrial structure in which the peripheral inner membrane is fundamentally different from the membrane of the cristae.

By contrast with the inner face of the inner membrane just discussed, the outer face appears to be smooth. The outer membrane is probably smooth on both its faces although occasional projections have been suggested for the outer surface.

(c) Freeze-fracture studies

The technique of freeze-fracturing has allowed an examination of the structure of the interior of the membrane while freeze-etching exposes the membrane surfaces. The technique involves freezing a specimen of tissue or a pellet of particles at low temperature (liquid nitrogen), fracturing the specimen (with a razor blade) under liquid nitrogen, producing a carbon replica of the surface and examining this carbon replica, after shadowing, under the electron microscope. The original fracture produced in the specimen will tend to follow lines of weakness in the material including the hydrophobic areas in the centre of the membrane (Fig. 4.3). Thus in the fracturing process, the membrane tends to be split along its middle rather like pulling the two halves of a sandwich apart. The carbon replica will show the form, in relief, of the fracture faces.

The cleavage of the inner membrane results in an uneven distribution of particles between the two exposed surfaces. The P face (facing towards the cytoplasm) carries a large number of particles, about $4000/\mu m^2$ while the E face (facing towards the matrix) carries many fewer particles, about $2000/\mu m^2$ (Fig. 4.4).

Freeze-etching of a fractured face allows an examination of the outer surfaces of the membranes (see Fig. 4.3). Replicas of the etched inner face of the inner membrane show a relatively smooth face without the stalked knobs found in negative staining. These results have led Wigglesworth *et al.* (1970) to conclude that the stalked knobs found in negatively stained preparations are not found

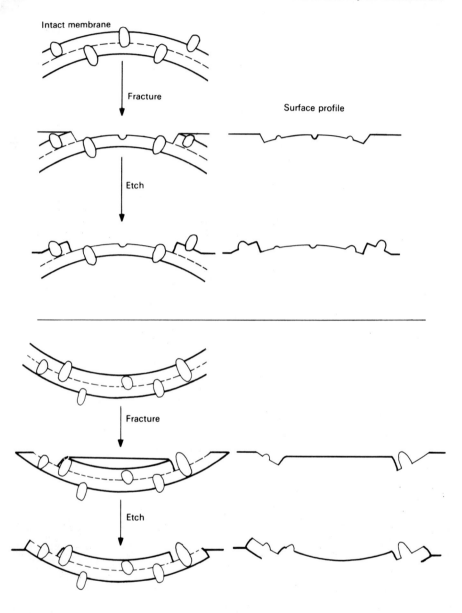

Fig. 4.3 *Freeze-fracture technique*. A membrane normally tends to fracture along the middle of the lipid bilayer. Embedded particles then appear as projections on the fracture face or alternatively leave depressions where they have been. A replica of such a surface shows the location of embedded particles. When membranes are fractured, the fracture faces are formed in such a way that the depressions in one face match the projections in the other.

If the ice in the preparation is partially removed by sublimation, the membrane surfaces are also exposed. Thus the replica of a 'freeze-etched' preparation provides information on both the fracture faces and the membrane surfaces.

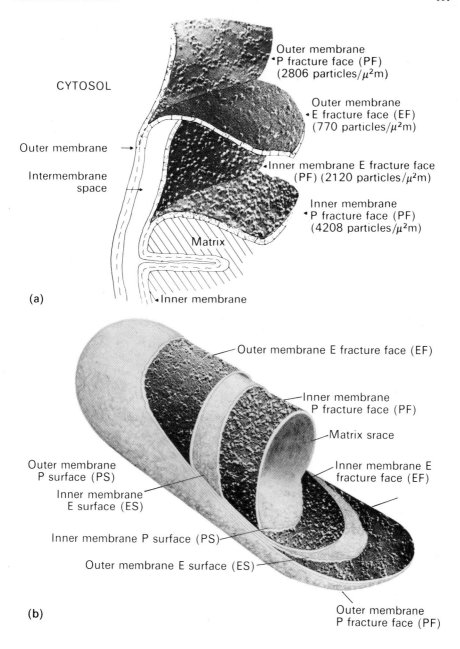

CYTOSOL

Outer membrane
• P fracture face (PF)
(2806 particles/μ^2m)

Outer membrane
• E fracture face (EF)
(770 particles/μ^2m)

Outer membrane →

Intermembrane
space →

• Inner membrane E fracture face
(PF) (2120 particles/μ^2m)

Inner membrane
• P fracture face (PF)
(4208 particles/μ^2m)

Matrix

(a)

• Inner membrane

Outer membrane E fracture face (EF)

Inner membrane
P fracture face (PF)

Matrix srace

Outer membrane
P surface (PS)

Inner membrane
E surface (ES)

Inner membrane E
fracture face (EF)

Inner membrane P surface (PS)

Outer membrane E surface (ES)

(b)

Outer membrane
P fracture face (PF)

Fig. 4.4 Freeze-fracture faces of mitochondrial membranes. Electron micrographs mounted in a collage of (a) mitochondrial inner and outer membranes (Packer 1973) and (b) of the organelle (from Packer and Worthington 1974, in *The Bioenergetics of Mitochondria* (Kroon, A. M., Saccone, C. eds) pp 537–540, Academic Press, New York.

In split membranes, the half nearest the matrix or cytoplasm is the 'protoplasmic' half (P) and the half nearest the intermembrane space is the 'endoplasmic' half (E). (See Branton *et al.*, 1975, for full definition.)

in vivo but are formed during preparation from existing structures buried in the membrane.[3]

(d) Membrane structure

The classical model (Fig. 4.5a) of membrane structure is that of Danielli and Davson, originally proposed in 1935. It consists of a double layer of lipid with polar groups outermost and hydrophobic fatty acid chains forming a hydrophobic region in the centre which must approach the state of a liquid hydrocarbon if the chains are in a disordered array. Membrane protein is attached to the structure by interaction with the polar regions of the lipid. Although current for a long period, this model has now been superseded. An example of a rather different approach to membrane structure is the model of Green and Perdue (1966). Noting the ease with which complexes of the respiratory chain can be removed from the inner membrane with bile salts (see Ch. 9), these workers have suggested a model consisting of lipoprotein particles of similar form and size, although functionally and chemically different, built into a two-particle-thick membrane (see Fig. 4. 5b). The particles which have the lipid chains directed into and around the hydrophobic regions of the protein, are bound to each other by forces weaker than those which bind the constituents of the particle together – these are hydrophobic interactions. The polar part of the phospholipid is primarily found at the membrane surfaces. Thus the particles can be dissociated from the membrane as discrete entities. A major difference between this and the previous model is that whereas, according to Danielli and Davson, the protein is on the outside of the membrane, here the protein has moved much more into the central regions.

Much evidence has accumulated which supports the idea of a lipid bilayer (as seen in the Danielli and Davson model) as a basic part of membrane structure, although protein distribution is probably much more complex. For example in the cristae of the mitochondrion, only 30–40 per cent of the proteins appear to be associated with the polar surface of the lipid bilayer, the remainder are totally or partly in the interior. Hence it is possible to consider the proteins of the membrane as extrinsic or peripheral and intrinsic or integral located in the lipid layers. The former require only mild treatment to remove them from the membrane and they may be held on the surface by non-covalent forces such as electrostatic interactions. The integral proteins form an intimate part of the membrane and require more drastic treatment for their release. Thus, in agreement with the findings of electron microscopists using freeze-fracturing techniques as described above, a model of a membrane must conform more with the type shown in Fig. 4.5c having protein complexes located in the membrane.

This concept has been further developed by Singer and Nicolson (1972) as the fluid mosaic model. This assumes a mosaic of alternating globular proteins and lipid bilayer in a dynamic state so that it may be thought of as a two-dimensional viscous solution. While specific protein complexes may occur in the mosaic in an ordered localised group (as would be necessary for the respiratory

---

[3] Note that Stiles and Crane (1966) found stalked particles on inner membranes fixed before negative staining.

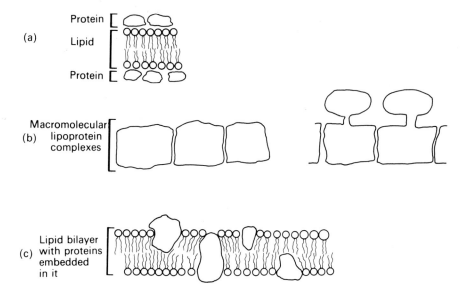

Fig. 4.5 Models of membranes. (a) Danielli–Davson model showing the lipid bilayer with polar heads directed towards the outer protein layer and the hydrophobic fatty acid chains at the middle of the bilayer. (b) Lipoprotein complex model. The lipoprotein complexes contain several proteins bonding to the lipid by hydrophobic interactions. Complexes found in the inner mitochondrial membrane may have stalked particles as seen in negatively stained preparations. (c) Mosaic model (fluid). The proteins are embedded in a lipid bilayer which may be fluid.

chain complexes), in general the arrangement of proteins over larger distances is random. Proteins or protein complexes are dispersed in a fluid two-dimensional system so that some proteins may even diffuse through the lipid layer. This latter point will assume importance in, for example, the case of the adenine nucleotide translocator, a protein thought to move and reorient itself within the membrane in order to bring about the translocation of nucleotides across the membrane. It has been shown that while phospholipids migrate readily in their own layer, they move from one side of the bilayer to the other at a very low rate. It should be noted, however, that Singer and Nicolson's view has been criticised on the ground that it puts too much emphasis on fluidity. The fluid regions in some membranes have been shown experimentally to be somewhat limited in size.

One of several other advantages of this model is that it provides a ready explanation of the freeze-fracture electron micrographs already discussed. It also is consistent with the demonstration by Packer, 1973, that the distribution of particles seen in freeze-fracture work changes with the metabolic state of the membrane.

4.3 Disruption of mitochondria

(a) Submitochondrial particles, sidedness of mitochondrial membranes

Several types of particles or fragments have been obtained by disruption of mitochondria under varying conditions. In general, these particles are derived principally from the inner membrane and carry the respiratory system. In almost all cases the products of disruption are fully enclosed vesicles. One of the first submitochondrial particles to become an object of intensive investigation was the ETP (electron transport particle) formed by homogenising beef heart mitochondria in a medium of ethanol and phosphate. ETP vesicles were shown to oxidise $NADH_2$ and succinate but not to possess the other tricarboxylic acid cycle dehydrogenases. Less drastic treatment gave phosphorylating ETP (PETP) which differed from ETP in being able to carry out oxidative phosphorylation and which also contained citric acid cycle dehydrogenases. ETPH particles are prepared by brief sonic treatment (30 s) of 'heavy' beef heart mitochondria. Several other similar particles were prepared from mitochondria in the late 1950s and early 1960s. Of special significance were particles which would carry out oxidative phosphorylation if missing factors, also prepared from mitochondria, were added. These experiments will be discussed in relation to oxidative phosphorylation.

However, two types of particles have been of particular interest in a variety of mitochondrial studies. These are digitonin particles and sonic particles (Fig. 4.6). Digitonin particles may be prepared by treatment of washed mitochondria with 1 per cent digitonin for 30 min at $0\,^{\circ}C$ and separation of the resulting particles by differential centrifugation between 37 000 g (15 min) and 105 000 g (40 min) (see Elliott and Haas, 1967). These particles oxidise succinate and $NADH_2$ but not other citric acid cycle intermediates, and carry out phosphorylation.

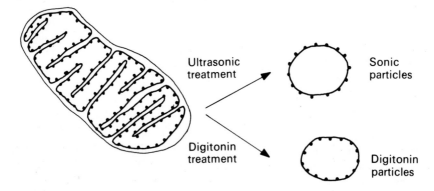

Fig. 4.6 Sonic and digitonin particles.

Sonic particles are prepared by treatment of a mitochondrial suspension in a hypotonic medium with a sonic oscillator (Gregg, 1967). The particles sediment between 25 000 g (20 min) and 144 000 g (30 min). They carry out phosphorylation with NADH or succinate as substrate. Studies of these two types of particles prepared from mammalian mitochondria suggest that while digitonin particles have a membrane orientation corresponding to that in the intact mitochondrion, sonic particles are mostly inverted.

However, the orientation in submitochondrial particles has proved a somewhat complex problem. The two faces of the inner membrane are distinguishable by a number of criteria. Firstly, the matrix-facing side of the inner membrane shows stalked spheres when negatively stained, although it is possible for some preparations to lose these structures. Secondly, cytochrome c is accessible from the outer face of the inner membrane when antibodies to cytochrome c are used to inhibit respiration. Nevertheless there is evidence that sonication results in a more random distribution of cytochrome c across the membrane. Thirdly, the dehydrogenases, succinate or NADH dehydrogenases, appear to be located on the inner face of the membrane in the intact mitochondrion. Fourthly, vectorial processes such as proton translocation, energy-dependent calcium transport and adenine nucleotide exchange, also indicate a sidedness to the membrane. For example, respiration in the intact mitochondrion is accompanied by a translocation of protons outwards across the mitochondrial membrane. The formation of sonic particles and digitonin particles is not entirely regular in that neither type of particle preparation normally shows 100 per cent uniform orientation. Thus it has been claimed that in some preparations of beef heart mitochondria at least 64 per cent have inverted membranes, while in digitonin particles from the same source, as few as 14 per cent of the particles may have inverted orientation (Malviya *et al.*, 1968). However, in the protozoan, *Tetrahymena*, 100 per cent inversion has been claimed for digitonin particles.

While the orientation of membranes in particles from ox heart or rat liver mitochondria appears to be predictable to the extent that in sonic particles the membranes are mostly inverted and in digitonin particles they are mostly not inverted, this is not a universal rule. The orientation of particle preparations from a specific source cannot necessarily be presumed without evidence.

(b) Separation of inner and outer membranes

From the foregoing account, it is clear that the mitochondrion is composed of two separate and morphologically distinct membranes. New possibilities for the study of mitochondria arose in the mid-1960s when several workers published methods for the preparation of inner and outer membrane suspensions. Parsons and coworkers (1966) described the first successful method dependent on the swelling of a mitochondrial preparation in 20 mM phosphate buffer (but no sucrose). The semipermeable membrane involved in the osmotic uptake of water is the inner membrane; swelling results in the loss of the cristae which appear to contribute to the limiting membrane of a simple spherical structure. This enlargement of the inner part of the mitochondrion disrupts the outer membrane releasing the enzymes from the intermembrane space. The outer membrane can then be separated from the inner part of the mitochondrion (mitoplast) by centrifugation (Fig. 4.7).

A refinement by Ernster's group (Sottocasa *et al.*, 1967), aimed at obtaining a better separation of inner and outer membranes, involves contracting the intact inner membrane particle after swelling and disrupting the outer membrane. This is achieved by adding a solution containing sucrose, ATP and Mg^{2+} (see Fig. 4.7). The preparation is then given a brief ultrasonic treatment

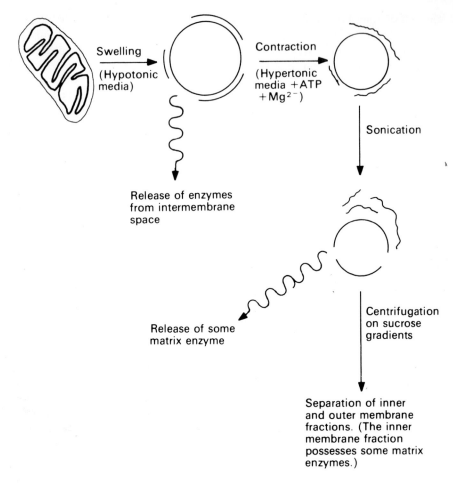

Swelling

(Hypotonic media)

Contraction

(Hypertonic media +ATP +Mg^{2-})

Sonication

Release of enzymes from intermembrane space

Release of some matrix enzyme

Centrifugation on sucrose gradients

Separation of inner and outer membrane fractions. (The inner membrane fraction possesses some matrix enzymes.)

Fig. 4.7 Separation of mitochondrial membranes.

and separated into three fractions on a sucrose gradient. The fractions, for liver mitochondria are: (1) a clear yellow soluble fraction; (2) a pinkish yellow light fraction of outer membranes; and (3) a dark brown heavy pellet of inner membranes with some matrix material.

4.4   Permeability of mitochondrial membranes

The examination of electron micrographs of mitochondria indicates the existence within the organelle of at least two independent spaces separated by the inner membrane, the matrix and the intermembrane space. The cristae also possess a central space but it will be assumed here that this is continuous with the intermembrane space and simply formed by the invagination of the inner membrane. The validity of this assumption will depend on the existence of an adequate opening between the cristal space and the intermembrane space at the junction of the cristae with the peripheral part of the inner membrane. Some workers have been inclined to the view that there is only a small pore between the two spaces, permitting only limited diffusion between them; the cristal space may hence be regarded as an independent region of the mitochondrion.

The experimental approach of several workers, including Werkheiser and Bartley (1957), Klingenberg and Pfaff (1966), has demonstrated the existence of two spaces in the mitochondrion on the basis of sucrose permeability (sucrose-permeable space and sucrose-impermeable space). The distribution of water in a pellet of mitochondria prepared by centrifugation in a medium containing a known concentration of sucrose (MW = 342) and radioactively labelled [^{14}C] polyglucose (MW = 50 000) is estimated. The water will be distributed between: (1) the space between the organelles which is represented by the polyglucose content of the pellet (the outer membrane is not permeable to polyglucose); (2) the sucrose-permeable space which may be calculated from the volume of water required to account for the sucrose in the pellet less the volume of water between the particles; and (3) the sucrose-impermeable space which can be determined from the total volume of water in the pellet less that associated with the sucrose. The estimates for the sucrose-permeable space and the sucrose-impermeable space can be compared with the mitochondrial spaces seen in electron micrographs. By varying the concentration of sucrose used, the relative sizes of the spaces can be varied. From such experiments it may be concluded that the sucrose-permeable space may be equated with the intermembrane space and the sucrose-impermeable space with the matrix.

This type of experiment also leads to conclusions about the permeability of the inner and outer membranes. After testing the penetration of various solutes, it has been concluded that the outer membrane is permeable to both charged and uncharged substances up to a molecular weight of the order of 10 000. This means that the intermembrane space is accessible to most metabolites in the cytoplasm but that enzymes and proteins in the inter-membrane space are kept apart from those in the cytoplasm. By contrast the inner membrane behaves as a semipermeable structure and an osmotic barrier which is not freely permeable to charged ions.

4.5   Properties of mitochondrial membranes

The general properties of mitochondrial membranes are summarised in Table 4.1 which is based on estimates made for mammalian mitochondria. It will be seen that a quarter of the mitochondrial protein is associated with membranes and that most of this is found in the inner membrane. Consequently the outer membrane contains a higher proportion of lipid and has a lower density. Lipid-depleted mitochondria lose the outer but retain the inner membrane. Most of the lipid (about 80 %) is phospholipid, the predominant components being phosphatidylcholine and phosphatidylethanolamine. Major differences between the two membranes are seen in relation to cholesterol content (primarily in the outer membrane) and cardiolipin (primarily in the inner membrane). Probably most of the phosphatidylinositol is in the outer membrane although workers disagree on this point. The phospholipids possess a high proportion of unsaturated fatty acids, particularly cardiolipin which has more than three-quarters of its fatty acids unsaturated. About half of the fatty acids in phosphatidylcholine are unsaturated. A similar lipid composition is found in plant mitochondria (see Moreau *et al.*, 1974).

Table 4.1 Properties of membranes of mammalian mitochondria

	Inner membrane	*Outer membrane*
Surface: outer	Smooth	Smooth but some projections have been reported
inner	Shows stalked spheres when negatively stained	Smooth
Thickness	70 Å (estimates vary 50–150 Å)	About 70 Å
Density	1.19–1.23	1.09–1.12
Protein content related to total mitochondrial protein content	21 %	4 %
Lipid content: phospholipid mg/mg protein	0.30	0.88
Cholesterol μg/mg protein	5.1	30.1
Phospholipid composition		
Phosphatidylcholine	42 %	54 %
Phosphatidylethanolamine	32 %	26 %
Phosphatidylinositol	4 %	14 %
Cardiolipin	19 %	4 %
Other phospholipid	3 %	2 %
Ubiquinone	Present	Absent

In *Neurospora* the presence of carotenoid (neurosporaxanthin) and the steroid ergosterol (rather than cholesterol) has been shown in the outer membrane which also contains a higher proportion of phospholipid than the inner membrane (Hallermayer and Neupert, 1974).

4.6   Mitochondrial enzymes

(a) Distribution

The methods described for the separation of the inner mitochondrial compartment (inner membrane + matrix) and inner and outer membranes have enabled biochemists to study the distribution of enzymes within the mitochondrial particle. A list of enzymes found in the mitochondrion, together with their probable location within the particle, is given in Table 4.2 (a fuller list will be found in Munn, 1974). When considering Table 4.2, a number of points should be borne in mind:

(i) A number of experimental problems are encountered when determining whether a particular enzyme found in a mitochondrial fraction is indeed associated with the mitochondrion itself *in vivo*. Thus there are a number of claims in the literature for the presence of enzymes in mitochondria which must be treated with caution. However, techniques have now advanced to the state where, in most cases, one may have considerable confidence in the results.

(ii) The list refers to the mitochondrion in general and no single organelle possesses all the enzymes shown, although a great many of them are found, for example, in the mammalian liver mitochondrion. Some, such as the ADP-dependent succinate thiokinase, are characteristically plant enzymes while others, such as some steroid-metabolising enzymes, are characteristic of mitochondria in specific tissues such as the adrenal cortex.

(iii) The distribution of a given enzyme may differ from tissue to tissue as, for example, the PEP carboxylase enzyme synthesising phosphoenolpyruvate (see sect. 3.5). Thus the inclusion of an enzyme in the list does not imply that it is necessarily found primarily in the mitochondrion in any particular tissue.

(iv) Some enzymes are found both inside and outside the mitochondrion. However, it is frequently found that the two enzymes differ in their properties. For example, in higher animals the intra- and extramitochondrial malate dehydrogenases, although they are both NAD-specific and convert malate to oxaloacetate, are nevertheless different in their kinetic properties and amino acid composition.

(v) Experimental problems also arise when attempts are made to determine the intramitochondrial location of enzymes. In the past a variety of methods were used to fractionate the mitochondrion and a variety of results were obtained. For example, the citric acid cycle was regarded by some as a matrix/inner membrane system and by others as associated more with the intermembrane space. Standardisation of methods has done much to overcome these problems, but it is still not clear whether some enzymes are loosely bound to membranes or free in the soluble portion of the organelle.

(vi) Considerable evidence has been put forward to suggest that mitochondria in a given tissue are not homogeneous with respect to enzyme content. Thus sampling of the mitochondrial fraction from a sucrose density gradient after centrifugation, shows a variation of activity of a specific enzyme with density suggesting that mitochondria of slightly different densities have different enzyme compositions. It now seems likely that mitochondria are not as heterogeneous enzymically as was originally thought. Apparent variations in enzyme content may simply reflect variations in the membrane contents of organelles of different densities. The specific activity of membrane-bound enzyme, unlike soluble enzymes of the matrix, will reflect the membrane content. Damage to organelles may, however, produce both loss of enzyme and change in density.

Examination of Table 4.2 shows that the major functions of the inner compartment of the mitochondrion (matrix and inner membrane) are the citric acid cycle, fatty acid oxidation (at least in vertebrates), some processes of nitrogen metabolism and the synthesis of proteins, nucleic acids and porphyrins (some reactions). The inner membrane itself is concerned with electron transport systems (including hydroxylation in some tissues) and phosphorylation. The

Table 4.2 Enzymes of mitochondria

The enzymes listed below are those which normally occur in mammalian mitochondria except where otherwise noted. The enzyme location within the mitochondrion is listed where it is known, i.e. matrix, i.m. (inner membrane), i.m.s. (intermembrane space) and o.m. (outer membrane).

	EC number	Location	Notes
I. Tricarboxylic acid cycle and related enzymes			
(a) Cycle reactions			
Pyruvate dehydrogenase complex	1.2.4.1	matrix (i.m.)	
	1.6.4.3		
	2.3.1.12		
Pyruvate dehydrogenase kinase	2.7.1.99	matrix	
Pyruvate dehydrogenase phosphatase	3.1.3.43	matrix	
Citrate synthase	4.1.3.7	matrix	
Aconitase (aconitate hydratase)	4.2.1.3	matrix	
Isocitrate dehydrogenase (NAD)	1.1.1.41	matrix	
Isocitrate dehydrogenase (NADP)	1.1.1.42	matrix	
Oxoglutarate dehydrogenase complex	1.2.4.2		
	2.3.1.61		
	1.6.4.3		
Succinate thiokinase (GTP) (succinyl CoA synthetase)	6.2.1.4	matrix	Animal mitochondria
Succinate thiokinase (ATP) (succinyl CoA synthetase)	6.2.1.5	matrix	Plant mitochondria
Succinate dehydrogenase	1.3.99.1	i.m.	
Fumarase (fumarate hydratase)	4.2.1.2	matrix	
Malate dehydrogenase	1.1.1.37	matrix (i.m.)	
(b) Carboxylation reactions			
Malic enzyme (NADP) (oxaloacetate decarboxylating)	1.1.1.40	matrix	Brain, not liver

Enzyme	EC no.	Location	Notes
Malic enzyme (NAD) (malate dehydrogenase decarboxylating)	1.1.1.39	—	?Mammalian mitochondria. Not oxaloacetate decarboxylating
Pyruvate carboxylase (ATP)	6.4.1.1	matrix	Animal mitochondria
Phosphoenolpyruvate carboxylase	4.1.1.31	—	Plant mitochondria
Phosphoenolpyruvate carboxykinase	4.1.1.32	matrix	Subcellular distribution varies with tissue in animals

II. Glycolysis (enzymes in this group attached to outer surface of mitochondria)

Enzyme	EC no.	Location	Notes
Hexokinase	2.7.1.1	o.m.	
Phosphofructokinase	2.7.1.11	o.m.	
Glyceraldehyde phosphate dehydrogenase	1.2.1.12	o.m.	
Pyruvate kinase	2.7.1.40	o.m.	Reported from *Tetrahymena*
L-lactate dehydrogenase	1.1.1.27	o.m.	Reported from rabbit muscle; Reported from *Tetrahymena* yeast

III. Fatty acid metabolism

(a) Fatty acid activation

Enzyme	EC no.	Location	Notes
Fatty acyl CoA synthetase (ATP)	6.2.1.3	o.m.	Activates long-chain fatty acids
Fatty acyl CoA synthetase (ATP)	6.2.1.2	matrix	Activates medium-chain fatty acids
Fatty acyl CoA synthetase (GTP)	6.2.1.10	matrix	
Acetyl CoA synthetase (ATP)	6.2.1.1	matrix	
Propionyl CoA synthetase (ATP)		matrix	
Carnitine acyltransferase	2.3.1.21	i.m.	

Table 4.2 (*Contd.*)

	EC number	Location	Notes
(b) β-Oxidation			
Acyl CoA dehydrogenase (long chain)	1.3.99.3		
Acyl CoA dehydrogenase (short chain) (Butyryl CoA dehydrogenase)	1.3.99.2		
Acyl CoA dehydrogenase (medium chain)	—	inner mitochondrial compartment : i.m. + matrix	
Enoyl CoA hydratase	4.2.1.17		
l-Hydroxyacyl CoA dehydrogenase	1.1.1.35		
3-Oxoacyl CoA thiolase (Acetyl CoA acyltransferase)	2.3.1.16		
cis-3, *trans*-2-Enoyl CoA isomerase	5.3.3.8		
3-Hydroxyacyl CoA epimerase	5.1.2.3		
(c) Propionyl CoA metabolism			
Propionyl CoA carboxylase	6.4.1.3	matrix (i.m.)	
Methylmalonyl CoA racemase	5.1.99.1	matrix	
Methylmalonyl CoA mutase	5.4.99.2	matrix	
(d) Ketone body metabolism			
Acetoacetyl CoA thiolase (acetyl CoA acetyltransferase)	2.3.1.9	matrix	
Hydroxymethylglutaryl CoA synthase	4.1.3.5	matrix	
Hydroxymethylglutaryl CoA lyase	4.1.3.4	matrix	
Hydroxybutyrate dehydrogenase	1.1.1.30	i.m.	
Succinyl CoA transferase	2.8.3.5	matrix	Peripheral tissues not liver
(e) Fatty acid elongation			
Enoyl CoA reductase (Acyl CoA dehydrogenase-NADP)	1.3.1.8	matrix	

IV. Oxidation–reduction systems

(a) Dehydrogenases

Enzyme	EC number	Location	Notes
NADH dehydrogenase	1.6.99.3	i.m. (inside)	An additional enzyme found on outside i.m. in plants and microorganisms
NAD(P) transhydrogenase	1.6.1.1	i.m. (inside)	
Cytochrome b_5 reductase	1.6.2.2	o.m.	
NADPH–cytochrome reductase	1.6.2.4	i.m.	Linked to cytochrome p450 (see VII below)
Succinate dehydrogenase	1.3.99.1	i.m. (inside)	
Glycerol phosphate dehydrogenase	1.1.99.5	i.m. (outside)	
Sarcosine dehydrogenase	1.5.99.1		
Dimethylglycine dehydrogenase	1.5.99.2	i.m.	Linked to respiratory chain through the electron transferring flavoprotein
Acyl CoA dehydrogenases (see III, (b) above)	1.3.99.2 / 1.3.99.3		
Choline dehydrogenase	1.1.99.1		
Lactate dehydrogenase (cytochrome b_2)	1.1.2.3	i.m.	Cytochrome-containing flavoprotein found in yeast mitochondria

(b) Electron-transferring systems

Enzyme	EC number	Location
Ubiquinone–cytochrome c reductase	1.10.2.2	i.m.
Cytochrome oxidase	1.9.3.1	i.m.
Sulphite dehydrogenase	1.8.2.1	i.m.s.

V. Phosphate metabolism

Enzyme	EC number	Location	Notes
ATPase	3.6.1.3	i.m. (inside)	Responsible for oxidative phosphorylation

THAMES POLYTECHNIC LIBRARY

Table 4.2 (*Contd.*)

	EC number	Location	Notes
GTP/AMP phosphotransferase (nucleoside triphosphate-adenylate kinase)	2.7.4.10	matrix	
Nucleoside diphosphokinase	2.7.4.6	i.m.s.	May be on outside of i.m.
Adenylate kinase	2.7.4.3	i.m.s.	
Adenylate cyclase	4.6.1.1	i.m.s.	Reported from rat testis
Creatine kinase	2.7.3.2	i.m.s.	Human, beef, rat heart; rat brain; rat muscle; not rat liver, kidney or testis.
Pyrophosphatase	3.6.1.1	i.m.	
VI. Amino acid and nitrogen metabolism			
Glutamate dehydrogenase	1.4.1.3	matrix	
Aspartate transaminase (aminotransferase)	2.6.1.1	matrix	
Alanine transaminase (aminotransferase)	2.6.1.2		
Glutaminase	3.5.1.2	matrix	
Carbamoyl phosphate synthetase	6.3.4.16	matrix	
Ornithine transcarbamylase	2.1.3.3	matrix	
Monoamine oxidase	1.4.3.4	o.m.	
Kynurenine hydroxylase (mono-oxygenase)	1.14.13.9	o.m.	
Arginase	3.5.3.1	i.m.	Plant mitochondria
Proline dehydrogenase	1.5.1.12		
Pyrroline carboxylate dehydrogenase		matrix	
Betaine aldehyde dehydrogenase	1.2.1.8		

Serine hydroxymethyl transferase	2.1.2.1	matrix
Glycine synthase	2.1.2.10	i.m.

VII. Steroid metabolism

Steroid 11β-mono-oxygenase	1.14.15.4	i.m. (inside)
Steroid Δ-isomerase	5.3.3.1	
β-Hydroxysteroid dehydrogenase	1.1.1.51	

Note: Many of the enzymes involved in steroid metabolism in the mitochondrion are poorly defined. The major pathways are shown in Fig. 4.9.

VIII. Phospholipid metabolism

Cholinephosphotransferase	2.7.8.2	o.m.
Glycerophosphate acyltransferase	2.3.1.15	o.m.
CDP-diglyceride pyrophosphorylase (phosphatidate cytidylyltransferase)	2.7.7.41	
Glycerolphosphate phosphatidyl transferase	2.7.8.5	
Phosphatidylglycerophosphatase	3.1.3.27	
Phosphatidate phosphatase	3.1.3.4	o.m.

IX. Porphyrin and haem synthesis

δ-Aminolaevulinate synthase	2.3.1.37	matrix
Coproporphyrinogen oxidase	1.3.3.3	
Ferrochelatase	4.99.1.1	i.m.

X. Nucleic acid and protein synthesis (see Ch. 5)

DNA polymerase	2.7.7.7	matrix
Deoxyribonuclease I	3.1.21.1	matrix

Table 4.2 (*Contd.*)

	EC number	Location	Notes
RNA polymerase	2.7.7.6	matrix	
Amino acid–tRNA synthetase	6.1.1.–	matrix	Many of the expected mitochondrial synthetases have now been found
XI. Miscellaneous enzyme activities			
(a) Detoxication enzymes			
Superoxide dismutase	1.15.1.1	matrix	Mn-protein similar to bacterial enzyme but dissimilar to enzyme in cytosol
(b) Isoprenoid synthesis			
Hydroxymethylglutaryl CoA reductase	1.1.1.34		Found in yeast mutant
Prenyl transferase	2.5.1.1		
Hydroxybenzoate polyprenyl transferase	—		
(c) Ethanol metabolism			
Alcohol dehydrogenase	1.1.1.1		
Acetaldehyde dehydrogenase	1.2.1.–	matrix	Yeast but not liver
(d) Other enzymes			
Mannosyl transferase	—	i.m.	

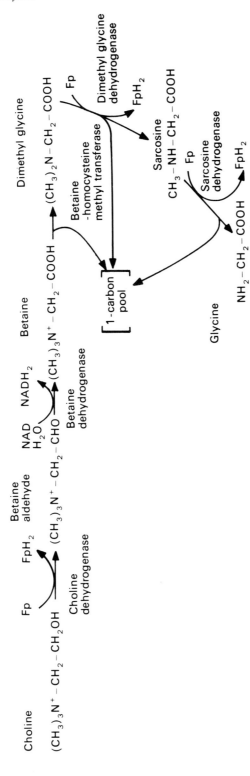

Fig. 4.8 Mitochondrial choline oxidation.

intermembrane space possesses few activities other than those concerned with phosphate metabolism, while the outer membrane carries a number of special activities, particularly phospholipid metabolism. The outside of the particle possesses binding sites for some enzymes which may be reversibly bound, with a consequent modification of activity in some cases. Thus brain hexokinase is mainly but not totally associated with the mitochondrion, although glucose-6-phosphate and ATP will solubilise the enzyme; sensitivity to inhibition by glucose-6-phosphate is decreased by binding.

(b) Metabolic functions of the mitochondrion (see Table 4.2)

I. Citric acid cycle. This is an almost invariable feature of the organelle associated with the matrix and inner membrane (inside).

II. Glycolysis. The extent of binding of the glycolytic enzymes to the outside surface of the mitochondrion is uncertain. However, in a number of animal tissues several glycolytic enzymes, particularly phosphofructokinase and hexokinase, are loosely bound to the surface of the organelle.

III. Fatty acid metabolism. The oxidation of long chain fatty acids is mainly mitochondrial in higher animals. In plants the major site of fatty acid oxidation is the peroxisome, but it has been suggested that mitochondria may also possess activity. In animals excess acetyl coenzyme A from fatty acid oxidation is channelled into ketone body formation, whereas in plants acetyl coenzyme A can be metabolised fully through the citric acid cycle and glyoxylate pathway.

IV. Oxidation–reduction systems. Mammalian mitochondria possess three electron transport systems. The major system is located in the inner membrane and linked to the oxidation of $NADH_2$, succinate, glycerol phosphate, acyl coenzyme A esters, sarcosine, dimethylglycine and choline (see Fig. 4.8). Oxidation of the first two substrates is a universal property of mitochondria but activity for the remainder will vary with the tissue. The oxidation system consists of dehydrogenases, ubiquinone and b-, c- and a-cytochromes terminating in the cytochrome oxidase.

The outer membrane possesses a simple oxidation–reduction system reducing cytochrome b_5

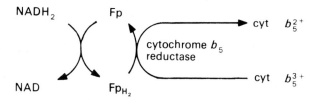

Although extensively studied from a structural point of view, the biological function of this cytochrome is not clear.

The third electron transport chain is found in the inner membrane of mitochondria from mammalian tissues, particularly the adrenal cortex and is responsible for steroid hydroxylations. It functions as a mono-oxygenase system using NADPH as substrate (Fig. 4.9a).

The intermembrane space contains a haem-protein which has sulphite dehydrogenase activity apparently utilising the cytochrome c on the outside of the inner membrane as electron acceptor. Oxidation of sulphite is coupled to ATP synthesis through one coupling site (P/O $= 1$ approx.). There is evidence to suggest that this enzyme is part of a mitochondrial system for sulphur metabolism.

V. Phosphate metabolism. The enzymes in this group are concerned primarily with relating ATP metabolically with the other nucleoside phosphates and will be discussed together with adenine nucleotide transport (Ch. 6).

VI. Amino acid and nitrogen metabolism. The role of glutamate in nitrogen metabolism was discussed in the last chapter. Although most amino acids are oxidised extramitochondrially, the proline dehydrogenase and pyrroline carboxylate dehydrogenase convert proline to glutamate in the mitochondrial matrix. The proline dehydrogenase is tightly bound to the inner membrane and probably linked to the respiratory chain through ubiquinone; the product of the reaction is pyrroline-5-carboxylate. The literature contains accounts of several other enzymes which appear to occur in mitochondria including several transaminases, enzymes in serine and glycine metabolism and enzymes for catabolism of branch-chain amino acids etc. Plant mitochondria may possess enzymes normally found in the cytosol in animal tissues, for example fungal mitochondria appear to contain the enzymes for the isoleucine–valine pathway.

Choline is oxidised by liver mitochondria (see Fig. 4.8) and is involved in mitochondrial one-carbon metabolism which is poorly understood, but probably includes the synthesis of methionine.

VII. Steroid metabolism. Cholesterol is synthesised extramitochondrially. Some modifications of the cholesterol molecule are associated with mitochondria in mammals. In adrenal cortex, testis, ovary and placenta, mitochondria play a role in hormone synthesis. The inner membrane possesses enzymes which cleave the side chain of cholesterol (Fig. 4.9b) by hydroxylation at C-20 and C-22 followed by lyase activity. These reactions involve a mixed function oxidase (mono-oxygenase) system (Fig. 4.9a) which consists of a flavoprotein dehydrogenase (NADP-specific), an iron–sulphur protein (adrenodoxin in adrenal mitochondria) and a CO-sensitive cytochrome, P450, which also acts as the hydroxylating enzyme. Two further inner membrane enzymes, a 3β-hydroxysteroid dehydrogenase (NAD(P)-linked) and a steroid isomerase catalyse the oxidation of $\Delta^5$3β-hydroxysteroids with 19 or 21 carbons to $\Delta^4$3-oxosteroids. These reactions also occur in the microsomal fraction.

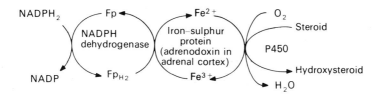

Fig. 4.9 (a) Steroid hydroxylation system for the inner mitochondrial membrane.

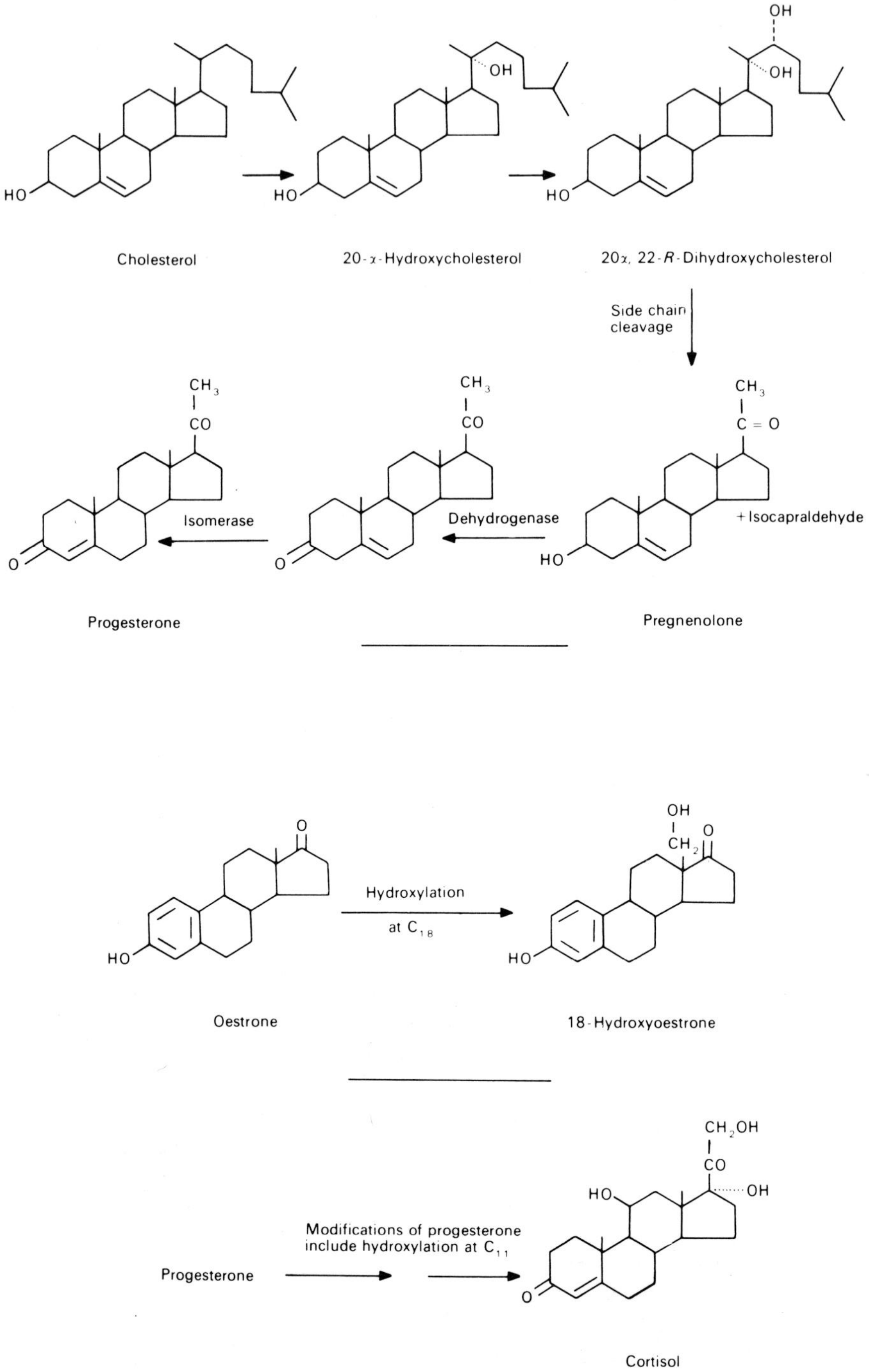

Fig. 4.9(b) Steroid metabolism in the mitochondrion: progesterone synthesis, C-18 hydroxylation and C-11 hydroxylation

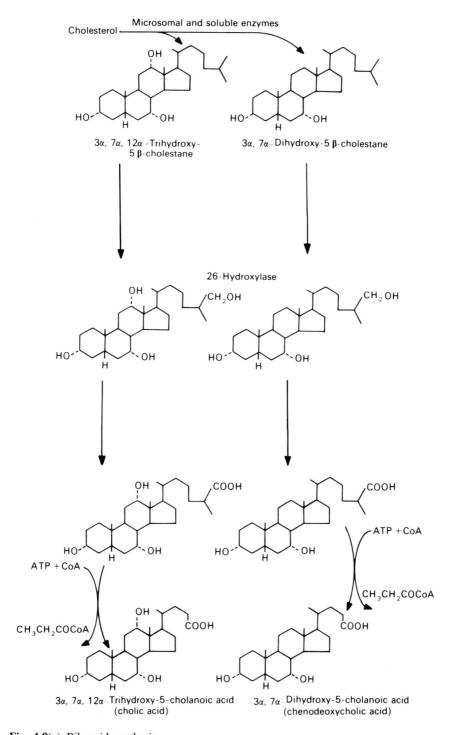

Fig. 4.9(c) Bile acid synthesis.

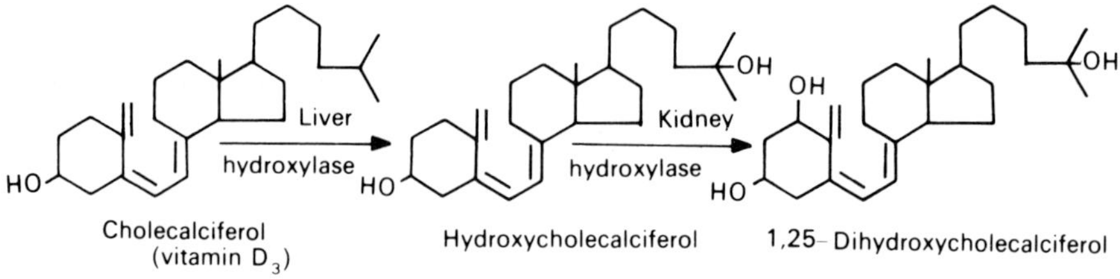

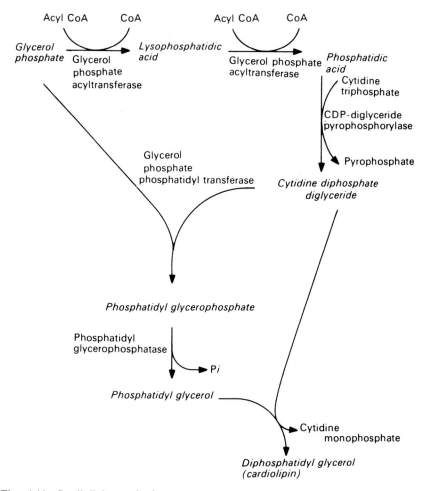

Fig. 4.9(d) Cholecalciferol metabolism.

Adrenal cortical mitochondria also catalyse C-18 hydroxylations in the synthesis of oestrogen and 18-hydroxyoestrone and a C-11 hydroxylation in corticosteroid synthesis (Fig. 4.9b). P450 is involved in all these hydroxylations. Although not fully understood, it is now clear, however, that there are several P450 enzymes each catalysing specific hydroxylations.

Fig. 4.10 Cardiolipin synthesis.

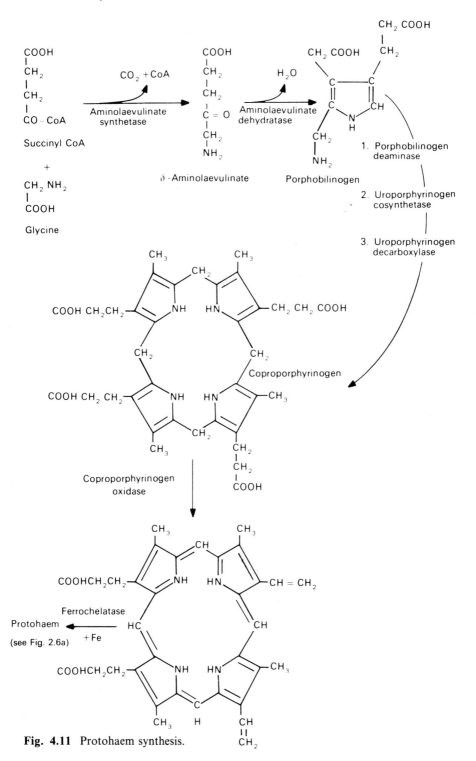

Fig. 4.11 Protohaem synthesis.

Vitamin D which has an important role in calcium metabolism, is metabolised to a biologically more active form initially in the liver to the 25-hydroxy derivative and subsequently in kidney to the 1,25-dihydroxy derivative (Fig. 4.9d). Enzymes for these reactions are located in the mitochondrion and appear to involve hydroxylation systems similar to those described above.

Cholesterol may be converted to bile acids in liver (Fig. 4.9c). Hydroxylations of the steroid nucleus occur in the microsomal fractions although some soluble enzymes are involved. Modification of the side chain occurs at the liver mitochondrial inner membrane.

VIII. Phospholipid metabolism. Most of the phospholipid in the mitochondrial membranes is synthesised in the endoplasmic reticulum and transferred to the mitochondrion. A probable exception to this is the cardiolipin which appears to be exclusively located in the inner mitochondrial membrane. The enzymes for cardiolipin synthesis (see Fig. 4.10) seem to be associated with the outer membrane. The existence of other enzymes of phospholipid metabolism in the mitochondrion appears to be a matter for debate.

IX. Porphyrin and haem synthesis. The mitochondrion contains a number of porphyrin-containing proteins, the cytochromes. Porphyrin synthesis occurs partly in the matrix and partly in the cytosol. The first step in the pathway (Fig. 4.11) takes place in the matrix and uses glycine and succinyl coenzyme A to form aminolaevulinate. The subsequent steps to coproporphyrinogen occur in the cytosol. Coproporphyrinogen oxidase and ferrochelatase are mitochondrial enzymes converting coproporphyrinogen to protohaem, the prosthetic groups of the *b*-cytochromes and haemoglobin.

X. Nucleic acid and protein synthesis. This is a property of the mitochondrial matrix and will be discussed in the next chapter.

XI. Miscellaneous enzyme activities. Ubiquinone found primarily in the inner mitochondrial membrane (although also reported in the nuclear membrane) appears to be synthesised in the mitochondrion. Hydroxymethylglutaryl coenzyme A is an intermediate in the synthesis of ketone bodies and of the isoprenoid side chain of ubiquinone. The enzyme reducing the coenzyme A ester to mevalonic acid, a major precursor of isoprenoids, is microsomal in liver. In yeast, where there is no ketone body formation, the reductase is mitochondrial. However, liver mitochondria appear able to synthesise ubiquinone from mevalonate and benzoate.

Evidence is also accumulating for the synthesis of NAD from nicotinamide by mitochondria.

Further Reading

Munn, E. A. (1974) *The Structure of Mitochondria.* Academic Press, London.

Chapter 5
Mitochondrial biogenesis

5.1 Early approaches to mitochondrial biogenesis

Since 1965, interest in the protein-synthesising and genetic systems of the mitochondrion has greatly increased. In this discussion two issues will be examined. Firstly, the evidence for, and the properties of, the protein-synthesising system in the mitochondrion will be considered. Secondly, the role of the mitochondrial genetic machinery in synthesising mitochondrial proteins will be assessed. The understanding of mitochondrial protein synthesis has been based on the general principles elucidated for protein and nucleic acid synthesis in bacteria and in the eukaryotic nucleus and cytoplasm. In a great many respects the mitochondrial system resembles that in bacteria much more closely than that in the cytoplasm of eukaryotes. It has become apparent that proteins found in the mitochondrion are products of two separate systems; while most of them are synthesised in the cytoplasm, some are synthesised in the mitochondrial matrix. Similarly, the genetic information is coded mainly in the nucleus but also in the mitochondrial genome. The proteins synthesised in the mitochondrion are found firmly bound to the inner membrane while the more soluble proteins are synthesised extramitochondrially.

Incorporation of radioactively labelled amino acids into protein by a mitochondrial preparation was demonstrated by McLean and others in 1958. The problem in these early experiments was whether the observed protein synthesis was due to activity of the mitochondrial particle or to contamination of the fraction. Mitochondrial protein synthesis could be shown to be sensitive to inhibitors such as chloramphenicol, but insensitive to cycloheximide, whereas the reverse is true of the cytoplasmic synthesis. Hence, the observed synthesis could not be attributed to cytoplasmic contamination of the mitochondrial fraction. However, sensitivity to chloramphenicol is characteristic of bacterial protein synthesis and this left the possibility that protein synthesis in the mitochondrial fraction was due to bacterial contamination. Mitochondrial preparations normally contain some bacteria but it was eventually shown that the level of protein synthesis bore no relation to the size of the bacterial population. Furthermore, mitochondria prepared aseptically still synthesised protein.

An additional problem arose from the failure to demonstrate incorporation of radioactively labelled amino acids into specific isolatable proteins such as malate dehydrogenase and cytochrome *c*. Radioactive amino acids were incorporated into insoluble components of the inner membrane. The demonstration of the synthesis of specific proteins by mitochondrial systems had to await the application of techniques for the isolation of firmly bound membrane proteins.

5.2   The nucleic acids of the mitochondrion

(a) Mitochondrial DNA

The existence of extranuclear genes and hence DNA has been realised for a long time. Interest in the possible presence of DNA in the mitochondrion was stimulated by the demonstration of DNA in chloroplasts (see sect. 12.7) from about 1960 onwards. Early claims for the presence of DNA in mitochondrial preparations had been dismissed as due to contamination by nuclei. More convincing evidence came from Nass and Nass (1963) who showed the presence of fibre-like inclusions in the matrix of chicken embryo mitochondria; these inclusions were specifically destroyed by DNase. Schatz and coworkers (1964), using purified yeast mitochondrial preparations, showed the presence of a small and constant amount of DNA associated with the mitochondria. In the same year Luck and Reich (1964) isolated DNA from *Neurospora*. By 1965, Nass and coworkers were able to conclude that 'DNA is an integral part of most and probably all mitochondria'.

Mitochondrial DNA (mtDNA) consists of a small duplex (Fig. 5.1) usually circular except in the case of some protozoa; it is probably bound to the inner mitochondrial membrane on the matrix face. The length of the duplex varies, being longer in plants and micro-organisms and shorter in higher animals (Table 5.1). The mtDNA of animals (length approx. $5\,\mu$m) has a molecular weight of about 10^7 and is composed of about 17 000 base-pairs. In petite mutants of yeasts (phenotypically characterised by small colonies) the mtDNA is smaller than in the wild-type strains or even absent and the mitochondrial structure is altered. The base composition of mtDNA is unrelated to that of nuclear DNA of the same species and the two DNAs can normally be separated by density gradient centrifugation. The guanine:cytosine (GC) content of mtDNA is particularly low in yeasts.

The DNA is replicated semiconservatively by a mitochondrial DNA polymerase which appears to be identical with the nuclear DNA polymerase-γ in mammalian cells (Bertazzoni *et al.*, 1977). Experiments using ^{15}N or bromodeoxyuridine to label new strands show that each duplex is composed of one old and one new strand. Isolated mitochondria incorporate deoxyribonucleotides into their endogenous DNA. Tritiated dATP requires dGTP, dCTP and dTTP for its incorporation into mtDNA. Actinomycin and mitomycin C inhibit the synthesis.

Electron micrographs of DNA released from mitochondria by osmotic shock show the presence of circular double-stranded DNA. However, other types of molecule are also found including the D-loop closed circular DNA discovered by Kasamatsu *et al.* (1971) in mammalian mtDNA. The D-loop is

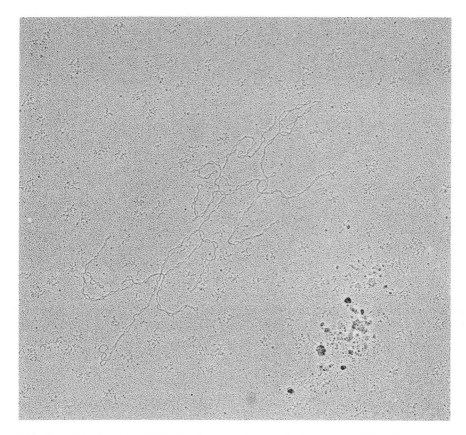

Fig. 5.1 Mitochondrial DNA. An electron micrograph of a circular DNA (26μ contour length) released from yeast (*Saccharomyces carlsbergensis*) by osmotic shock × 30 600. (From Hollenberg *et al.*, 1969, *Biochim. Biophys. Acta* **186**, 417–19).

formed by separation of the strands of the duplex in a limited region and the synthesis of a small piece of DNA complementary to the L^1 (light) strand (Fig. 5.2b). This additional piece of DNA can be separated from the rest of the duplex at elevated temperatures as 7S DNA; it has a variable length with a mean close to 550 nucleotides in mouse liver, a little longer in human liver (Gillum and Clayton, 1978). Synthesis of 7S DNA is rapid and the variable lengths suggest several termination sites. The replication of the circular duplex proceeds by extension of the 7S fragment unidirectionally as shown in Fig. 5.2.

Mitochondrial nucleases, a DNA ligase and a system for DNA repair have also been detected.

(b) Mitochondrial RNA

The first isolation of relatively undegraded mtRNA was carried out by Wintersberger (1966) from yeast. Three species were obtained with sedimen-

---

[1] Separation of the two strands of the duplex gives strands of different buoyant densities (H, heavy and L, light) which are separable on gradients. The different densities reflect different base compositions.

Table 5.1 Size and structure of mtDNAs

Species	Structure	Contour length(μm)	Molecular weight ($\times 10^{-6}$)
Protozoa			
Tetrahymena	linear	15	35
Acanthamoeba	circular	13	
Plasmodium	circular	10.3	18
Paramecium	linear	14	30–35
Algae			
Chlamydomonas	circular	4–5.4	9.8
Fungi			
Saccharomyces cerevisiae	circular	25	50
Schizosaccharomyces pombe	circular		17
Kluyveromyces	circular	11	24
Neurospora	circular	19–26	40
Aspergillus	circular	11	20–26
Higher plants			
Pisum (Pea)	circular	30	
Phaseolus (Red bean)	circular	20	
Platyhelminthes			
Hymenolepis	circular	4.8	
Nematoda			
Ascaris	circular	4.8	
Annelida			
Urechis	circular	5.9	
Arthropoda			
Musca (house fly)	circular	5.2	
Drosophila	circular	6.2	12.4
Echinodermata			
Echinoidea	circular	4.6–4.9	
Chordata			
Fish	circular	5.4	
Amphibia	circular	4.9–5.8	
Birds	circular	5.1–5.4	
Mammals	circular	4.7–5.6	
Human	circular	4.7	9.8

tation constants of 23S, 16S and 4S, the two larger being attributed to ribosomal RNA and the smaller to tRNA. Isolation of mitochondrial ribosomes was achieved by Kuntzel and Noll (1967) and Rifkin *et al.* (1967) from *Neurospora* and by O'Brien and Kalf (1967) from rat liver. In each case it was shown that the ribosomes clearly differed from those in the cytoplasm. The sedimentation constants for ribosomes from a variety of sources are shown in Table 5.2. Only those from *Tetrahymena* have the same values as cytoplasmic ribosomes but even here the mitochondrial ribosomes dissociate to give different sized subunits. In

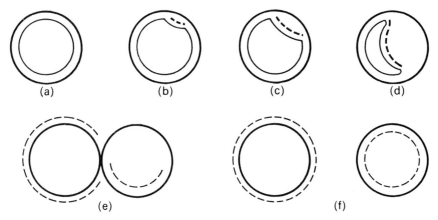

Fig. 5.2 Proposed scheme for replication of mtDNA:
(a) shows circular double-stranded mtDNA molecule; (b) shows the insertion of a D-loop; (c) and (d) the extension of the new DNA strand unidirectionally; (e) the completion of one double stranded molecule and replication of the second strand; and (f) completion of replication to form two double-stranded molecules.

Table 5.2 Sedimentation constants of ribosomes and rRNA

	Mitochondrion			Cytoplasm		
	Ribosome	Subunits	rRNA	Ribosome	Subunits	
Tetrahymena	80	55	21 14	80	60	40
Euglena	72	50 32	21 16	87	67	46
Yeast	72	50 38	23 16	80	60	38
Neurospora	73	51 37	23 16	77	60	37
Higher plants	77–78	60 44	26 18	80	60	40
Locust	60	40 25	16 12	80	60	40
Mammals	55	39 28	16 12	80	60	40

the photosynthetic micro-organism, *Euglena*, three types of ribosome have been distinguished, 72S (mitochondria), 87S (cytoplasm) and 69S (chloroplast).

In general, there are two sizes of ribosomes found in mitochondria, the smaller 55–60S type found in higher plants and animals and the larger 70–80S type found in lower organisms. Like the cytoplasmic particles, the mitochondrial ribosomes dissociate to give two subunits. In rat liver, for example, the 55S particle ($MW = 2 \cdot 5 \times 10^6$) dissociates to give 39S ($MW = 1 \cdot 6 \times 10^6$) and 28S ($MW = 0 \cdot 86 \times 10^6$) subunits. The large subunit possesses 16S rRNA ($MW = 0 \cdot 53 \times 10^6$) and the smaller subunit 12S rRNA ($MW = 0 \cdot 30 \times 10^6$). About 40 proteins, equivalent to a molecular weight of $1 \cdot 1 \times 10^6$ for the large and $0 \cdot 56 \times 10^6$ for the small subunit, are associated with each of the subunits. Although these proteins are synthesised mainly in the cytoplasm, they are different from those found in the cytoplasmic ribosome. Like bacterial ribosomes but unlike cytoplasmic ones, mitochondrial ribosomes dissociate at low Mg^{2+} concentrations. The base composition shows a small proportion of methylated bases. rRNA will hybridise[2] with mtDNA, suggesting that the sequence of bases matches that in a segment of the mtDNA and was synthesised with the DNA as template.

Although 5S rRNA is found in bacterial and also in choloroplast ribosomes, this RNA species is missing from mitochondrial ribosomes of animals and micro-organisms, although it has been reported in higher plants. Several electron microscopic studies have suggested that the ribosomes of mitochondria are attached to the inner membrane (matrix face) of the mitochondrion. Such a location would be consistent with their role in the synthesis of hydrophobic membrane proteins. The attachment of cytoplasmic ribosomes to the outside of the mitochondrion has also been demonstrated.

As noted above, mitochondria possess a 4S RNA fraction which is attributed to tRNA. In the cytoplasm more than one tRNA is found for each amino acid; it now appears that a comparable situation is found in the mitochondrial matrix. Three forms of $tRNA^{leu}$ have been found and more than one tRNA for some other amino acids. At least 25 different species of tRNA can be distinguished (de Vries *et al.*, 1978). They are distinct from those found in the cytoplasm and many of them have been shown to hybridise with mtDNA, suggesting that they are coded in the mitochondrial genome. Mitochondrial tRNA contains methylated bases, but the degree of methylation is not as great as in cytoplasmic tRNA. A mitochondrial methylase has been isolated and is distinct from the cytoplasmic one. Similarly the aminoacyl tRNA synthetases of the mitochondrion are also found to be distinct from those in the cytoplasm.

5.3 Protein synthesis

(a) Transcription of mtDNA: synthesis of mtRNA

Isolated mitochondria will incorporate labelled uridine into an unstable RNA fraction suggesting that the DNA duplex of the mitochondrion is transcribed. DNA-dependent RNA polymerase has been isolated from mito-chondria of several species. The enzyme, which is synthesised in the cytoplasm, has a subunit molecular weight of 67 000 in rat liver and differs substantially from other eukaryotic polymerases. It is inhibited by rifampicin which inhibits many bacterial polymerases but not by amanitin which inhibits several eukaryotic nuclear RNA polymerases.

In microbial and animal tissue culture cells it is possible to demonstrate the synthesis of labelled mitochondrial rRNAs if cells are incubated with ^{32}P and pulse-labelled with [^3H] uracil: this synthesis unlike that in the nucleus is sensitive to ethidium bromide. The isolation of the mitochondrial polymerase made

---

[2] When DNA is denatured to form separate strands and then allowed to renature in the presence of RNA synthesised from homologous DNA, the RNA will form hybrid molecules by hydrogen-bonding to the DNA. Such a hybrid will only form if there are nucleotide sequences in the DNA complementary to those in the RNA.

possible the demonstration of mitochondrial transcription *in vitro*. The polymerase incubated with mtDNA gave two high molecular weight RNAs identified as ribosomal RNA and smaller molecules of 4–5S type. Perlman *et al.* (1973) used labelled human cells (HeLa cells) to obtain transcribed mtRNA. The 4S fraction was found to contain heterogeneous RNAs with covalently attached poly(A). Since poly(A) is found in cytoplasmic mRNA, this fraction was assumed to be the mitochondrial mRNA. The poly(A) segment of RNA from higher animals is composed of about 50 to 70 nucleotides which are probably added after transcription. In yeast the poly(A) segment is much smaller, about 20–30 nucleotides. A poly(A) polymerase has been separated from the RNA polymerase (Gallerani *et al.*, 1976).

An alternative approach to the study of mRNA has been based on the isolation of polysomes, that is complexes of mRNA and ribosomes. Separation of the mRNA gives a fraction which is composed of both poly(A)-containing and poly(A)-lacking mRNA (Lewis *et al.*, 1976). By this means at least eight species of mRNA can be isolated and these hybridise with mtDNA.

The RNA destined to become rRNA appears to undergo modification after transcription, including a reduction in molecular weight.

(b) Translation of mRNA: synthesis of protein

The mRNA fraction from mitochondria can be translated in bacterial (*Escherichia coli*) protein-synthesising systems. The products have been analysed and include polypeptides, such as are found in the cytochrome oxidase complex, which are known from other experiments to be synthesised in mitochondria (Padmanaban *et al.*, 1975). Initiation of protein synthesis requires formylated methionyl tRNA as in bacteria.

A simpler and widely adopted approach to determine which proteins are synthesised in the mitochondrion makes use of the difference in sensitivity to inhibitors of mitochondrial and cytoplasmic protein synthesis. The latter is inhibited by cycloheximide whereas the former, like bacterial protein synthesis, is inhibited by chloramphenicol (also erythromycin and lincomycin). Incubation of cells with labelled amino acids results in labelling of mitochondrial proteins. Those labelled in the presence of chloramphenicol are assumed to be synthesised in the cytoplasm, whereas those labelled in the presence of cycloheximide and not in the presence of chloramphenicol are presumed to be synthesised in the mitochondrion. Isolated mitochondria will also incorporate labelled amino acids into mitochondrial polypeptides and this process is inhibited by chloramphenicol.

The role of the mitochondrial biosynthetic system may also be studied in petite mutants of yeast. A few petite colonies can be seen in normal yeast cultures, but they can also be induced by treatment with acriflavin or ultraviolet light. Petite mutants lack all or part of their mtDNA, have a deficient mitochondrial protein-synthesising system, and are usually deficient in respiration due to deficiencies in the cytochrome oxidase and the $b–c_1$ region of the chain. The cristae are generally few and may be disorganised. Examination of the mitochondrial morphology shows no detectable changes in the outer membrane, but small changes in the inner membrane.

The analysis of petite strains of yeast, the effects of inhibitors on the synthesis of mitochondrial protein and studies on the transcription and translation of mtDNA show that some polypeptides in the cytochrome oxidase complex, in the ATPase complex and in the cytochrome $b-c_1$ complex are synthesised in the mitochondrion. Almost all the remaining mitochondrial protein, more than 90 per cent of the total, is synthesised in the cytoplasm on the basis of information coded in nuclear genes. Mitochondria do not appear to be able to import mRNA from the cytoplasm as once proposed; hence all proteins synthesised in the mitochondrion must be coded in mitochondrial genes located in the circular duplex of mtDNA.

Until very recently, it has been assumed that the genetic code which specifies the amino acid sequence in proteins is universal. However, it now seems that in mammalian mitochondria, some codons may differ from those of the classical genetic code. UGA for example codes for tryptophan rather than acting as a termination codon (Barrell *et al.*, 1979).

(c) Cytochrome oxidase

In *Neurospora*, yeast and mammalian mitochondria, this enzyme can be isolated from the inner mitochondrial membrane and shown to be composed of seven polypeptides of varying molecular weights: 42 000 (I) to 5,000 (VII) in yeast (see sect. 9.8). The synthesis of active enzyme is inhibited by both cycloheximide and chloramphenicol. However, in experiments with labelled amino acids, the synthesis of the three largest polypeptides, I, II and III is inhibited by chloramphenicol and not by cycloheximide (Mason and Schatz, 1973; Rubin and Tzagoloff, 1973). These large polypeptides, two of which are deeply embedded in the membrane in yeast, are therefore synthesised on mitochondrial ribosomes. The synthesis of the smaller polypeptides occurs in the cytoplasm. In petite mutants of yeast, mitochondria possess loosely bound polypeptides IV–VI, but lack I–III.

(d) The ATPase

This enzyme consists of at least ten polypeptides (see sect. 9.1). Of these, five are associated with the cold-labile ATPase which is equivalent to F_1, one is concerned with oligomycin sensitivity and the remaining four are hydrophobic. F_1-ATPase activity is present in cytoplasmic petite mutants of yeast and in cells grown in the presence of chloramphenicol. It is synthesised in the cytoplasm. However, the four hydrophobic polypeptides are synthesised in the mitochondrion (Tzagoloff and Meagher, 1972).

(e) Cytochrome b

There are at least two b-cytochromes in animal and microbial mitochondria. The apoproteins of these cytochromes are either similar or identical and have a molecular weight of about 25 000. The apoprotein is synthesised in the mitochondrion. The synthesis in the mitochondrion of one or more additional polypeptides of the cytochrome $b-c_1$ complex has also been proposed.

(f) Cytoplasmic synthesis of mitochondrial protein

Most of the protein in the mitochondrion is synthesised in the cytoplasm and hence must be transferred to the mitochondrion. Although little is known about this process, some recent studies throw some light on the process. For example, two different aspartate aminotransferases (transaminases) are found in eukaryotic cells, one is cytoplasmic and the other is in the mitochondrial matrix. If mitochondria are incubated with these enzymes, only the mitochondrial one is transferred to the matrix (Marra *et al.*, 1978). In the case of cytochrome *c*, the apoprotein is synthesised on cytoplasmic ribosomes, but the functional cytochrome appears to be formed on the inner membrane where the prosthetic group, protohaem, is covalently attached to the protein, the haem being synthesised in the matrix (Korb and Neupert, 1978). The synthesis in the cytoplasm of precursors of the three large subunits of the F_1-ATPase has been demonstrated. These precursor polypeptides are larger than the mature subunits and undergo a reduction in size before incorporation into F_1 (Maccecchini *et al.*, 1979).

(g) Regulation of the synthesis of mitochondrial protein

From the foregoing it is clear that several important components of the mitochondrion are formed by the co-operative activity of the cytoplasmic and mitochondrial protein-synthesising systems. This applies to the ATPase, the cytochrome oxidase, the cytochrome $b-c_1$ complex and also to the ribosomes where the rRNA is coded in the mtDNA but most, although perhaps not all, ribosomal proteins are synthesised outside the mitochondrion. Similarly, while most or all of the mitochondrial tRNA is transcribed from mtDNA, the aminoacyl tRNA synthetases required to form the charged tRNA are synthesised extramitochondrially. It would therefore be reasonable to expect regulatory mechanisms to operate between the mitochondrial and nuclear-cytoplasmic systems. A number of experiments have suggested that such regulation exists. For example, cycloheximide which inhibits cytoplasmic synthesis also appears to reduce mitochondrial synthesis *in vivo*, suggesting that the mitochondrial system requires cytoplasmic products for its proper function. Evidence for a mitochondrial pool of cytoplasmically synthesised protein necessary for mitochondrial protein synthesis has also been produced. Barath and Kuntzel (1972) showed in *Neurospora* that chloramphenicol treatment stimulated the synthesis of a mitochondrial protein (RNA polymerase) formed in the cytoplasm. This was interpreted as inhibition of mitochondrial formation of a repressor for cytoplasmic synthesis. However, there is little evidence at present for regulation of the synthesis of mitochondrial proteins.

A rather different regulatory problem arises in the transition from anaerobically grown yeast with only non-functional promitochondria to aerobic cells with fully functional organelles. The promitochondria possess the normal mtDNA. The presence of oxygen is necessary for the formation of at least two of the mitochondrially synthesised subunits of cytochrome oxidase.

5.4   Mitochondrial genetics: mapping the mitochondrial DNA

This discussion will be restricted to the mitochondrial genome of yeast, *Saccharomyces cerevisiae*, where most progress has been made. The DNA duplex is 25 μm long, has a molecular weight around 50 000 and contains 68–75 kb (kilobases). Mitochondrial mutants resistant to antibiotics such as oligomycin, erythromycin, paromomycin and chloramphenicol, were used in early studies. These attempted to carry out a transmission and recombination analysis by crossing haploid strains to form diploid zygotes which initially possessed mitochondria from both parents. Evidence of recombination was obtained both from the appearance of new combinations of markers and from the formation of DNA of a different buoyant density to that of either parent. More recent evidence of recombination (Fonty *et al.*, 1978) has been based on analysis with restriction endonucleases (see below). It appears that several rounds of recombination occur and that, after a period, mitochondria possess a uniform genome. Analysis of crosses produced little information since most markers were not linked. However, it did prove possible to determine the order of a few markers on the genome.

A more successful approach came from the hybridisation studies with *petite* mutants. A range of petites can be isolated with reduced mtDNA in which there is a deletion. If spontaneous petites are used (as opposed to those induced by acridines, etc.) the number of secondary deletions will be minimised. The analysis of the petites proceeds by DNA–DNA hybridisation in which the mtDNA from the petite is denatured and bound in excess to a filter. Labelled mtDNA from the *grande* parent strain is also denatured and the fraction of the grande mtDNA not hybridising with the petite DNA on the filter is measured. This fraction represents the magnitude of the deletion. The analysis is extended by testing hybridisation between mtDNAs of different petites. Where the mtDNA of one petite overlaps with that of another, hybridisation will occur. If labelled mtDNA is used, the minimum size of the overlap can be estimated. Information from such studies is combined with a knowledge of the presence of antibiotic resistance factors in specific petites. This type of study forms the basis for a simple genetic map (Fig.5.3).

Restriction endonucleases (such as *Eco*R1 or *Hin*dIII) have also proved a powerful tool in the analysis of mtDNA. The enzyme cleaves the mtDNA at several specific sites giving a uniform series of fragments which can be separated electrophoretically. These fragments can be mapped by the DNA-hybridisation technique outlined above.

The ability of rRNA to hybridise with some segments of mtDNA and not with others enables these genes to be located. A number of methods have been used to tag the progressively increasing number of known tRNAs which (in a denatured form) can be hybridised with the mtDNA and the hybridisation site mapped (Morimoto *et al.*, 1978).

Recent studies (see for example Moorman *et al.*, 1978) demonstrate the possibility of mapping the genes for specific polypeptides. Fractionation of mtRNA gives mRNA species which can be transcribed by bacterial (*E. coli*)

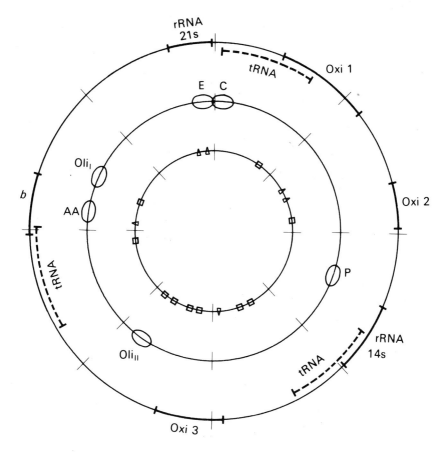

Fig. 5.3 The genetic map of the yeast mitochondrial genome.
The approximate locations are shown of the sites of action of two restriction endonucleases (inner circle, *Eco*R1 □ and *Hind*III △), the sites of antibiotic resistance markers (middle circle, erythromycin E, chloramphenicol C, paromomycin P, oligomycin Oli$_I$ and Oli$_{II}$ and antimycin AA) and the genes for rRNA, tRNA, cytochrome *b* and cytochrome oxidase (Oxi 1–3). (See Morimoto *et al.*, 1978 and Sanders *et al.*, 1977.)

systems and which can be hybridised with mtDNA. Thus the RNA-hybridising segment can be correlated with specific polypeptides. Direct transcription–translation studies with bacterial systems have so far been less successful.

A tentative map of the mtDNA of yeast is shown in Fig. 5.3. A comparable map of the mammalian mtDNA (rat liver) can be seen in Kroon *et al.* (1977) and Barrell *et al.* (1979). In the case of the mtDNA from mammals with a molecular weight of about 10^7, it is estimated that the known mitochondrial functions of the mammalian mitochondrial genome account for about 70 per cent of the total DNA. The function of the remaining mtDNA in mammals and particularly in yeast where the genome is much longer and known genes account for 20–25 per cent of the mtDNA, remains to be understood.

Note added in proof

The human mitochondrial genome has been sequenced and shown to consist of 16569 base pairs. Among the interesting aspects of this work is the discovery that the genetic code is not universal. In human mitochondria, UGA codes for tryptophan rather than 'stop' while AGA and AGG code for 'stop' rather than arginine. AUA codes for methionine while AUA and AUU code for initiation in preference to the normal AUG. In yeast mitochondria, the normal code applies except that UGA codes for tryptophan and CTn for leucine rather than threonine. Also of great interest is the demonstration in human mtDNA of eight unassigned reading frames presumed to code for eight unidentified polypeptides. Human mtDNA codes for 22 tRNA together with two tRNAs (1559 and 954 base pairs as opposed to 3200 and 1660 in yeast). The mRNA appears to lack leader and trailer sequences (which would not be transcribed); indeed the whole genome shows extreme economy with very few non-coding bases.

For a map of the mitochondrial genome see Borst, P. and Grivell, L. A. (1981) *Nature* **290** 443–4.

See also: Anderson, S. *et al.* (1981) *Nature* **290** 457–65; Montoya, J. *et al.* (1981) *Nature* **290** 465–70; Ojala, D. *et al.* (1981) *Nature* **290** 470–4.

Further Reading

Nagley, P., Sriprakash, K. S. and Linnane, A. W. (1977) Structure, synthesis and genetics of yeast mitochondrial DNA. *Adv. Microbial Physiol.*, **16**, 158–277.

Schatz, G. and Mason, T. L. (1974) The biosynthesis of mitochondrial proteins *Ann. Rev. Biochem.*, **43**, 51–87.

Tzagoloff, A. and Macino, G. (1979) Mitochondrial genes and translation products. *Ann. Rev. Biochem.*, **48**, 419–41.

Chapter 6

Mitochondrial water movement and substrate transport

6.1   Compartmentalisation

Compartmentalisation of cellular metabolism is an important aspect of the evolution of eukaryotic cells. The mitochondrion, bounded by membranes of limited permeability, provides the cell with compartments within which certain processes are confined. The outer membrane, with a permeability limit equivalent to a molecular weight of 5000–10 000, is permeable to most substrates and ions but not to proteins. The inner membrane is selectively permeable and is readily permeated by water and lipophilic substances only. It does possess permeation systems for a number of metabolites including adenine nucleotides, phosphate, pyruvate, acylcarnitine esters, dicarboxylic and tricarboxylic acids and some amino acids.

6.2   Water movement: mitochondrial swelling

As far back as the nineteenth century, it was realised that mitochondria showed osmotic behaviour such as swelling in dilute solutions. In the 1950s, studies showed mitochondria to behave as simple osmometers when suspended in solutions of non-permeating solutes such as mannitol. They underwent rapid and usually reversible volume changes according to the relationship:

$$V = k/\pi + V_{\pi \to \infty}$$

where V is the volume of the organelle, k is a constant, π the osmotic pressure and $V_{\pi \to \infty}$ is the osmotic dead space which is not involved in osmosis. The volume of the osmotically sensitive part of the mitochondrion is inversely proportional to the osmotic pressure of the surrounding solution. At least the major part of the osmotic dead space is the sucrose-permeable space, i.e. the intermembrane space including the cristal space. Osmotic swelling of mitochondria is seen when the particles are suspended in hypotonic solutions or in solutions containing permeant ions. This type of large-amplitude swelling leads to disruption of the outer membrane. The inner membrane will contract, extruding water from the matrix, if ATP and Mg^{2+} are added (Parsons *et al.*, 1966), but this is not necessarily the reverse of swelling.

In addition to volume changes attributable to the simple osmotic effects described above, small-amplitude changes in mitochondrial volume associated with energetic aspects of metabolism have also been observed. Packer (1960–62) measured changes in light scattering of mitochondrial suspensions. It was assumed that the increase in light scattering was due to low-amplitude swelling and the decrease to contraction. The light-scattering properties were correlated with the metabolic state (see Table 2.2) of the mitochondria which were swollen in states 1 and 4 and contracted in states 2, 3 and 5. Hence mitochondria with ADP-limited respiration contracted on addition of ADP (state 4 to state 3 transition). Swelling could also be induced by oligomycin-sensitive ATP hydrolysis and was reversed by uncouplers. Thus the creation of an energised state by ATP hydrolysis or by respiration in the absence of ADP leads to swelling, while dissipation of the energised state results in contraction.

Light-scattering changes were found to be due to internal morphological changes rather more than with total swelling and shrinking of the organelles. Hackenbrock (1966) found that mitochondria in state 3 were in the 'condensed mode' (Fig. 6.1b). The orthodox mode (Fig. 6.1a) was characterised by small cristal and intermembrane spaces while mitochondria in the condensed mode had expanded cristal spaces and a dense matrix. Mitochondria *in situ* are seen to be in the orthodox mode.

Although there may be some contraction in the transition from state 4 to state 3, it is clear that there is a loss of water from the matrix and also an increase in the external (intermembrane) mitochondrial space. Two views have been expressed on the morphological changes associated with metabolic state. Some authors have seen changes in conformation of the membrane as an expression of the intermediate high-energy state and detailed interpretations have been based on this view (Penniston *et al.*, 1968; Harris *et al.*, 1968). The changes are held responsible for driving water from the matrix. An alternative view, widely held, is that the volume changes reflect movement of water in response to a redistribution of ions associated with change of metabolic state.

6.3   Permeability to substrates

(a) Factors influencing substrate transport

Since many of the mitochondrion's metabolic processes are intimately linked to those in the extramitochondrial region of the cell, it is clear that the inner membrane must possess mechanisms which enable metabolites to enter and leave the inner mitochondrial compartment. Uncharged molecules are frequently able to permeate the inner membrane readily. However, charged molecules require carrier systems. Thus the membrane is impermeable to NH_4^+ but permeable to NH_3 (Table 6.1). In considering the movement of ions across the membrane, several points must be considered.

(i) There is a need to maintain electrical neutrality. The movement of charged ions in one direction may be balanced either by movement of an ion of the same sign in the opposite direction or movement of an ion of opposite sign in the same direction.

(ii) The movement of an ion across a membrane will create a potential across the membrane. For example the movement of a positively charged ion into the mitochondrion will result in a potential across the inner membrane, positive

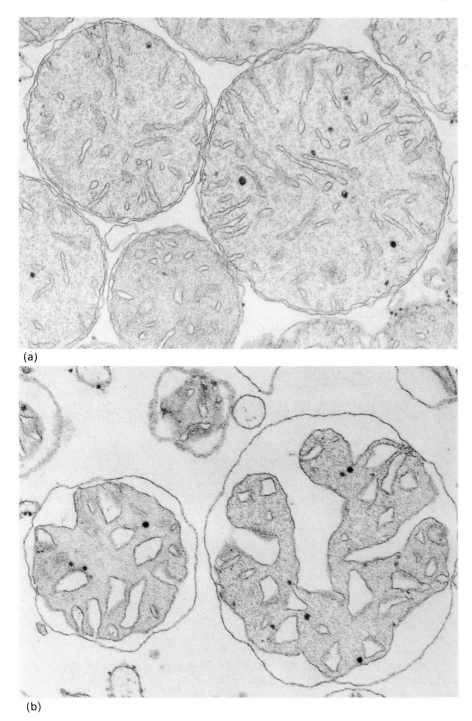

(a)

(b)

Fig. 6.1 Structure of the mitochondrion under various metabolic conditions × 70 000 (Hackenbrock, 1966).
(a) Orthodox mode characteristic of state 4 (ADP-limited) and also of mitochondria *in situ*. (b) Condensed mode characteristic of state 3.

Table 6.1 Permeability of the mitochondrion to anions and metabolites

Substances to which the membrane is permeable (no translocation mechanisms involved)	Substances for which there is a translocation system in the inner membrane	Substances which do not readily permeate the inner membrane
acetate	phosphate (arsenate)	Cl^-, NO_3^-, Br^-, NH_4^+
formate	ADP, ATP	fumarate, (oxaloacetate)
propionate	pyruvate	NAD, $NADH_2$, NADP,
butyrate	hydroxybutyrate,	$NADPH_2$
H_2O, O_2, CO_2, NH_3	acetoacetate	acetyl CoA, succinyl CoA
	malate, succinate	coenzyme A, acyl CoA esters
	(oxaloacetate)	sucrose, mannitol, etc.
	citrate, isocitrate, aconitate	AMP
	glutamate, aspartate	GMP, GDP, GTP, etc.
	proline	
	acyl carnitines	
	phosphoenolpyruvate	
	sulphate, sulphite	
	ornithine, citrulline	

inside and negative outside. Once formed, such a potential will encourage the entry of anions and exit of cations while discouraging entry of further cations and exit of anions. The movement of ions in response to a potential will tend to dissipate that potential.

(iii) The mitochondrial respiratory chain, when oxidising substrates in the presence of oxygen, pumps protons outwards across the inner membrane. It appears that ATP hydrolysis by the membrane-bound ATPase also pumps protons outwards, while ATP synthesis is associated with the return of the protons to the matrix. Thus an intact inner membrane of a mitochondrion will normally have a potential across the membrane (positive outside) and a pH gradient, the external pH being lower than the internal. The pH gradient will promote systems involving transport of H^+ inwards or OH^- outwards.

(iv) Where the transport of a specific ion is under consideration, it is necessary to determine whether it is transported in the ionised or un-ionised state. The entry of positively charged ions and exit of negatively charged ions (electrogenic transport) will be promoted in coupled mitochondria.

(v) In studying ion uptake, it is necessary to distinguish between binding of the ion to the membranes and net transport into the mitochondrial matrix.

(b) Early studies of substrate transport

Little was known about the permeability of mitochondria to metabolites until the mid-1960s. However, Pfaff and coworkers (1965) demonstrated that exogenous labelled adenine nucleotides would exchange with endogenous mitochondrial nucleotides. This exchange was inhibited by atractyloside, previously regarded as an inhibitor of oxidative phosphorylation itself. This specific exchange of labelled nucleotide discredited the hypothesis that high-energy phosphate (~P) was transported outwards across the inner membrane to phosphorylate ADP externally.These experiments laid the foundation for an understanding of mitochondrial translocation systems (translocases) which operate by way of reciprocal exchange of anions between matrix and intermembrane space (cytosol).

At about the same time Chappell and Crofts (1966) showed that while small ions such as Cl^-, Br^- and NO_3^- (2.3 to 2.7 Å hydrated ion diameter) do not enter the mitochondrion, phosphate (hydrated diameter 5.2 Å for $H_2PO_4^-$ or 6.4 Å for HPO_4^-) penetrates readily suggesting the existence of a carrier mechanism. Penetration was measured as swelling in the presence of a permeant cation. The cations used were either NH_4^+ which relies on the formation of the readily permeant NH_3 or K^+ which will penetrate the mitochondrion if valinomycin is added (Fig.6.2). These experiments lead to the conclusion that phosphate exchanges for OH^-.

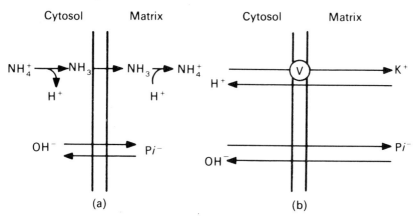

Fig. 6.2 Demonstration of entry of phosphate into mitochondria.
(a) Phosphate enters with ammonia. (b) Phosphate enters with potassium in the presence of valinomycin (V) (see sect. 7.4). The formation of the phosphates in the matrix leads to osmotic swelling.

Using the ammonium swelling technique, Chappell and Haarhoff (1967) showed little or no swelling when succinate was added to intact mitochondria. By contrast, butyrate, propionate and acetate did enter mitochondria. However, it was found that swelling with succinate could be induced if small amounts of phosphate were added. Several other anions, D- and L-malate, and malonate would also enter mitochondria if phosphate were present. These results suggest the presence of a translocase[1] which exchanged phosphate for malate, succinate or similar dicarboxylates. Following these early experiments, the existence of a number of translocase systems was demonstrated in mitochondria (Table 6.2). Because of their specificity, it was assumed and it has now been demonstrated in several cases that translocases are proteins.

[1] The term 'translocase' implies a stereospecific protein which transports substrate across lipid membranes. It is probably a misnomer in that a reaction involving making and breaking of covalent bonds, is not involved.

Table 6.2 Translocation systems

Translocator	Substrates	Inhibitors
Phosphate	Inorganic phosphate, arsenate, OH^-	*N*-Ethylmaleimide, organic mercurials
Dicarboxylate	Inorganic phosphate, D-malate, L-malate, malonate, succinate, *meso*-tartrate, sulphate sulphite	*N*-Butylmalonate organic mercurials
Tricarboxylate	Citrate, isocitrate, L-malate, isomalate, *cis*-aconitate, phosphoenolpyruvate	Benzene-1,2,3-tricarboxylate
Oxoglutarate	Oxoglutarate, L-malate, malonate, succinate (weak), oxaloacetate	
Glutamate	Glutamate	4-Hydroxyglutamate 2-aminoadipate, *threo*hydroxyaspartate avenaciolide
Glutamate/aspartate	Glutamate, aspartate	
Adenine nucleotide	ADP, ATP, phosphoenol-pyruvate	Atractyloside, bongkrekate carboxyatractyloside
Pyruvate	Pyruvate	α-Cyano-4-hydroxycinnamic acid Phenylpyruvate 4-Ketoisocaproate
Carnitine	Carnitine, acyl-carnitines	
Ornithine	Ornithine	
Citrulline	Citrulline	

6.4   Mitochondrial translocation systems

(a) Phosphate

There are probably two independent carrier systems involved in the transport of phosphate, although it has been suggested that they share common components. These two systems may be differentiated by the use of inhibitors (Johnson and Chappell, 1973). The phosphate carrier (P*i*/OH$^-$ exchange) is readily inhibited by reagents reacting with thiol groups such as organic mercurials (e.g. mersalyl) and *N*-ethylmaleimide but it is insensitive to *n*-butylmalonate. In contrast, phosphate transport by the dicarboxylate carrier (P*i*/dicarboxylate exchange) is inhibited by butylmalonate and organic mercurials but not by *N*-ethylmaleimide. Transport by either carrier may be studied in the presence of an inhibitor of the other. Both translocases will transport phosphate into or out of the mitochondrion.

The transport of inorganic phosphate into the matrix of the mitochondrion is an essential part of oxidative phosphorylation. The phosphate translocator probably represents the major route in mammalian tissues and the only one in insect flight muscle since the dicarboxylate system is absent in that tissue. Both carriers are present in plant mitochondria. The phosphate translocator has been isolated from several sources. Banerjee *et al.* (1977) incorporated the beef heart protein into liposomes,[2] where it was shown to catalyse a Pi/Pi and Pi/OH$^-$ exchange. The latter exchange was influenced by an artificially imposed proton gradient.

Although phosphate is usually regarded as exchanging with OH$^-$ when translocated by the phosphate carrier, it is not possible to distinguish between this mechanism and entry of phosphate accompanied by a proton. The existence of a proton gradient in coupled mitochondria under most conditions will promote a Pi/OH$^-$ exchange, resulting in a higher phosphate concentration in the matrix.

(b) Dicarboxylate and tricarboxylate transport

The realisation that entry of succinate into the mitochondrion required phosphate, led to the idea of a carrier mediating phosphate–succinate or phosphate–malate exchange. In intact mitochondria, *n*-butylmalonate was found to inhibit L-malate oxidation while not affecting any of the enzymes required for that oxidation in broken mitochondria. The carrier mediates exchanges of dicarboxylates (succinate or malate) or dicarboxylates and phosphate and probably possesses separate sites for Pi and dicarboxylates. Fumarate is not transported and oxaloacetate only weakly.

Using the ammonium sulphate swelling technique, oxoglutarate, citrate, *cis*-aconitate and isocitrate were shown to require both phosphate and malate for entry. The sensitivity of the malate-dependent citrate and oxoglutarate entry to butylmalonate supports the view that malate must first enter the matrix by the dicarboxylate carrier before being exchanged for tricarboxylate or oxoglutarate (Fig.6.3). Assay (spectrophotometric) of the reduction of intramitochondrial pyridine nucleotides shows, furthermore, that citrate oxidation requires catalytic amounts of malate. These experiments lead to the conclusion that oxoglutarate and tricarboxylates such as citrate enter the mitochondrion in exchange for malate. Consequently phosphate, malate, tricarboxylate and oxoglutarate transport are related (see Fig. 6.3).

There is good evidence for two carrier systems, one for tricarboxylates and the other for oxoglutarate. Firstly, heart mitochondria, which possess an active oxoglutarate translocase, lack significant tricarboxylate translocase activity. Secondly, the range of compounds which will replace malate as the exchanging anion is different for oxoglutarate and the tricarboxylates (Table 6.2). Succinate, a good substrate for the dicarboxylate translocase, is a poor substrate for oxoglutarate and not transported by the tricarboxylate carrier. Most of the low level of oxaloacetate transport is probably due to the oxoglutarate carrier.

Most anion exchange systems can mediate transport in both directions. Chappell and Robinson (1968) showed an exchange of isocitrate for citrate. A suspension of intact mitochondria was incubated with isocitrate and the formation of extramitochondrial citrate measured. Since the aconitase for converting isocitrate is intramitochondrial, the formation of citrate can only

[2] Liposomes are enclosed vesicles formed artificially from phospholipids.

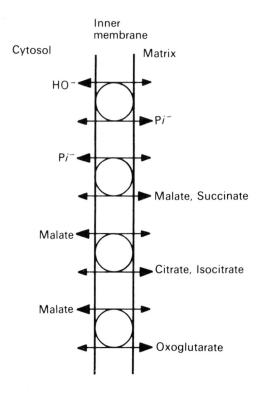

Fig. 6.3 Transport of phosphate, dicarboxylates and tricarboxylates across the inner membrane.

occur through entry of the isocitrate in exchange for citrate.

A major function of the tricarboxylate carrier in lipogenic tissues is the transfer of acetyl groups from the matrix where they are formed, to the cytosol where they are required for lipid synthesis. The membrane is impermeable to acetyl CoA, but this may be converted to citrate which is transported by the tricarboxylate system to the cytosol (Watson and Lowenstein, 1970). In the cytosol it is cleaved by the citrate lyase to acetyl CoA (Fig. 6.4). The tricarboxylate translocase appears to be inhibited by acyl CoA esters; this would provide a form of feedback inhibition of fatty acid synthesis.

(c) Pyruvate (monocarboxylate) transport

Although pyruvate is an important substrate for mitochondrial metabolism, the mechanism of entry into the matrix has been a matter for debate. Earlier, opinions were divided as to whether it was necessary to postulate a carrier-mediated process or whether non-specific diffusion would be adequate to explain the observed transport. However, several workers (Halestrap and Denton, 1974; Mowbray, 1974; Land and Clark, 1974), using mitochondria from various mammalian tissues, demonstrated the existence of a specific translocase which could be inhibited by α-cyano-4-hydroxy-cinnamate, phenylpyruvate and α-ketoisocaproate. The system is generally regarded as catalysing an exchange

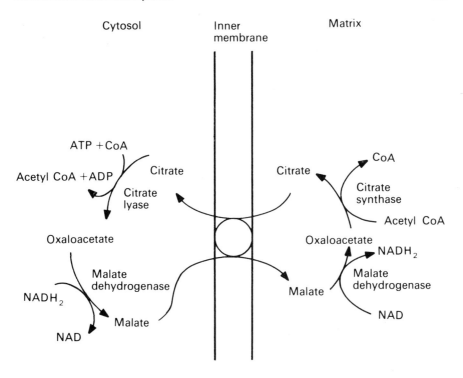

Fig. 6.4 The malate–citrate shuttle. Note that the citrate synthase and citrate lyase reactions are irreversible.

with OH$^-$. In a re-examination of the system, Pande and Parvin (1978) have concluded that only the carrier-mediated transport would be adequate for pyruvate metabolism at physiological concentrations of substrate in the cytosol. At higher concentrations, non-specific diffusion would be adequate to account for pyruvate metabolism by mitochondrial suspensions.

The carrier can be regarded as a monocarboxylate system. In mammalian mitochondria it probably transports acetoacetate and possibly hydroxybutyrate, while in yeast, lactate appears to be transported in addition to pyruvate.

(d) Amino acid transport

Azzi and coworkers (1967) found that glutamate metabolism (measured as intramitochondrial reduction of NAD(P)) was inhibited by glutamate analogues, e.g. 4-hydroxyglutamate in intact mitochondria, but not in broken particles. These observations in mammalian mitochondria led to the elucidation of a glutamate translocase system in which glutamate (like phosphate) enters the mitochondrion in exchange for OH$^-$ (or is co-transported with a proton). The exchange appears to be electroneutral, with no net movement of charge. The system is sensitive to lipid-soluble thiol reagents such as N-ethylmaleimide and to avenaciolide.

A second system for glutamate translocation involves exchange for aspartate. Aspartate metabolism by intact mitochondria requires the presence of glutamate or some of its analogues. The exchange of extramitochondrial glutamate for intramitochondrial aspartate involves the co-transport inwards of a proton. Thus the exchange (glutamate$_{in}$/aspartate$_{out}$) will be promoted by the proton gradient and membrane potential in coupled mitochondria under most conditions (see Tischler *et al.*, 1976). Unlike the translocases discussed previously, this system can be expected to function irreversibly. A glutamate-binding proteolipid has been isolated, which may be the glutamate/aspartate translocase (Julliard and Gautheron, 1978).

In liver mitochondria, where the entry of glutamate is important for nitrogen metabolism through the urea cycle (see Fig. 3.16), both translocases are functional. Brain mitochondria also possess both translocases, but heart probably possesses only the glutamate/aspartate system. In kidney cortex mitochondria, which in mammals have high glutaminase or glutamate metabolism forming ammonia, both translocases are probably present. In rat and pig kidney where mitochondrial glutaminase I is important, the glutamate formed probably leaves the mitochondrion by means of the glutamate translocase, although a glutamine/glutamate translocase has also been proposed. In guinea pig and rabbit, glutamate, which is the major source of ammonia, enters the mitochondrion by either translocase (Bryla and Dzik, 1978).

The urea cycle requires the transport of ornithine into, and citrulline out of the mitochondrion (see Fig. 6.5). Translocase systems for these amino acids have been described; two separate systems are probably involved (Gamble and Lehninger, 1973; Bryla and Harris, 1976). The full nature of the systems has not been elucidated. Meijer and coworkers (1975) found that when liver cells metabolised ammonia, urea synthesis was dependent on malate translocation. They proposed the system shown in Fig. 6.5.

Translocase systems have been proposed for several other amino acids including proline (Meyer, 1977), which is oxidised in the mitochondrion; these have not been studied in detail.

(e) Adenine nucleotide translocation

Oxidative phosphorylation requires the transport of ADP into, and ATP out of the mitochondrion. It has been shown that labelled adenine nucleotides are exchanged across the membrane in a 1 : 1 manner. The system is specific for ADP and ATP; other nucleotides, including AMP and the guanine nucleotides, are not translocated in significant amounts. There is, however, some translocation of deoxyribonucleotides, dATP and dADP. Studies of the kinetics of translocation showed that, with mitochondria in the coupled state, the rate of entry of ATP relative to ADP is low. In uncoupled particles, the rate of ATP entry is the same as that of ADP. When the exit of nucleotides is considered, ATP leaves in preference to ADP in coupled but not in uncoupled particles.

Two approaches have been made to understanding the control of adenine nucleotide transport by the energy state of the mitochondrion. Klingenberg (1970) has considered the effect of the proton gradient and potential across the inner membrane. The effect of the pH gradient and the potential will depend on the degree of ionisation of the additional phosphate in ATP when it is exchanged

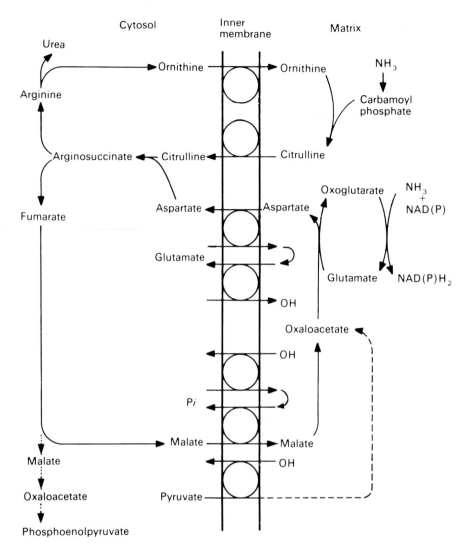

Fig. 6.5 Synthesis of urea from ammonia in liver. Pathways for the formation of aspartate are shown from fumarate and malate and from pyruvate (dotted line).

with ADP. Thus the transport may be considered as electrogenic (Fig. 6.6a) or electroneutral but proton translocating (Fig. 6.6b). The existence of the membrane potential, positive outside and negative inside, will discourage the entry of ATP in the ionised state more than that of ADP. In contrast, the proton gradient would have the reverse effect. The preference of the adenine nucleotide translocase for extramitochondrial ADP in coupled mitochondria could be explained if the translocation were electrogenic. Early measurements suggested that up to 0.5 or more protons per ATP were ejected when external ADP was

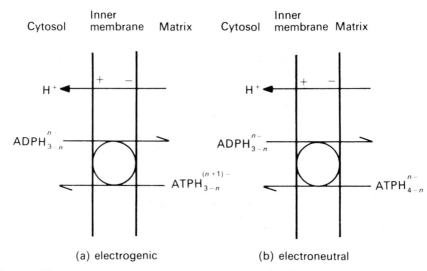

Fig. 6.6 Electrogenic and electroneutral adenine nucleotide exchange. In the electrogenic exchange, the ATP carries an additional negative charge when it exchanges for ADP, while in the electroneutral exchange the additional phosphate in ATP is un-ionised and therefore transports a proton.

exchanged for internal ATP. The electrogenic effects of ADP/ATP exchange can be determined by studying the associated movements of K^+ in the presence of valinomycin. However, LaNoue *et al.* (1978) have concluded that the adenine nucleotide translocase is fully electrogenic, that is the ADP/ATP exchange can be fully compensated by K^+ movement in the same direction as the ATP. The proton movements observed earlier are attributed to independent ion redistribution, compensating for the electrogenic exchange. If this interpretation is correct, then some cation (probably proton) redistribution consequent on adenine nucleotide transport will need to be taken into account when the stoicheiometric relationship between oxidative phosphorylation and proton transport is assessed (sect. 9.11(b)). The movement of negative charge outwards will dissipate the membrane potential formed by the proton pump.

   A second approach to the asymmetric behaviour of the translocase has considered the possibility that the conformations of the translocase are different on the inside and the outside of the membrane (cf. Fig. 6.7). On the outer surface the carrier is seen as having a greater affinity for ADP, while on the inner face it binds ATP preferentially. The translocation of the carrier–nucleotide complex across the membrane will be influenced by the membrane potential favouring the ATP complex in the outward direction and the ADP complex in the inward direction. The conformation of the carrier is denoted as c-form on the outside (cytosol) and m-form on the inside (matrix).

   The study of the translocase has been facilitated by the use of inhibitors, bongkrekate, atractyloside and the related carboxyatractyloside. Atractyloside, originally known as an inhibitor of oxidative phosphorylation, but later shown to block phosphorylation by its effect on translocation, is competitive with

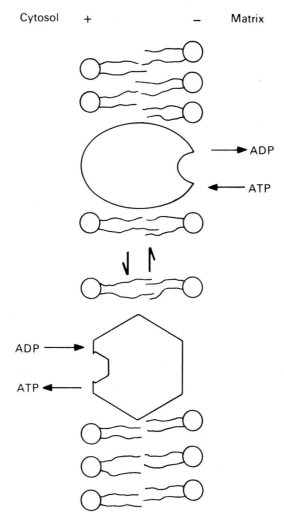

Cytosol   +   −   Matrix

Fig. 6.7 Transport of adenine nucleotides across the coupled inner membrane. It is assumed that the carrier rotates in the membrane so that the binding site is always at the surface. Transfer of the binding site from one face of the membrane to the other only occurs if an adenine nucleotide is bound and the reorientation is associated with a conformational change of the translocase.

adenine nucleotides. Bongkrekate, a branched unsaturated fatty acid with three carboxyl groups, is a non-competitive inhibitor.

Evidence for carrier movement in the membrane has been provided by studies with inhibitors. Atractyloside and carboxyatractyloside act competitively with external adenine nucleotide and bind to the c-form of the translocase. In contrast, bongkrekate, which is able to penetrate the membrane, appears to bind on the inside. This may be deduced from the fact that ADP promotes the binding of inhibitor by promoting the passage of the carrier to the inside of the membrane. The 1 : 1 exchange requires that the carrier will not cross the membrane in the absence of bound nucleotide.

In a different approach to adenine nucleotide transport, a gated pore has been proposed. The translocase binds ADP at the membrane outer surface and then undergoes a conformational change which creates a pore within the protein leading to the matrix; the ADP then migrates through the pore. In this conformation ATP is bound from the matrix side. A change to the original conformation releases ATP on the cytosol side. This model, which does not require carrier migration across the membrane, is consistent with the evidence from the inhibitor studies (see Klingenberg 1981, ref. cit. p. 156).

Evidence for conformational changes comes partly from the asymmetry of carrier inhibition and partly from the finding that the translocase is inhibited by permeant thiol reagents; these act only when the carrier is in the m-form and the inhibition is stimulated by small amounts of ADP. When in the c-form, the thiol groups are not available for attack.

The translocase has been isolated by several methods as the bongkrekate complex and the carboxyatractyloside complex (see Aquila *et al.*, 1978; Klingenberg *et al.*, 1978). It is composed of two subunits, each having a molecular weight of about 30 000 in mammals and 37 000 in yeast (Lauquin *et al.*, 1978). The carboxyatractyloside protein and the bongkrekate protein complexes differ in their reactivity with thiol reagents, presumably reflecting the different conformations of the c-form and the m-form respectively. The bongkrekate protein can be converted into the carboxyatractyloside protein in the presence of adenine nucleotides. The isolated translocase has been incorporated into phospholipid vesicles and its ability to translocate ADP and ATP specifically across the vesicle membrane has been demonstrated (Kramer and Klingenberg, 1977). Translocation is sensitive to bongkrekate and atractyloside (Shertzer and Racker, 1976). It has been estimated that the adenine nucleotide translocase constitutes 12 per cent of the mitochondrial membrane protein and is therefore the major protein in the inner mitochondrial membrane.

Regulation of the carrier is probably exerted by long-chain acyl CoA esters such as palmitoyl CoA which are inhibitory. An immediate effect of high acyl CoA content in liver after starvation will be an increase in the mitochondrial ATP/ADP ratio and a decrease of the cytosolic ratio.

Plant mitochondria possess a similar translocase to that found in animals; the plant system is, however, much less sensitive to atractyloside. The adenine nucleotide concentration of plant mitochondria is also substantially lower.

The fundamental role of the adenine nucleotide translocase is to make the product of mitochondrial oxidation, ATP, available to the extramitochondrial ATP-consuming reactions. Some of these give rise to AMP to which the inner membrane is impermeable. The adenylate kinase thus fulfils an important role in the intermembrane space between the regions of cellular ATP synthesis and ATP utilisation (Fig. 6.8). The 1 : 1 exchange ensures that the size of the intramitochondrial adenine nucleotide pool remains constant.

Before leaving the subject of adenine nucleotide translocation, we should consider the energetic implications of this system. We have noted that 'non-

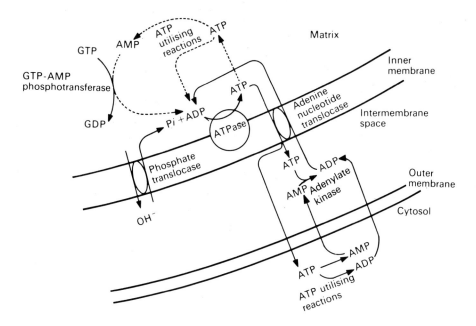

Fig. 6.8 Adenine nucleotide metabolism.

energised' mitochondria show no preference for ADP or ATP and consequently the ATP/ADP ratio will be the same inside the mitochondrion as outside. However, in the energised particle, ADP entry and ATP exit are favoured, resulting in a lower ATP/ADP ratio in the matrix than outside.

Since the energy required for ATP synthesis can be represented by:

$$\Delta G = \Delta G^{\oplus} + RT \ln \frac{[\text{ATP}]}{[\text{ADP}][\text{P}i]}$$

the effects of changes in ATP/ADP ratios can readily be calculated. The energy necessary for the synthesis of ATP from ADP in the intact mitochondrion can now be regarded as the sum of the energy involved in translocation and the energy required for the ATP synthetase (ATPase) reaction inside the mitochondrion. Experiments with isolated coupled liver mitochondria show that the ATP/ADP ratio can be 15–20 fold higher in the suspending medium than in the matrix, while the phosphate concentration may be 6–8 fold lower. Thus, the phosphate potential,

$$\frac{[\text{ATP}]}{[\text{ADP}][\text{P}i]}$$

is much lower inside the mitochondrion than outside. The difference in phosphorylation potential between the cytosol and the matrix is probably about 8–12 kJ mol^{-1}, a significant proportion of the total energy required for ATP synthesis (about 50 kJ).

(f) Carriers for other substrates

Some mitochondria synthesise phosphoenolpyruvate, which is then transported to the cytosol for conversion to glycogen, etc. In liver, the tricarboxylate carrier transports most of the phosphoenolpyruvate, while some may be transported by the adenine nucleotide carrier. In heart, where the tricarboxylate carrier is very weak, the major route is through the adenine nucleotide carrier.

A translocase exchanging carnitine and acyl carnitines across the inner mitochondrial membrane has been described by Pand · (1975). This system forms an essential part of the oxidation of fatty acids which are activated to their coenzyme A esters extramitochondrially and then converted to the acyl carnitine ester before being transported across the inner mitochondrial membrane.

Sulphite, which is formed intramitochondrially from cysteine, is transported by the dicarboxylate carrier (Crompton *et al.*, 1974). Sulphate and thiosulphate appear to be transported similarly.

6.5 Transfer of reducing equivalents across the inner membrane

(a) Maintenance of $NAD/NADH_2$ ratios

As demonstrated by Lehninger (1951) in his early studies of oxidation of NADH, pyridine nucleotides do not penetrate the inner mitochondrial membrane. Nevertheless, the mitochondrial matrix is a major locus of production of reducing equivalents and the matrix side of the inner membrane provides the main site for NADH oxidation, the NADH dehydrogenase. Metabolic processes in the cytosol may require reducing equivalents from the mitochondrion, for example in gluconeogenesis, from pyruvate. Alternatively, the cytosol may generate an excess of reducing equivalents (e.g. in glycolysis) which must be reoxidised in order to maintain the appropriate $NAD/NADH_2$ ratios. Several shuttle systems have been proposed to account for the fact that reducing equivalents are transferred across the inner mitochondrial membrane.

There is a second aspect of the maintenance of $NAD/NADH_2$ ratios to be considered, which arises from the difference in ratio across the mitochondrial membrane. In general the ratio is 100 times more electronegative inside the mitochondrion than outside (for rat liver cells the $NAD/NADH_2$ ratio is 300 to 1000 for the cytosol and about 10 for the mitochondrial matrix, see Gumaa *et al.*, 1971), a difference of the order of 50 mV in terms of redox potential (see Tischler *et al.*, 1977). The maintenance of such a difference requires energy (~ 9.6 kJ/mol) and the energetic aspects of the translocation systems will therefore need to be taken into account.

(b) Malate–aspartate shuttle

This system proposed by Borst in 1963 (see Safer and Williamson, 1973; Meijer and van Dam, 1974) involves intra- and extramitochondrial malate dehydrogenases and transaminases (aminotransferases) together with the glutamate–aspartate and oxoglutarate translocases (Fig. 6.9). The operation of the cycle results in the transfer of two reducing equivalents from the cytosol to the mitochondrion, the equivalents being transferred as malate. Since oxaloacetate is

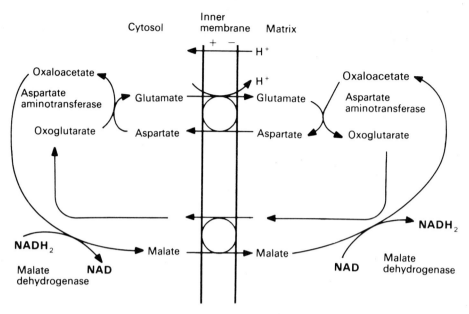

Fig. 6.9 Malate–aspartate shuttle.

not a good substrate for the dicarboxylate carrier systems, it is exported in the form of aspartate. Although the shuttle appears to be symmetrical, it is probable that it is primarily concerned with the transfer of reducing equivalents into the mitochondrion since in the energised particle, aspartate is normally exchanged outwards only. This shuttle will therefore tend to produce a difference in $NAD/NADH_2$ ratio across the inner membrane, making this high outside and low inside.

Evidence for the shuttle has been obtained in many tissues, especially liver and heart. For example, Safer and Williamson (1973) found a rapid exchange of oxoglutarate for malate associated with the oxidation of cytosolic $NADH_2$. Other workers have used transaminase inhibitors to show the importance of transamination in the oxidation of substrates in the cytosol.

(c) α-Glycerophosphate shuttle

This system proposed by Estabrook and Sactor (1958) for insect flight muscle requires the cytosolic NAD-linked α-glycerophosphate dehydrogenase and a flavoprotein α-glycerophosphate dehydrogenase associated with the outside of the inner membrane and linked directly to the respiratory chain (Fig. 6.10). Thus transport of metabolites across the inner membrane is not required. Energetically the system favours a high $NAD/NADH_2$ ratio in the cytosol, since the oxidation of $NADH_2$ gives a P/O ratio of 3 whereas the mitochondrial oxidation of α-glycerophosphate gives a P/O ratio of only 2. There is evidence that this shuttle functions in mammalian cells, at least to a limited extent. It probably plays a significant role in insects, which do not possess the dicarboxylate transport systems required for the malate–aspartate shuttle. This system

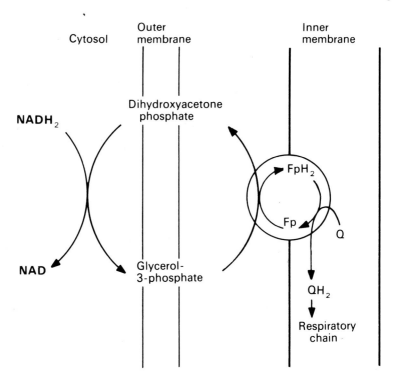

Fig. 6.10 α-Glycerophosphate shuttle.
The membrane-bound flavoprotein, α-glycerophosphate dehydrogenase, is linked to the respiratory chain.

probably acts unidirectionally, importing reducing equivalents into the mitochondrion.

(d) Fatty acid cycle

This system proposed by Whereat *et al.* (1969) relies on the oxidation of fatty acids intramitochondrially and their elongation outside the inner membrane (see Ch. 3); the latter process oxidises reduced pyridine nucleotides. Since β-oxidation is irreversible, this shuttle will be concerned only with the import of reducing equivalents into the mitochondrion. Evidence for the operation of the shuttle *in vivo* is weak.

(e) Shuttles concerned with the export of reducing equivalents

All the above systems are concerned primarily with the transfer of reducing equivalents into the mitochondrion and are to some extent energy-linked. A number of systems for export of reducing equivalents have been proposed and, in contrast to those above, there should be no energy requirement since transport will be down a potential gradient.

Fig. 6.11 The export of reducing equivalents from the mitochondrial matrix.
(a) Export of reducing equivalents during gluconeogenesis from pyruvate. (b) A modified version of (a) to give a shuttle in which malate is exchanged for pyruvate.

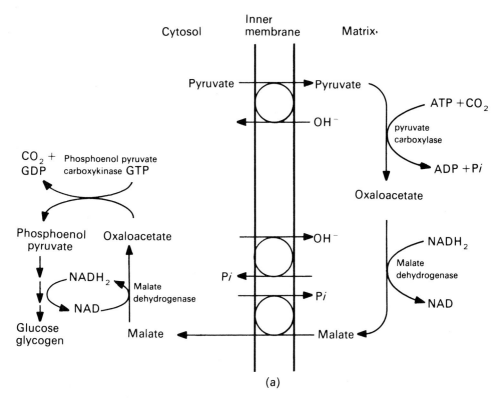

(a)

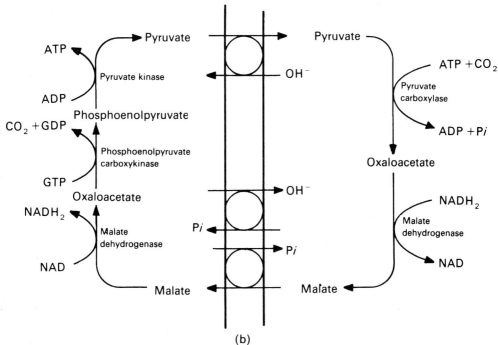

(b)

The pyruvate–malate shuttle for gluconeogenesis operates under conditions where pyruvate is converted to malate intramitochondrially and phosphoenolpyruvate is synthesised extramitochondrially (Fig. 6.11a). A simplified version of this shuttle has been proposed for the export of reducing equivalents (Fig. 6.11b). Both systems are energy-consuming.

(f) Other shuttles

In sperm cells, where there is a lactate dehydrogenase (NAD-dependent) in the mitochondrial matrix, a pyruvate/lactate shuttle has been proposed for the transfer of reducing equivalents across the membrane.

In plant and microbial mitochondria, there appear to be two NADH dehydrogenases, one on either side of the inner membrane. Thus in these systems both intramitochondrial and extramitochondrial NADH can be oxidised directly by the mitochondria.

Further Reading

Klingenberg, M. (1970) Mitochondrial metabolite transport. *FEBS. Lett.*, **6**, 145–54.

La Noue, K. F. and Schoolworth, A. C. (1979) Metabolite transport in mitochondria. *Ann. Rev. Biochem.*, **48**, 871–922.

Meijer, A. J. and van Dam, K. (1974) The metabolic significance of anion transport in mitochondria. *Biochim. Biophys. Acta*, **346**, 213–44.

Vignais, P. V. (1976) Molecular and physiological aspects of adenine nucleotide transport in mitochondria. *Biochim. Biophys. Acta*, **456**, 1–38.

Also:

Klingenberg, M. (1981) Membrane protein oligomeric structure and transport function. *Nature* **290**, 449–54.

Chapter 7

Mitochondrial cation transport

7.1 Mitochondrial cations

The transport of cations in the mitochondrion has attracted attention primarily because of the way in which such studies have extended our knowledge of oxidative phosphorylation. One of the earliest demonstrations of the importance of ions in mitochondrial metabolism was the discovery by Lehninger (1949) that calcium ions uncouple oxidative phosphorylation, stimulating respiration and inhibiting ATP synthesis. In the period following this discovery, calcium was classed as an uncoupler along with dinitrophenol. Later it was shown that mammalian mitochondria possessed a strong system for calcium uptake and accumulation. Some other divalent cations, such as manganese and strontium and also probably iron, are taken up in a similar manner to calcium. Intraperitoneal injection of radioactive Ca^{2+}, Sr^{2+} or Mn^{2+} results in an accumulation of these ions in mitochondria of tissues such as liver.

In the early 1950s, several workers showed that isolated mitochondria retained K^+ and Mg^{2+} as long as respiration continued, but that these ions leaked out when respiration ceased. However, unlike calcium, net accumulation was small or negligible. The transport of K^+ has been studied by using ionophores, such as valinomycin, which increase membrane permeability to specific ions.

Before proceeding to examine ion uptake systems in detail, it is useful to note the approximate ion composition of freshly isolated mammalian mitochondria (Table 7.1). Potassium is the most abundant ion, with magnesium also present as a major component. Despite the strong calcium pump, calcium is not present in high concentration. The anions associated with these cations are primarily phosphate, adenine nucleotides and possibly other phosphate esters, together with organic acids such as citrate.

7.2 Calcium uptake

(a) 'Massive loading' experiments

Although a number of observations had been made in the 1950s,[1] a systematic study of calcium uptake was not undertaken until 1961–62.

---

[1] For references to this work see Vasington and Murphy (1962).

Table 7.1 Metal cation* content of rat liver mitochondria

	nmol/mg mitochondrial protein
Potassium	270
Magnesium	70
Sodium	15
Calcium	10
Iron	6
Zinc	0.8
Copper	0.4
Manganese	0.3

* Since a proportion of these cations will be bound, the intramitochondrial concentration of free cation is not known (see Thiers and Vallee, 1957).

Vasington and Murphy (1962) and De Luca and Engstrom (1961) examined the uptake of ^{45}Ca by suspensions of rat kidney mitochondria. Under suitable conditions the endogenous calcium level rose to as much as 200 times the initial value. Such an enormous uptake damages the mitochondrion but significant conclusions were drawn from such experiments. This uptake of calcium occurred in only 5 to 10 minutes at $37°C$, but required other ions including phosphate, adenine nucleotides and Mg^{2+}. There was an energy requirement which could be met either by respiration or ATP hydrolysis. Calcium uptake driven by respiratory activity was sensitive to amytal, azide, antimycin and cyanide. When driven by ATP hydrolysis, uptake was inhibited by oligomycin. Uncouplers such as dinitrophenol inhibit the uptake. Calcium uptake appears to take precedence over oxidative phosphorylation, in that ATP synthesis does not occur during calcium accumulation.

Calcium is deposited inside the mitochondrion as calcium hydroxyapatite $[Ca_3(PO_4)_2]_3Ca(OH)_2$. This insoluble deposit can be seen in electron micrographs as dense granules in the matrix. The molar ratio of calcium to phosphate taken up is 1.67:1, a value consistent with calcium hydroxyapatite formation.

Studies by Rossi and Lehninger (1963) showed that calcium uptake, like ATP synthesis, is linked to each coupling site. They obtained values for a Ca^{2+}/O ratio (μmol Ca^{2+} accumulated per μg atom oxygen consumed, analogous to the P/O ratio) of 4.9 for hydroxybutyrate, 2.7 for succinate and 1.8 for ascorbate–cytochrome c oxidation. They concluded from studies with mitochondria from several tissues that 1.67 Ca^{2+} and 1.0 phosphate ions were taken up for each pair of electrons passing through a coupling site.

(b) 'Limited loading' experiments

Early experiments had shown that calcium stimulates respiration and that the oxygen uptake was proportional to the calcium uptake. Chance (1965) showed that addition of small amounts of Ca^{2+} to mitochondria in state 4 resulted in a shift of the oxidation–reduction state of the components of the respiratory chain. Repetition of cross-over experiments (sect. 2.8) using small amounts of Ca^{2+} in place of ADP gave a cross-over point between cytochromes

b and c_1. The similarity between effects of ADP and Ca^{2+} on the oxidation–reduction state of the carriers of the respiratory chain led to the conclusion that Ca^{2+} transport was coupled to respiration at the same sites as ATP synthesis. However, the stimulation of respiration by Ca^{2+} is greater than that obtained by ADP addition and comparable to that of dinitrophenol.

Measuring the oxygen uptake associated with the addition of a small amount of Ca^{2+} in the absence of phosphate to mitochondria in state 4, provided a better experimental approach. Under these conditions almost all the Ca^{2+} is taken up by the end of the 'oxygen jump'. With this type of experiment, about 2 Ca^{2+} are taken up for each pair of electrons passing through a coupling site ($Ca^{2+}/O = 6$). Considerably greater values for the Ca^{2+}/O ratio have been obtained for short periods following addition of calcium (referred to as superstoicheiometry). These high ratios are attributed to additional uptake driven by hydrolysis of endogenous ATP.

With experiments of this type, it is possible to demonstrate a limited uptake of calcium in the absence of phosphate. Small amounts of phosphate added subsequently are taken up, showing that it is the calcium transport which is energy-driven and that phosphate follows to compensate the charge. More recent studies with phosphate-depleted mitochondria have also shown that calcium uptake can occur in the absence of phosphate (Crompton *et al.*, 1978). Greater amounts of Ca^{2+} may be accumulated in the presence of phosphate than in its absence and even greater amounts still if adenine nucleotides are present. Phosphate may be replaced by several other permeant anions such as acetate, propionate or bicarbonate $+ CO_2$. With acetate the mitochondria swell and lyse since calcium acetate, unlike calcium phosphates, is highly soluble, resulting in an increase in osmotic pressure of the matrix.

Although calcium accumulation takes precedence over oxidative phosphorylation, ATP synthesis continues after calcium uptake is complete.

The addition of a small amount of Ca^{2+} to mitochondria incubated with ATP results in increased ATP hydrolysis associated with calcium uptake. This ATP hydrolysis 'jump' is similar in principle to the oxygen uptake jump, but it is sensitive to oligomycin. For each ATP hydrolysed, approximately 2 Ca^{2+} and about 1 phosphate are taken up.

(c) Relationship between calcium transport, phosphorylation and proton transport.

Calcium uptake can be driven by each coupling site ($Ca^{2+}/ \sim = 2)^2$ in the respiratory chain but, since it is insensitive to oligomycin, ATP synthesis and the membrane-bound ATPase are not involved. On the other hand ion transport driven by ATP hydrolysis is oligomycin-sensitive, but it is not sensitive to inhibitors of respiration. The simplest interpretation is that ion uptake must be driven by the high-energy intermediate of oxidative phosphorylation (Fig. 7.1).

In addition to phosphate, adenine nucleotides appear to have a role in massive accumulation of calcium. Thus 60–100 nmol/mg mitochondrial protein is accumulated in the absence of phosphate. This figure rises to 100–200 nmol in

---

[2] The symbol $\sim$ is used to represent an energised state of the mitochondrion resulting from the passage of two electrons through one coupling site.

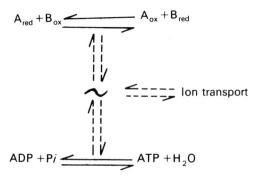

Fig. 7.1 Oxidative phosphorylation and ion transport. The intermediate high-energy state ($\sim$) generated from oxidative phosphorylation or from ATP hydrolysis may be used to drive ion transport. Under suitable conditions, ion gradients may be used to drive ATP synthesis (see sect. 7.5(c)).

the presence of a permeant anion such as phosphate and up to 3000 nmol in the presence of phosphate, adenine nucleotides and Mg^{2+}. The role of adenine nucleotides is not clear, but it appears to be independent of oxidative phosphorylation.

Phosphate uptake normally accompanies calcium uptake, a fact which has led Moyle and Mitchell (1977) to propose a calcium–phosphate symport[3], a transport system translocating calcium and phosphate with one positive charge per calcium (Fig. 7.2b). However, as already noted, small calcium movements can occur independently of phosphate or other anions. It is believed that calcium is translocated electrophoretically across the membrane as a positively charged ion. Transport is thus driven by the membrane potential. The active uptake of calcium, driven either by respiration or ATP hydrolysis, is coupled to the extrusion of protons from the mitochondrion. As noted above, one calcium ion exchanges for one proton. Since calcium normally carries two positive charges, this would lead to an imbalance of charge. The problem is solved if the true H^+/Ca^{2+} ratio is 2 and negatively charged phosphate is exchanged for a hydroxyl ion by the phosphate translocase (Fig. 7.2a). Under these circumstances the observed H^+/Ca^{2+} ratio in the presence of phosphate would be 1 since the second proton would be neutralised by the phosphate/OH^- exchange (Fig. 7.2). Lehninger's group (Vercesi *et al.*, 1978) have observed a H^+/Ca^{2+} ratio of 2 under conditions where the phosphate translocase was inhibited. The mechanism requires that the accompanying anion should be proton-translocating. Lehninger (1974) found that not all permeant anions promoted calcium transport, but only those which donated a proton to the mitochondrial matrix such as phosphate, acetate, etc.

If two calcium ions are taken up for every pair of electrons passing through a coupling site or for each ATP hydrolysed, it might be deduced that each pair of electrons passing through a coupling site results in the ejection of four protons from the mitochondrion. However, this figure is not agreed and will be discussed later (sect. 7.5). It should be noted that the calcium–phosphate

[3] The terms uniport, symport and antiport have been used by Mitchell to describe transport of a single species, linked transport of two species in the same direction and linked transport of two species in opposite directions, respectively.

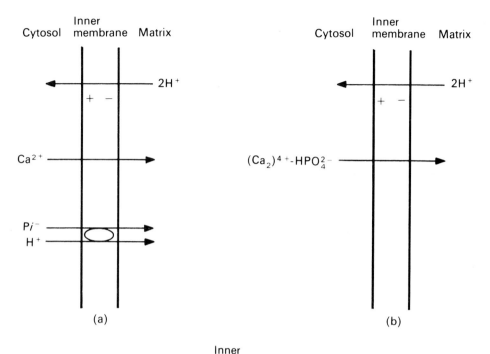

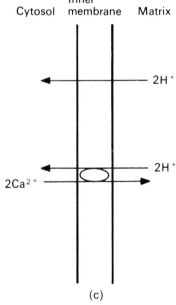

Fig. 7.2 Proposed systems for active uptake of calcium.
(a) An electrogenic calcium transporter for Ca^{2+}. For simplicity, the phosphate translocase is shown as translocating a proton with phosphate rather than OH^- in exchange for phosphate. (b) The calcium–phosphate symport of Mitchell and Moyle showing a net transfer of charge of calcium as Ca^+. (c) A partially charge-compensated transport of calcium in which one Ca^{2+} exchanges for one H^+. All three systems account for an observed net H^+/Ca ratio of 1.

symport (and calcium–monocarboxylate symport) proposed by Moyle and Mitchell (1977) assumes that calcium will be transported with phosphate carrying a single charge and that the H^+/Ca^{2+} ratio will be 1 (Fig. 7.2b). Alternatively, if the calcium translocase functions as a partially charge-compensated exchange (by a Ca^{2+}/H^+ exchange) then the overall H^+/Ca^{2+} ratio will be 2 but only one proton would be ejected by the respiratory system (Fig. 7.2c).

(d) Calcium-binding proteins and the calcium translocase

The transport of calcium is inhibited competitively by strontium and manganese and by low concentrations of ruthenium red (ammoniated ruthenium oxychloride, $Ru_2(OH)_2Cl_47NH_33H_2O$) and lanthanides (rare-earth cations including lanthanum, La^{3+}) without directly affecting respiration. Lehninger (see Carafoli and Lehninger, 1971) distinguished two types of site responsible for the binding of calcium to mitochondria. The non-specific low-affinity sites had a K_m for calcium of about 50 μM, whereas the high-affinity sites had a K_m of about 1 μM. The specific inhibition of calcium transport by lanthanides and ruthenium red, together with the studies of high-affinity binding sites, supported the notion that calcium is transported by a translocase. This view was strengthened by the fact that high-affinity sites were found only in those mitochondria capable of calcium accumulation.

High-affinity binding sites are lost by ageing, lysis or detergent treatment of mitochondria, suggesting the solubilisation of the calcium-binding protein. Several isolations of binding proteins have been made. Gomez-Puyou et al. (1972) isolated a glycoprotein ($MW = 67\,000$) which contained about 27 per cent phospholipid, and bound calcium; the binding was inhibited by lanthanides and ruthenium red. A similar but smaller glycoprotein ($MW = 33\,000$) was isolated by Sottocasa et al. (1972), but subsequent attempts to demonstrate calcium transport with the protein failed. Recently, a small calcium-binding protein (MW: 3000 approx.) has been isolated which has the ability to extract calcium into an organic phase, the extraction being sensitive to lanthanides and ruthenium red. The protein will also transport calcium through a hydrophobic layer in response to an applied potential (Jeng et al., 1978). Whether such a protein is the calcium translocase remains to be tested.

(e) Calcium efflux

Early studies on calcium transport assumed that efflux involved the calcium translocase. However, more recently it has been shown that the addition of Na^+ induces Ca^{2+} efflux, which is insensitive to ruthenium red. A translocase for the exchange of Na^+ for Ca^{2+} has been proposed. Heart muscle, skeletal muscle, brain and adrenal mitochondria possess this system, which appears to be absent from liver and kidney. It is further suggested that sodium can also exchange for a proton – see Fig. 7.3. (Crompton and Heid, 1978).

(f) Physiological significance of calcium transport

The reason why mitochondria, which normally contain only small amounts of calcium, should possess a strong active uptake of this ion, is only now beginning to become clear. The calcium concentration appears to have a specific

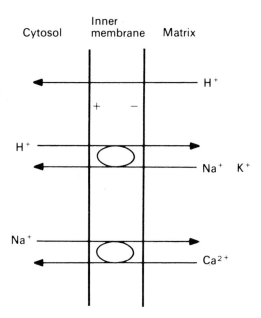

Fig. 7.3 Proposed cation-exchange systems.

role as a regulator of cell function. Thus a number of enzyme systems are regulated by calcium; pyruvate dehydrogenase and phosphorylase *b* kinase, for example, are activated by calcium while pyruvate kinase is inactivated. Calcium also has a role in hormone regulation, various membrane functions and in several contractile and motile systems. The full significance of mitochondrial calcium influx and efflux for cell metabolism has still to be elucidated.

It should be noted that the energy-dependent calcium transport system described here is found in vertebrates, but seems to be absent from plant and microbial mitochondria. High-affinity calcium-binding sites and calcium-stimulated respiration are found only in the organelle from higher animals. In blowfly flight muscle mitochondria, calcium enters only slowly and does not affect respiration. In plants, some calcium uptake has been recorded, but opinion is divided on the ability of the plant organelle to transport calcium.

7.3 Magnesium and iron transport

In heart mitochondria, an active accumulation of magnesium, which is respiration-dependent, has been observed. Under different conditions it is possible to observe a respiration-dependent efflux. In neither case does ATP hydrolysis function as effectively as respiration in driving transport.

An energy-dependent accumulation of iron has been demonstrated in rat liver mitochondria although uptake of iron is negligible in heart. The iron is accumulated in the matrix and high- as well as low-affinity binding sites have been identified; the high-affinity sites are associated with an energy-dependent uptake. Uptake is competitively inhibited by Ca^{2+} and inhibited by lanthanides and ruthenium red. It is possible that the calcium translocase is also responsible for iron uptake.

7.4   Monovalent cation transport

(a) Non-induced transport

From early work it was concluded that the inner mitochondrial membrane has very low permeability to potassium. It was observed that mitochondria retained their potassium as long as respiration occurred, but this ion leaked into the medium when respiration ceased. Treatment with uncoupling agents also led to potassium leakage. Little net accumulation was obtained and labelled ^{42}K exchanged only slowly with potassium in the matrix. Other studies showed a doubling of intramitochondrial potassium in respiring mitochondria and, in potassium-depleted mitochondria, some active uptake.

More recently evidence has accumulated which suggests that there may be regulated transport of cations across the inner membrane. For example, respiring heart mitochondria readily swell in the presence of acetate (which permeates the inner membrane) and potassium or sodium, while non-respiring (non-energised) mitochondria swell with sodium acetate but not with potassium acetate. Such observations have led to the proposal for a Na^+/H^+ exchange system and a translocase for monovalent cations comparable with that for calcium, namely a carrier conveying K^+ or Na^+ into the matrix in response to the membrane potential. Some evidence for regulation of transport by Mg^{2+} and phosphate has also been obtained (Brierley *et al.*, 1978).

The postulated cation/H^+ exchange system has also been invoked to explain the efflux of ions and contraction of the matrix which occurs when mitochondria, previously swollen in salt solutions, are energised (Brierley *et al.*, 1977). As noted earlier the cation/H^+ exchange has also been allotted a role in calcium efflux (Fig. 7.3).

The energy-dependent influx and efflux of cations, if allowed to proceed unregulated, would dissipate substantial amounts of metabolic energy by futile cycles of ions. It thus seems likely that some parts of the closely interrelated cation transport systems will only be observed under conditions where they are released from regulation.

(b) Ionophore-induced transport

Valinomycin is an antibiotic which was initially found to uncouple oxidative phosphorylation and stimulate ATPase activity. Moore and Pressman (1964) found that small additions of valinomycin to rat liver mitochondria in the presence of K^+ and a respiratory substrate, caused an increase in the respiratory rate, uptake of K^+ and ejection of H^+. Potassium uptake can also be activated by ATP hydrolysis. This valinomycin-induced transport resembles that of calcium in its general properties. It differs in that the mitochondrial K^+ reaches a new steady-state level related to the rates of influx and efflux of potassium; respiration also reaches a higher steady-state level after initial stimulation, since there is a cycling of potassium in and out of the matrix. Valinomycin itself acts by rendering the inner membrane permeable to potassium which then migrates in response to the membrane potential. Permeant anions such as acetate and phosphate promote potassium accumulation. Another antibiotic, gramicidin, also induces potassium transport; the overall energetic properties are similar to those of valinomycin.

The active uptake of K^+ in the presence of valinomycin is understood as a response to the membrane potential created by the respiration-driven extrusion of protons. This uptake has been quantified as the $K^+/\sim$ ratio, that is the number of potassium ions translocated per ATP hydrolysed or per pair of electrons passing through a coupling site. Azzone and Massari (1973) concluded that the $K^+/\sim$ ratio appoached a value of 4, a figure confirmed by Reynafarje and Lehninger (1978). Thus the charge$/\sim$ ratio of 4 appears to apply to both calcium and potassium transport.

(c) Role of ionophores

Valinomycin (Fig. 7.4a) is a 12-residue ring structure composed of three repeating sequences of four residues – D-valine, L-valine, D-hydroxyvalerate and L-lactate. The preferred conformation involves internal hydrogen bonding associated with the amide groups. The formation of the valinomycin–K^+ complex involves a conformational change so that the hydrophilic groups are directed inwards and the hydrophobic groups outwards (Fig. 7.4b). In this form the positively charged complex is lipid-soluble and will migrate through the membrane. The potassium ion is an appropriate size to fit the middle of the ring where it is protected from solvent and anions by ester groups, hydrogen bonds and hydrocarbon side chains. Valinomycin has been shown to confer selective permeability in a number of natural and artificial membranes to K^+, Rb^+ and Cs^+. Protons are not transported and Na^+ and Li^+ only slightly.

Valinomycin added to a potassium-containing organelle induces the efflux of potassium and the establishment of a membrane potential, negative inside and positive outside. A potential of reverse sign can be obtained by using valinomycin and higher external potassium concentrations, resulting in influx of this cation.

Gramicidin (Fig. 7.4c) is a linear polypeptide which appears to act by forming channels or pores through the membrane. It thus differs significantly from valinomycin. Two molecules of gramicidin co-operate to form a channel which facilitates the passage of alkali cations, including sodium, across the membrane. There is also significant transport of H^+.

Nigericin (Fig. 7.4d) is also capable of forming a ring structure and carrying K^+ in a manner similar to that of valinomycin. It induces cation/cation and cation/H^+ exchanges in mitochondria or artificial membranes. Although nigericin transports alkali cations, it does not induce an energy-dependent uptake of K^+ against a concentration gradient in mitochondria. In this respect it differs from gramicidin and valinomycin.

The action of some ionophores is represented diagrammatically in Fig. 7.5, together with their effects on ion transport in intact coupled mitochondria and inverted submitochondrial particles.

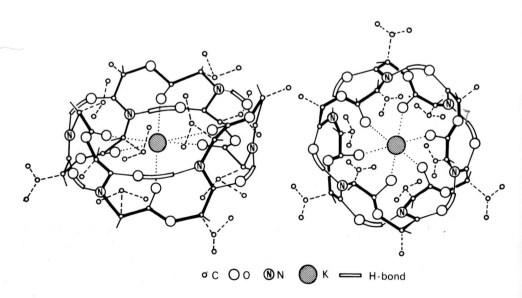

Fig. 7.4(a) Ionophores: valinomycin.

Fig. 7.4(b) Ionophores: valinomycin−K⁺ complex (after Ovchinnikov, 1972).

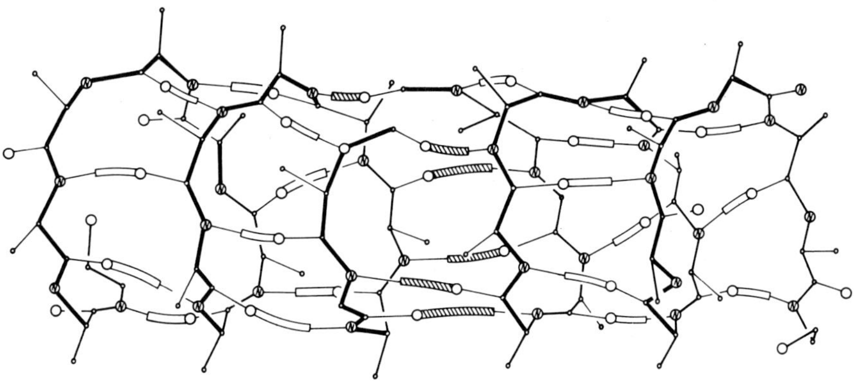

CHO—L-val—gly—L-ala—D-leu—L-ala—D-val—L-val—D-val—L-try—D-leu—L-try— D-leu—

—L-try—D-leu—L-try-NH—CH$_2$—CH$_2$OH

Fig. 7.4(c) Ionophores: a possible helical structure for gramicidin showing the formation of a channel across the membrane composed of two molecules of gramicidin (after Ovchinnikov, 1972).

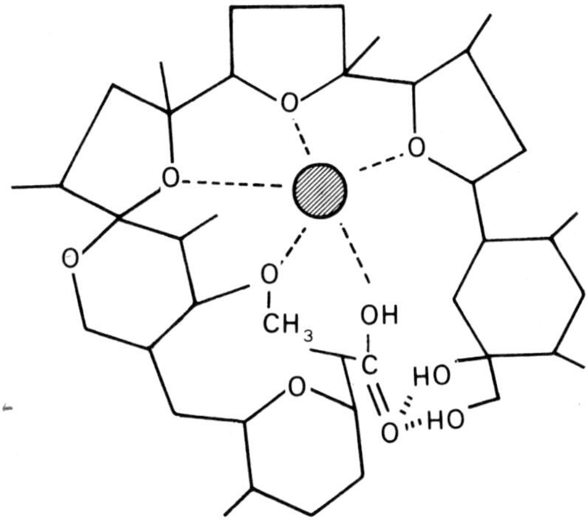

Fig. 7.4(d) Ionophores: nigericin.

7.5   Proton translocation

(a) Demonstration of proton transport

As noted earlier, respiring mitochondria expel protons. The ability of mitochondria to acidify the suspending medium has been noticed by several workers (see Bartley and Davies, 1954, for example). Proton movement has become a central issue, not only in ion transport but in the energy transducing systems of mitochondria, chloroplasts and bacteria.

The early studies of Mitchell and Moyle (1967) used a glass electrode

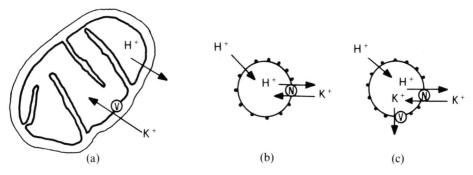

Fig. 7.5 Action of ionophores.
(a) Respiration-dependent potassium uptake in the presence of valinomycin (V).
(b) Respiration-dependent potassium uptake in inverted submitochondrial particles in the presence of nigericin (N). (c) Uncoupling of inverted submitochondrial particles in the presence of nigericin, valinomycin and potassium ions.

linked to a fast reacting chart recorder to measure rapid changes in pH. They measured the ejection of protons from a mitochondrial suspension incubated with substrate when a pulse of oxygen was added (as a small volume of KCl saturated with air). With this method it was demonstrated that protons were released very rapidly (in less than 1 s) lowering the pH. They found that the proton gradient thus formed disappeared slowly with a $t_{1/2}$ of about 1.5 min. The vectorial transport of protons could be demonstrated using submitochondrial particles. Thus digitonin particles which have the same membrane orientation as intact mitochondria eject protons when respiring. In contrast, sonic particles with mainly inverted membrane orientation translocate protons inwards.

The dependence on respiration can be demonstrated with inhibitors. Thus with β-hydroxybutyrate as substrate, rotenone inhibits both oxidation and proton translocation but is without effect when succinate is substrate.

Injection of small amounts of ATP into mitochondrial suspensions also gives a rapid expulsion of protons in the absence of respiratory substrate. Proton translocation here is insensitive to respiratory inhibitors but sensitive to ATPase inhibitors.

Uncouplers such as dinitrophenol (DNP) and carbonylcyanide-*p*-trifluoromethoxyphenylhydrazone (FCCP) greatly enhance the rate of decay of proton gradients but do not inhibit proton expulsion itself. Conventional uncouplers are usually lipid-soluble organic acids, and include phenolic substances (DNP, pentachlorophenol and dicoumarol), derivatives of carbonylcyanide phenylhydrazone (CCCP, carbonylcyanide-*m*-chlorophenylhydrazone) and derivatives of salicylanilide (see Fig. 9.3). Such uncouplers rapidly dissipate a proton gradient induced by respiration or by ATP hydrolysis. They also render artificial phospholipid membranes permeable to protons. In general there appears to be a relationship between uncoupling capacity and proton conductivity (Liberman *et al.*, 1969). Thus uncouplers such as DNP, FCCP, etc. can be seen as acting as proton conductors analogous to valinomycin in its potassium-conducting properties.

(b) Stoicheiometry

Mitchell and Moyle (1967) measured the H^+/O ratio for various substrates under a variety of conditions. Oxidation of β-hydroxybutyrate and succinate gave H^+/O ratios of 6 and 4 respectively. They concluded that two protons were ejected for every pair of electrons passing through a coupling site $(H^+/\sim = 2)$. Hydrolysis of ATP also gave an $H^+/P = 2$.

More recently these values have been criticised, mainly on the grounds that the protons ejected are partly compensated by a very rapid phosphate–hydroxyl ion exchange mediated by the phosphate translocase. Thus when phosphate transport was inhibited by N-ethylmaleimide or when phosphate-depleted mitochondria were used, an $H^+/\sim$ ratio of about 4 has been obtained (Reynafarje and Lehninger, 1978). Using succinate or β-hydroxybutyrate as substrates, H^+/O ratios of 8 and 12 respectively have been obtained when compensatory movements of Ca^{2+} or K^+ (in the presence of valinomycin) were allowed to occur (Alexandre *et al.*, 1978). As noted earlier, much work on calcium and potassium transport is consistent with an $H^+/\sim$ ratio of 4. Moyle and Mitchell (1978) have confirmed that the oxygen pulse experiments do give an $H^+/\sim$ ratio of 2 when phosphate movements are controlled. They also assert that calcium is transported as Ca^+, and they have proposed a calcium–phosphate symport (Mitchell and Moyle, 1977). Thus the stoicheiometry of proton translocation in relation to respiration remains a matter for further elucidation. However, the H^+/P ratio for hydrolysis of ATP estimated at 2 by Mitchell and Moyle (1973) may be 3 according to Lehninger's group (Alexandre *et al.*, 1978).

(c) Reversible relationship between ion transport and the ATPase reaction

The hydrolysis of ATP by mitochondria is associated with the formation of a proton gradient and a membrane potential. It has proved possible to demonstrate the reverse reaction, namely the synthesis of ATP brought about by the creation of a pH gradient and a membrane potential, positive outside. Thus, subjecting mitochondria to an acid wash gives rise to limited ATP synthesis. The process is enhanced if a membrane potential is also set up by the addition of valinomycin which allows potassium ions to migrate outwards across the membrane, creating a potential positive on the outside, negative inside.

A discussion of the mechanisms of proton transport requires a detailed discussion of the ATPase on the one hand and the respiratory chain on the other. This will be left until Chapter 9.

7.6 Models of cation transport

There have been three main models for cation transport as shown in Fig. 7.6. Early studies on calcium transport prompted the view that protons were ejected in response to a hypothetical calcium pump driven by the high-energy intermediate of oxidative phosphorylation (Fig. 7.6b). This view lost favour when it was shown that calcium transport was not essential for proton transport. Further, protons were ejected in association with potassium transport artificially

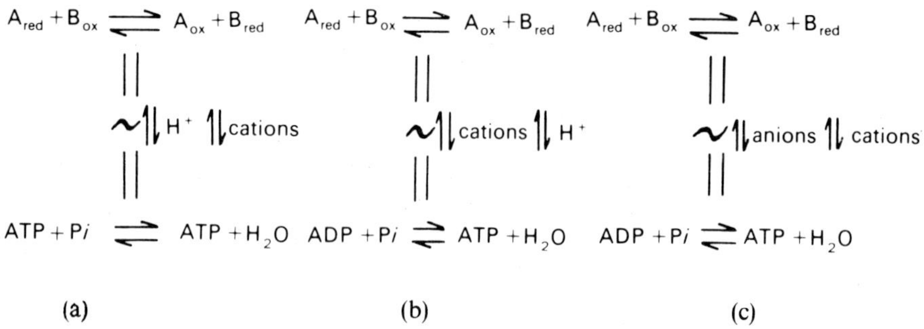

$$A_{red} + B_{ox} \rightleftharpoons A_{ox} + B_{red} \qquad A_{red} + B_{ox} \rightleftharpoons A_{ox} + B_{red} \qquad A_{red} + B_{ox} \rightleftharpoons A_{ox} + B_{red}$$

$$\parallel \qquad\qquad\qquad\qquad \parallel \qquad\qquad\qquad\qquad \parallel$$

$\sim \!\! \Downarrow H^+ \ \Uparrow$ cations $\qquad \sim \!\! \Downarrow$ cations $\Uparrow H^+ \qquad \sim \!\! \Downarrow$ anions $\Uparrow$ cations

$$\parallel \qquad\qquad\qquad\qquad \parallel \qquad\qquad\qquad\qquad \parallel$$

$$ATP + Pi \rightleftharpoons ATP + H_2O \quad ADP + Pi \rightleftharpoons ATP + H_2O \quad ADP + Pi \rightleftharpoons ATP + H_2O$$

(a)   (b)   (c)

Fig. 7.6 Three alternative views of the energetics of cation transport: (a) cations migrate in response to proton movements; (b) there is a primary cation pump and protons migrate in response to cation transport; (c) cations migrate in response to anion movements.

induced by valinomycin, although no potassium pump was detectable. It is now clear that proton movements are primary to the movement of other cations (Fig. 7.6a). Indeed, the movements of cations must now be seen as a response to the membrane potential and proton gradient set up by respiratory activity or ATP hydrolysis. As noted earlier, cation transport itself appears to be independent of anion movements (thus the scheme in Fig. 7.6c is rejected).

Further Reading

Carafoli, E. and Crompton, M. (1978) The regulation of intracellular calcium. *Current Topics in Membranes and Transport*, **10**, 151–216.

Chance, B. and Montal, M. (1971) Ion translocation in energy-conserving membrane systems. *Current Topics in Membranes and Transport*, **2**, 99–156.

Lehninger, A. L., Carafoli, E. and Rossi, C.S. (1967) Energy-linked ion movements in mitochondrial systems. *Adv. Enzymol.*, **29**, 259–320. (See also Azzone and Massari, 1973.)

Chapter 8
Theories of phosphorylation

8.1 The high-energy intermediate

The model of oxidative phosphorylation used in previous chapters has included the concept of a high-energy intermediate which may be used for ion transport, reverse-flow electron transport or ATP synthesis. The high-energy intermediate may be formed by oxidation reactions of the respiratory chain, ATP hydrolysis or the creation of ion gradients (Fig. 7.1). Many attempts to provide a plausible mechanism for these processes have been made. Most of these fall within one of three major theories and it is these three which will be considered here. In each case it is assumed that essentially the same mechanism will be involved in oxidative phosphorylation in the mitochondrion, phosphorylation associated with the bacterial plasma membrane and photophosphorylation in chloroplasts and bacterial chromatophores.

8.2 The chemical theory

The earliest theory of phosphorylation regarded the high-energy intermediate as a chemical substance. An early version of the chemical theory was proposed by Lipmann (1946) and these ideas were substantially developed by Slater (1953a). The basis on which Slater worked was that oxidative phosphorylation would not differ in principle from substrate-level phosphorylation. It was noted that, in uncoupled systems, phosphate was not necessary for respiration and that phosphorylation itself must be independent of the oxidation–reduction reactions of the respiratory chain. Thus phosphorylated intermediates of components of the respiratory chain as proposed by Lipmann were abandoned by Slater.

Substrate-level phosphorylation such as the triose phosphate dehydrogenase reaction and the conversion of oxoglutarate to succinate were used as models. In these reactions, the oxidation–reduction steps occur first and separately from the phosphorylation as, for example, in the triose phosphate dehydrogenase and phosphoglycerokinase reactions.

Oxidation: $E–SH + NAD + RCHO = E–S \sim \underset{\underset{O}{\|}}{C} – R + NADH_2$

Phosphorylation: $E–S \sim \underset{\underset{O}{\|}}{C}–R + Pi = E–SH + R–\underset{\underset{O}{\|}}{C}–O(P)$

$$R-\underset{\underset{O}{\|}}{C}-O(P)+ADP \qquad RCOOH+ATP$$

Slater's scheme which resembles these reactions is as follows:

$$A_{red}+B_{ox}+XOH+YH \;\rightleftharpoons\; A_{ox}+B_{red}+X\sim Y+H_2O$$
$$X\sim Y+H_3PO_4 \;\rightleftharpoons\; XO\sim PO_3H_2+YH$$
$$XO\sim PO_3H_2+ADP \;\rightleftharpoons\; ATP+XOH$$

where A and B are members of the respiratory chain at a coupling site and $X\sim Y$ is the high-energy intermediate. The scheme as shown here incorporates a modification by Racker (1970) to account for the requirement for adenine nucleotides in reactions exchanging oxygen in $H_2{}^{18}O$ with Pi (see sect. 9.1).

The role for uncouplers such as dinitrophenol was originally seen as promoting an alternative reaction:

$$X\sim Y+H_2O \rightarrow XOH+YH$$

and thus breaking the link between redox reactions and oxidative phosphorylation. Alternatively, since most uncouplers render the membrane permeable to protons or other ions, the phosphorylation process might be uncoupled by dissipation of the high-energy intermediate in continually pumping ions across a leaky membrane.

One of the obvious predictions of the chemical theory is the existence of a phosphorylated intermediate ($XO\sim PO_3H_2$)formed from inorganic phosphate; such an intermediate would have a high rate of turnover. The theory stimulated studies into phosphorylated intermediates, a number of which were found (such as quinol phosphates, phosphorylated proteins etc.) but their turnover rates and properties were not consistent with a role in oxidative phosphorylation. Interest in this theory has tended to fade after the failure to find the predicted chemical intermediate and after the rise of interest in alternative theories which are more amenable to experimental investigation. Nevertheless the recent studies of Griffiths (see Griffiths *et al.*, 1977) in which lipoic acid is seen as a coupling factor linking respiration and ATP synthesis represent a revival of this approach to oxidative phosphorylation.

8.3   The chemiosmotic theory

The basis of the chemiosmotic hypothesis proposed by Mitchell (1961) is that the high-energy intermediate is an electrochemical gradient of protons across the inner mitochondrial membrane. Respiration pumps protons outwards from the mitochondrion setting up the gradient. The gradient drives the synthesis of ATP by the mitochondrial ATPase (ATP synthase), which tends to dissipate the gradient by translocating protons inwards. As noted earlier, the transport of charged ions across a membrane results in a charge difference across the membrane. The energy stored can be seen as the sum of the membrane potential and the hydrogen ion concentration gradient:

$$\Delta p = \Delta \psi - \frac{2.303\, RT \cdot \Delta pH}{F}$$

where Δp is the proton motive force (PMF), $\Delta \psi$ the membrane potential and ΔpH the proton gradient (pH difference across the membrane). It is obviously essential that the membrane has a low permeability to protons and hydroxyl ions. The function of uncoupling agents is seen as rendering the membrane permeable to protons and thus dissipating the proton gradient. The classical uncouplers are known to render membranes permeable to protons.

Mitochondrial ATP synthesis is catalysed by a proton-translocating ATPase. The overall reaction can be represented as:

$$ADP + Pi + 2H_c^+ = ATP + H_2O + 2H_m^+$$

where H_c^+ and H_m^+ refer to protons outside and inside the mitochondrion respectively (c, cytoplasm; m, matrix). For this purpose the outer membrane can be neglected and the intermembrane space is taken as being in equilibrium with the cytoplasm. Thus the membrane-bound ATPase reaction obligatorily involves protons as set out in the above equation (see Fig. 8.3). In simple terms the translocation ($H_c^+ \rightarrow H_m^+$) of two protons through the ATPase drives the synthesis of one molecule of ATP. The driving force is the proton motive force, Δp. The reaction is reversible and ATP hydrolysis will generate a proton gradient.

How does transport of protons drive the synthesis of ATP? The answer to such a question must lie in the realm of reaction mechanisms for the ATPase which are not understood. One mechanism which has been proposed is shown in Fig. 8.1. A negatively charged ADP together with a magnesium–phosphate complex approach the active site. The magnesium–phosphate complex carries two negative charges on the oxygen which is attacked by the two protons coming from the c-side to form water and MgATP; this latter is translocated back to the m-side of the membrane. Protons will be released on the m-side in the formation of the negatively charged $ADP + Mg-PO_4^-$. Additional negative charges on the ADP and ATP which would be dependent on the pH, are not shown.

The function of inhibitors is seen as follows. Aurovertin binds the F_1-ATPase and inhibits the enzyme reaction. Oligomycin does not bind F_1 in the absence of the protein known as the oligomycin sensitivity-conferring factor. Oligomycin, or N,N-dicyclohexylcarbodiimide which acts similarly to oligomycin, is regarded as blocking the passage of protons through F_0 (the basal part of the ATPase complex embedded deeply in the membrane). It has been demonstrated that the removal of F_1 from the membrane renders the membrane permeable to protons and also uncouples it. The addition of dicyclohexylcarbodiimide decreases the permeability and restores coupled activities.

The proton motive force responsible for driving the ATP synthesis is generated by the operation of the respiratory chain. The chain is seen as being composed of three proton-translocating loops (Fig. 8.2). In each loop, oxidation–reduction involves the movement of electrons and protons in the outward direction (m→c) and electrons in the inward direction (c→m). This results in

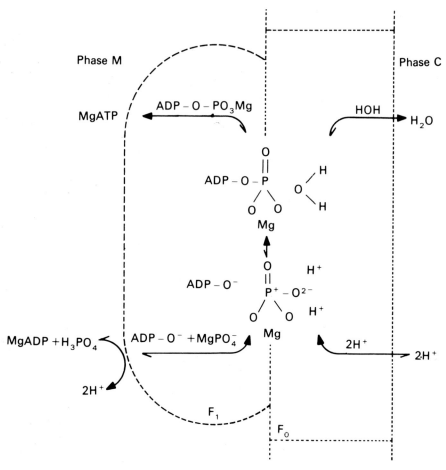

Fig. 8.1 A hypothetical scheme for the proton-translocating ATPase (from Mitchell, 1976).

proton transport across the membrane and can be represented as:

$$A_{red} + B_{ox} + 2H_m^+ \longrightarrow A_{ox} + B_{red} + 2H_c^+$$

Thus the NADH dehydrogenase, which is located with its catalytic site on the m-side of the membrane, might well catalyse the reactions:

(i) transfer of electrons and protons across membrane

$$NADH_2 + [Fe-S] \longrightarrow NAD + [Fe-S]^{2-} + 2H_c^+$$

(ii) completion of the loop by transfer of electrons inwards with the uptake of protons

$$[Fe-S]^{2-} + Fp_{FMN} + 2H_m^+ \rightleftharpoons [Fe-S] + Fp_{FMNH_2}$$

The three loops are taken to represent the three segments of the chain originally

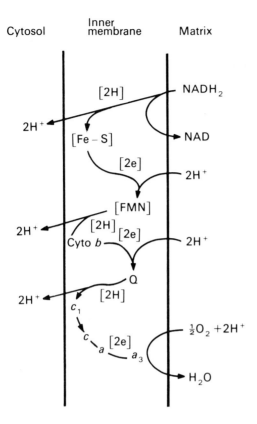

Fig. 8.2 A hypothetical scheme for the respiratory chain. Note that the order of components of the chain is not the same as that discussed in Chapter 2. The proton translocation, in association with the oxidation–reduction reactions, is examined further in Chapter 9 (Mitchell 1976)

associated with phosphorylation. Thus the process of oxidation–reduction in the respiratory chain comprises not only redox reactions but also obligatory reactions involving uptake of protons from the m-side and their release on the c-side. It should be noted that the notion of proton-translocating loops described here is not the only mechanism for proton translocation in biological membranes. The purple protein of the plasma membrane of *Halobacterium halobium* (see sect. 15.9), like the ATPase, apparently pumps protons without oxidation–reduction loops. Further, there is evidence that the cytochrome oxidase complex may also be able to translocate protons without an oxidation–reduction loop.

Values for the H^+/O ratio have already been discussed; Mitchell proposes a ratio of 6 for NADH oxidation and 4 for succinate oxidation although these figures are not agreed. The H^+/P ratio is 2 for the ATPase and this gives the expected P/O ratio of 3 for NADH and 2 for succinate oxidation. Although this arithmetic is satisfying, it does not take account of proton movements associated with phosphate transport and possibly ADP/ATP exchange (see Fig. 8.3).

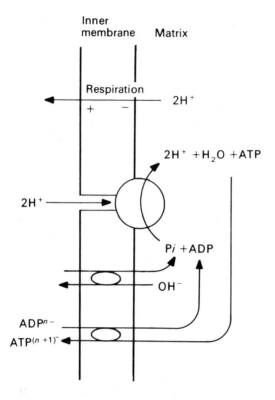

Fig. 8.3 Transport and phosphorylation.
Note that the scheme as shown requires further ion movements to achieve overall neutrality.

Since respiration is obligatorily coupled to proton translocation, the oxidation–reduction reactions in a coupling region will proceed until the electromotive force (difference in reduction potentials of the redox carriers, e.g. A and B above) balances the proton motive force (PMF). ATP synthesis by the proton-translocating ATPase will tend to dissipate the PMF and promote respiration. Anything which dissipates the PMF will promote respiration, such as uncoupling agents, membrane damage or decrease of the ATP/ADP ratio.

   The attractions of the chemiosmotic hypothesis lie in the fact that it readily explains the link between phosphorylation and ion transport, that uncoupling activity is clearly explicable and that intact vesicular structures are necessary for phosphorylation. Further, proton transport in association with ion transport, ATP hydrolysis and respiration are all readily observable phenomena and can be manipulated experimentally. This does not in itself necessarily mean that the proton gradient and the accompanying membrane potential are the driving force for oxidative phosphorylation. Nevertheless, in recent years a growing number of workers have found the ideas of the chemiosmotic theory a valuable means of understanding the bioenergetics of the mitochondrion, chloroplast and bacterial membranes. The award of a Nobel prize to Dr Mitchell in 1979 underlines the importance attributed to the chemiosmotic theory in cell bioenergetics.

8.4   Conformational theories

The failure to obtain support for the chemical theory also led to the development of conformational ideas by Boyer (see Boyer *et al.*, 1966; Boyer, 1975) and others in parallel with the development of the chemiosmotic theory. Several theories based on the storage of energy in conformational states of proteins have been evolved, but a simple outline only will be given here.

Respiration, according to these theories, creates a conformationally strained state of components or complexes in the membrane. This strained state also involves the ATPase which returns to its original conformation by synthesising ATP. The transfer of energy can be represented as follows:

Redox energy $\rightleftharpoons$ conformational energy in cytochrome proteins etc. $\rightleftharpoons$ conformational energy in other membrane proteins (and membranes) $\updownarrow$ energy of ion gradients $\rightleftharpoons$ conformational energy in ATPase $\rightleftharpoons$ ATP-associated energy

The passage of electrons through any of the three coupling regions of the chain would induce a change in the conformation of individual redox proteins. This change in secondary and tertiary structure of the protein would be associated with an alteration of weak bonding (hydrogen bonds, hydrophobic bonds, etc.) and would be accompanied by the storage of energy from respiration as conformational energy. The altered structure of the respiratory components would in turn induce similar strained states in other membrane proteins and in the membrane ATPase. The strained state (marked* below) would be relaxed by a reaction involving formation of free ATP. The reaction can be represented:

$$A_{red} + B_{ox} \rightleftharpoons A^*_{ox} + B_{red}$$

$$A^*_{ox} + F_1 \rightleftharpoons F^*_1 + A_{ox}$$

$$F^*_1 + ADP + Pi \rightleftharpoons F_1 + ATP + H_2O$$

where F_1 is the ATPase. The last reaction again raises the question of reaction mechanism and in particular of how conformational energy can be used to drive a chemical reaction. An example of a proposed mechanism is shown in Fig. 8.4. Bound ATP is synthesised from ADP and Pi in a reaction regarded as independent of respiratory energy. This is followed by an energy-dependent release of ATP. The ATPase has been shown to undergo conformational changes and also to bind adenine nucleotides. Further conformational changes have been demonstrated at the coupling sites. In particular, a *b*-type cytochrome, b_T, has been shown to undergo a change of redox potential in association with ATP

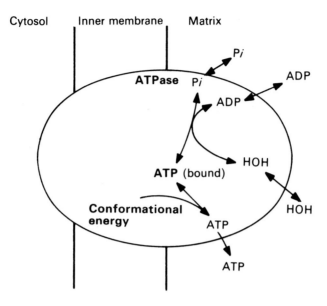

Cytosol   Inner membrane   Matrix

Fig. 8.4 A hypothetical scheme for ATP synthesis at the expense of conformational energy. This scheme can be further elaborated for two catalytic sites as in Fig. 9.2.

hydrolysis by the ATPase, which suggests a link between a conformational state of the cytochrome and the activity of the ATPase.

In a membrane with a high protein content, such as the mitochondrial inner membrane, changes in conformation under extreme conditions would be expected to alter the structural appearance of the particle. Structural changes in association with ATP hydrolysis and with respiration are observed and have formed the basis for some conformational theories (cf. Green and Baum, 1970).

8.5   Which theory?

The chemical, conformational and chemiosmotic theories represent three very different approaches to the problem of the mechanism of oxidative phosphorylation. In addition there are several variants of each theory. As a broad generalisation, the validity of scientific theories can be approached from the point of falsification rather than verification. Thus theories survive (in logical terms) until they are shown to be false, or until they cease to arouse the interest of the scientific community. The chemical theory still stands as a possible mechanism for oxidative phosphorylation and is still invoked by some workers. However, since it predicts the existence of an intermediate for which extensive searches have been made without significant success, its standing among biochemists is not high and indeed it is dying for lack of interest.

By contrast, the alternatives, the chemiosmotic and conformational theories, are consistent with a substantial body of experimental observations and attempts to disprove either view convincingly have so far failed. The chemiosmotic theory, however, allows for considerable experimentation, since proton movement may be readily measured. It has gained considerably in popularity not least because of its usefulness in the interpretation of experimental results.

It has been necessary for each theory to take account of the experimental data used in support of the other. Thus it has been argued that conformational changes in components of the oxidative phosphorylation system are consistent with the chemiosmotic viewpoint, since conformational changes accompany the interaction of many enzymes with their substrates or with ligands such as allosteric effectors. Boyer (1975), for instance, has argued that the uptake and release of protons may be intimately associated with the conformational changes involved in oxidative phosphorylation. Thus a change in conformation may expose on one side of the membrane, groups which become protonated resulting in net proton uptake while protons are released from the opposite side of the membrane. Such a relationship between protons and the conformational processes of oxidative phosphorylation could be used for a net synthesis of ATP, if a gradient of protons was applied and allowed to drive the conformational changes.

As already noted, the explanation of phosphorylation and transport given for the mitochondrion must also apply to photophosphorylation, to phosphorylation and to many transport systems and phosphorylation in bacteria. The properties of all these systems are broadly similar. In bioenergetics of bacteria and chloroplasts, the chemiosmotic theory has enjoyed wide popularity.

Chapter 9

Resolution of the respiratory chain and oxidative phosphorylation

9.1 The mitochondrial adenosine triphosphatase (ATP synthase)

(a) General properties and exchange reactions

The system for oxidative phosphorylation has been resolved into five complexes. Of these, four are components of the respiratory chain, while the fifth has ATPase activity. This latter enzyme is responsible for ATP synthesis (ATP synthase), although in general it is referred to as the ATPase. As already noted, ATP hydrolysis is associated with proton translocation; this property is most clearly demonstrated in experiments with artificially formed vesicles (see sect. 9.1(d) below). The enzyme requires Mg^{2+} for activity, is sensitive to a number of inhibitors including oligomycin, aurovertin, rutamycin and dicyclohexylcarbodiimide and, in a coupled system, catalyses the reaction

$$ADP + Pi \longrightarrow ATP + H_2O$$

and various exchange reactions. Since the equilibrium of the reaction lies very much to the left, ATP synthesis is only observed when it is coupled to electron transport or a proton gradient.

For many purposes, coupled activity (as opposed to uncoupled) can also be observed with exchange reactions such as the ATP/Pi exchange, which only occurs in a coupled system and is sensitive to ATPase inhibitors (see also sect. 2.9(b)). Incubation of ^{32}Pi and ATP with coupled mitochondria or submitochondrial particles leads to labelling of the γ-phosphate of ATP (Boyer et al., 1954). The exchange is not dependent on electron transport and is inhibited by uncoupling agents and oligomycin. Early interest in such reactions was stimulated by the search for evidence supporting the chemical theory of oxidative phosphorylation, but more recently by interest in the mechanism of the ATPase and the conformational ideas of Boyer (see Boyer et al., 1973). Exchange reactions such as ATP/Pi also demonstrate the existence of a high-energy intermediate state in oxidative phosphorylation.

Rather similar to the ATP/Pi system is the oligomycin- and uncoupler-sensitive ADP/ATP exchange. This can be assayed in coupled mitochondria and submitochondrial particles using ^{14}C-labelled nucleotides. An H_2O/Pi exchange was first discovered by Cohn (1953). Using the heavy isotope of oxygen in water $(H_2^{18}O)$, it has been possible to demonstrate transfer of label between the oxygens of phosphate and water. This shows a requirement for adenine nucleotides. A further oxygen exchange between the oxygen in the γ-phosphate group of ATP and the oxygen of water was described by Cohn and Drysdale (1955). At first sight, it is difficult to reconcile these observations since the oxygen of water might be expected to exchange with either ADP or Pi but not ATP. More complex reaction mechanisms involving intermediates which exchange their oxygen with that of water will however allow for a water/ATP reaction (see Boyer *et al.*, 1973). The Pi/H$_2$O system differs from the water/ATP reaction in being insensitive to uncouplers but inhibited by oligomycin. The ATP/H$_2$O system is sensitive to uncouplers; only the latter reaction is found in the chloroplast ATPase.

(b) Coupling factors

A new phase of study in phosphorylation was initiated by the isolation of a soluble dinitrophenol-stimulated ATPase from mechanically disrupted beef heart mitochondria in Racker's laboratory (Pullman *et al.*, 1960). This soluble ATPase could restore the ability of submitochondrial particles to carry out phosphorylation coupled to electron transport. The term 'coupling factor' was applied to the soluble ATPase and it was subsequently designated F$_1$. It has a diameter of 9 nm (90 Å), a molecular weight of about 360 000 and catalyses an oligomycin-insensitive hydrolysis of ATP but not the ATP/Pi exchange reaction. After recombination with the membrane, the enzyme becomes oligomycin-sensitive. Curiously the soluble enzyme is cold-labile due to partial dissociation of the subunits, although it is cold-stable when membrane-bound. Following the initial isolation of this coupling factor, other factors were isolated in Racker's laboratory and elsewhere. None of these had enzyme activity. Submitochondrial particles were treated by physical or chemical means to give preparations capable of oxidation of substrates but incapable of coupled phosphorylation. Addition to these particles of various membrane proteins (such as F$_1$ and other 'coupling factors') restored the ability to carry out phosphorylation or to catalyse an ATP/Pi exchange. Proteins F$_2$, F$_3$, F$_5$ and F$_c$ (F$_4$ = F$_2$ + F$_3$) simply assisted in restoring normal phosphorylation activity, although they had no activity on their own.

As noted above, the F$_1$ ATPase, unlike the membrane-bound enzyme, is insensitive to oligomycin and also rutamycin. A protein which confers oligomycin (and rutamycin) sensitivity to the reconstituted ATPase has been purified and has a molecular weight of about 18 000. Known as the OSCP (oligomycin sensitivity-conferring protein) or F$_c$, it is probably the active component in most of the other F factors (F$_3$, F$_4$, F$_5$). F$_0$ is the membrane-bound protein complex to which the ATPase attaches.

Sanadi and coworkers have also resolved the phosphorylating system into various fractions. Their factor A is a soluble oligomycin-insensitive ATPase similar to F$_1$.

A further heat-stable protein, originally isolated from beef heart mitochondria, inhibits ATPase activity of F$_1$ and has a molecular weight of about 10 000. It is readily inactivated by trypsin. In appropriate conditions in the presence of ATP and Mg^{2+} the protein binds to, and forms a stable complex with, F_1. Normal preparations of F_1 contain a variable amount of inhibitor protein. Although inhibiting ATP hydrolysis, the protein does not inhibit oxidative phosphorylation by submitochondrial particles (Ferguson *et al.*, 1977).

An oligomycin-sensitive cold-stable ATPase can be prepared by extract-ing mitochondrial preparations with detergent to give a lipoprotein preparation insoluble in the absence of detergent. Such preparations include the CF_0F_1 complex of Kagawa and Racker (1966). F_0 is a lipid-depleted form of CF_0. The lipoprotein, CF_0, constitutes the membrane sector of the ATPase to which F_1 is attached (see Fig. 9.1). From a study of oligomycin-sensitive ATPases, it is possible to distinguish those proteins which are neither part of F_1 nor OSCP and are therefore a part of the membrane 'base piece' of the enzyme. In yeast, Tzagoloff and Meagher (1971) found four hydrophobic proteins of molecular weight 29 000, 22 000, 12 000 and 8000. The nature of the 'base piece' of the mitochondrial ATPase is not yet fully understood.

The ATPase can be seen in electron micrographs of negatively stained inner membrane preparations as a stalked particle. F_1 corresponds to the particle. If, after removal of F_1 with urea (or NaBr), the membrane is further extracted with dilute ammonia, the OSCP is removed. It can then be shown that this protein is necessary for binding F_1 to the membrane. It has therefore been assumed that OSCP constitutes the stalk seen in negatively stained preparations[1]. If CF_0 and OSCP are incorporated into liposomes, the proton permeability of the liposome is increased. Permeability of F_1-depleted sub-mitochondrial particles to protons is decreased by addition of oligomycin or by $F_1 + OSCP$ (Hinkle and Horstman, 1971). A variety of experiments of this type have suggested that the proteolipid components of the ATPase constitute a proton-conducting pore in the inner membrane, or possibly a proton-conducting ionophore. In either case, the protons outside the mitochondrial inner membrane have access to the F_1 component of the ATPase.

(c) F_1 ATPase

Although insensitive to oligomycin, rutamycin, DCCD or mercurials (all of which inhibit the bound enzyme), the isolated F_1 ATPase binds and is inhibited by aurovertin. Activity is stimulated by dinitrophenol. The enzyme has a molecular weight of about 360 000 and can be dissociated into five main subunits (excluding the inhibitory protein discussed above). These are the α, β, γ, δ and ε subunits as shown in Table 9.1 (Senior, 1973; Tzagoloff and Meagher, 1971). Earlier studies suggested a stoicheiometry of α_3, β_3, γ, δ, ε, but more recent studies have tended to favour α_2, β_2, γ_2, δ_2, ε_2. The γ and δ subunits appear to be necessary for binding to OSCP (Fig. 9.1). Aurovertin binds specifically to the β-subunit. The substrate analogue, 8-azido-ATP, which is slowly hydrolysed by the enzyme, can also be shown to bind to the β-subunit, a maximum of two molecules of the ATP analogue per F_1 being bound. Freshly isolated, F_1 may possess up to three molecules of tightly bound ATP and up to two molecules of tightly bound

---

[1] It has also been proposed that a second protein F_{c2} or F_6 is also necessary for F_1 attachment.

Table 9.1 Polypeptides of the mammalian oxidative phosphorylation system

ATPase		NADH–ubiquinone reductase		Succinate–ubiquinone reductase		Ubiquinol–cytochrome c reductase		Cytochrome oxidase	
MW	designation	MW	designation	MW	designation	MW	designation	MW	designation
53 000 (58 500)	F_1 α-subunit	75 000*	Fe–S protein	70 000	flavoprotein	43 000		35 400 (40 000)	I, cytochrome
50 000 (54 000)	F_1 β-subunit	53 000	Fe–S protein	37 000	Fe–S protein	40 000		24 100 (33 000)	II
33 000 (38 500)	F_1 γ-subunit	53 000*				29 000	cyt c_1	21 000 (22 000)	III
17 000 (31 000)	F_1 δ-subunit	42 000				28 000		16 800 (14 500)	IV
7 500 (12 000)	F_1 ε-subunit	39 000				24 000	Fe–S protein	12 400 (12 700)	V
10 000	F_1 inhibitor	33 000				12 500		8 200 (12 700)	VI
18 000	OSCP = F_{C1} = F_5	29 000	Fe–S protein			8 000		4 400 (4 600)	VII
8 000	F_{C2} = F_6	26 000*				5 000			
29 000	⎱ 'base piece'	25 000							
22 000	⎰ membrane	23 500							
12 000	⎱ sector	22 000							
8 000		20 500							
		18 000							
		15 500							
		8 000							
		5 000							
Yeast subunits shown in brackets								Yeast subunits shown in brackets	

* Polypeptides associated with the type II dehydrogenase.

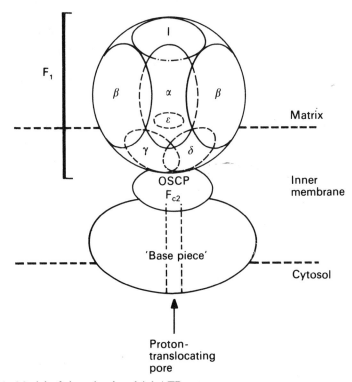

Fig. 9.1 Model of the mitochondrial ATPase.
Since there is inadequate evidence to support an agreed model of the ATPase, this model involves several assumptions. (I represents the inhibitor protein.)

ADP. Estimations of bound adenine nucleotide have, however, produced variable results. The role of this bound adenine nucleotide is not yet clear.

A recently discovered inhibitor of the ATPase is efrapeptin. This peptide antibiotic appears to act at the catalytic site of the ATPase (Cross and Kohlbrenner, 1978). One molecule of antibiotic is bound per F_1. More detailed studies with this inhibitor support the developing view that there are two alternating catalytic sites in the ATPase (Kayalar *et al.*, 1977; see Fig. 9.2).

Alternation of catalytic sites implies conformational changes in the ATPase. Such changes have been known for some time. For example aurovertin, which inhibits both ATPase activity and exchange reactions, forms a fluorescent complex with the ATPase. In coupled submitochondrial particles treated with aurovertin, energisation of the membrane enhances fluorescence, suggesting that the ATPase undergoes conformational changes associated with phosphorylation activity.

(d) Reconstitution studies

As already noted, it is possible to reconstitute phosphorylating sub-mitochondrial particles which have been depleted of the ATPase, F_1. A more refined constitution of the ATPase was carried out by Racker's group (Kagawa *et al.*, 1973) using phospholipids, hydrophobic mitochondrial proteins, F_1 and other coupling factors. The phospholipids and hydrophobic proteins were dissolved in the ionic detergent, cholate, and the cholate slowly dialysed away to give phospholipid vesicles. Coupling factors could then be added to the vesicles. Phosphatidylcholine and phosphatidylethanolamine were both required for the reconstitution, but the best activity was obtained with a three- to fourfold excess of the latter. However, if equimolar amounts of the two phospholipids were used, activity was then stimulated by the inclusion of cardiolipin. The reconstituted system gave an ATP/Pi exchange sensitive to ATPase inhibitors and uncouplers; it would also couple ATP hydrolysis to the inward pumping of protons. Using this type of system, it is possible to reconstitute an oxidative phosphorylation system with cytochrome oxidase, F_1 and additional factors. A simpler rapid method of forming active vesicles involves suspension of dried phospholipids and membrane proteins in salt solution.

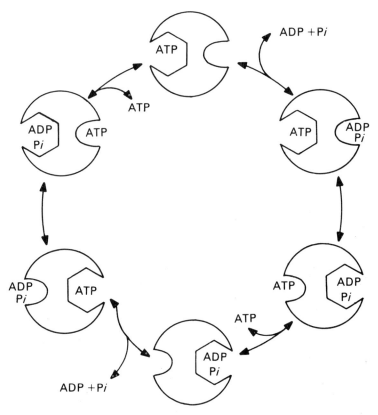

Fig. 9.2 An alternating site model for the ATPase. (Based on Kayalar *et al.*, 1977.)

These experiments clearly demonstrate the requirement of the coupled ATPase for phospholipids and the ability of the enzyme to pump protons when asymmetrically bound to a membrane.

9.2 Coupling and uncoupling

(a) Properties of coupled mitochondria

Fresh mitochondria from most sources are coupled, i.e. respiratory activity is obligatorily and stoicheiometrically linked to the synthesis of ATP or ion transport; hydrolysis of ATP may also be linked to reverse electron flow or ion transport. Uncoupling may be experimentally demonstrated in terms of reduced P/O ratio, inhibition of exchange reactions such as ATP/^{32}Pi, release from respiratory control (increase of respiratory activity in state 4) and release of respiration from inhibition by ATPase inhibitors such as oligomycin. However, it should be noted that loosely coupled mitochondria and submitochondrial particles may lack respiratory control while retaining ability to phosphorylate.

Coupling may be broken in a number of ways:

(i) Damage to mitochondria or submitochondrial particles affecting the integrity of the inner membrane. In addition to mechanical disruption, phospholipase and detergents also uncouple by virtue of their disruptive effect on the inner membrane.

(ii) Classical uncouplers, which include a large number of substances, many of which are of the phenolic or anilinic type (Fig. 9.3). These aromatic anionic substances have two properties either or both of which may be the basis for their uncoupling activity. Firstly, they are known to increase the electrical conductance of artificial phospholipid membranes; they can be shown to increase membrane permeability to protons by swelling experiments with vesicles. This activity correlates fairly well with their uncoupling activity (see Cunarro and Weiner, 1975). Secondly, many uncouplers have been found to bind to the inner mitochondrial membrane, although not to other membranes such as that of the erythrocyte. Since many uncouplers are lipophilic and tend to pass unspecifically into the membrane, specific binding has been studied with hydrophilic uncouplers such as 4-azido-2-nitrophenol or 2-nitro-4-azidocarbonylcyanide phenylhydrazone. A specific high-affinity binding site has been demonstrated (Katre and Wilson, 1977). A curious uncoupling activity is found with picrate (2,4,6-trinitrophenol), which is relatively unable to penetrate the inner membrane and is therefore ineffective in whole mitochondria. However, inverted phosphorylating submitochondrial particles are uncoupled by picrate suggesting action at a binding site on the inner face of the inner membrane.

There are two explanations for the action of uncoupling agents. They are seen as rendering the membrane permeable to protons and hence dissipating the proton gradient and the membrane potential (chemiosmotic) theory. Alternatively they bind to the membrane and dissipate some high-energy intermediate in the membrane.

(iii) Ionophores which render the membrane permeable to cations such as K^+. These uncouplers dissipate the energy available from respiration or ATP hydrolysis by ion transport (see Ch. 7).

(b) Brown adipose tissue mitochondria

In most cells, mitochondria appear to be tightly coupled. However, some tissues, particularly the brown fat of hibernating and new-born mammals have a

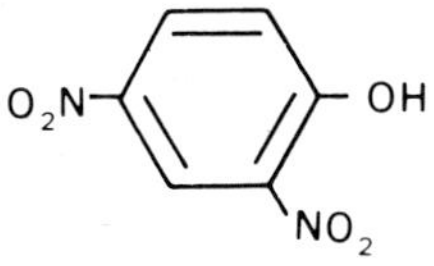

2,4-Dinitrophenol 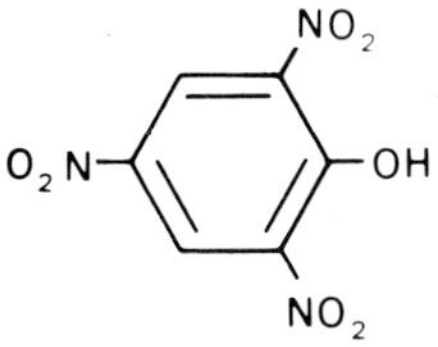   Trinitrophenol 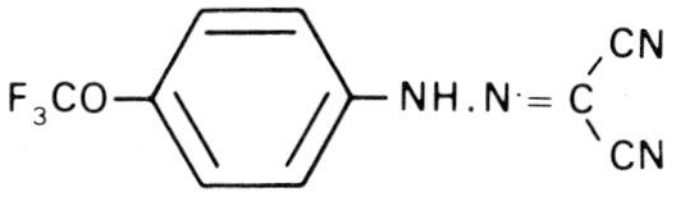   Carbonylcyanide-*p*-
                                             trifluoromethoxyphenyhydrazone (FCCP)

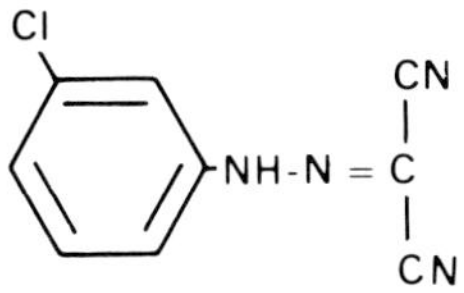

Carbonylcyanide-*m*-   Trifluoromethyl   5-Chloro-2-mercaptobenzothiazole
chlorophenylhydrazone   tetrachlorobenzimidazole
(CCCP)

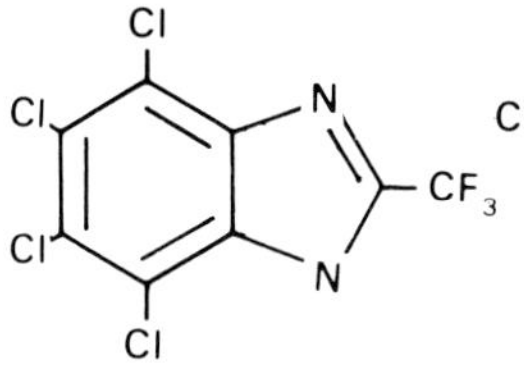

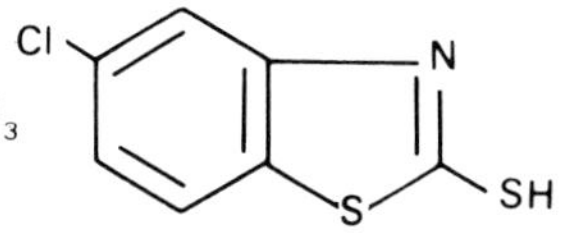

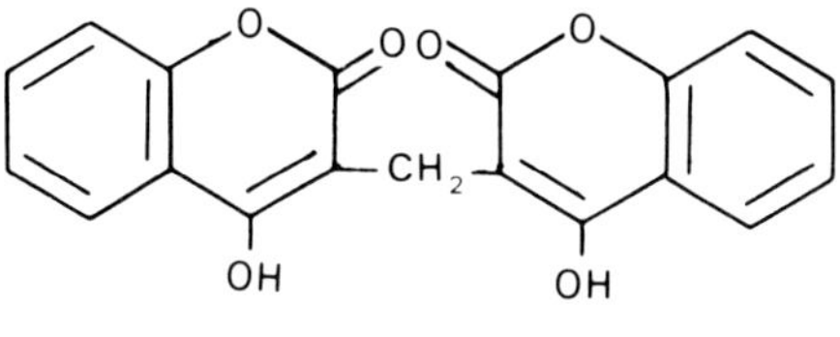

Dicoumarol 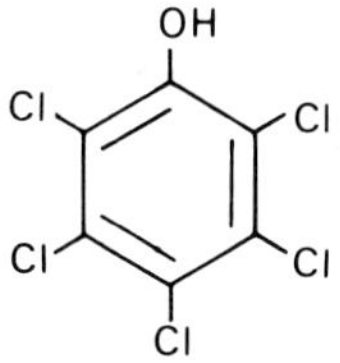   Pentachlorophenol

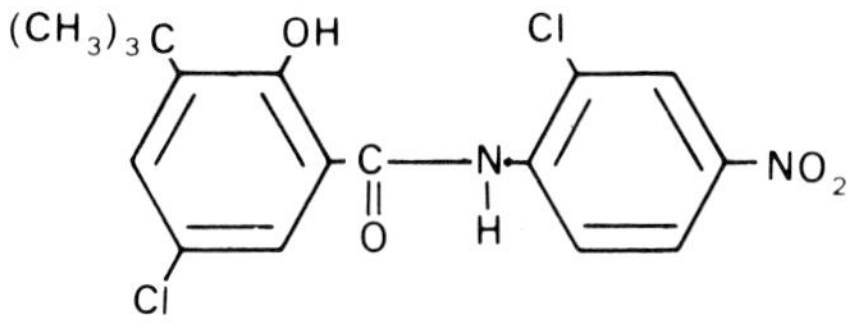

5-Chloro-3-*t*-butyl-2'chloro 4'-nitro salicylanilide

Fig. 9.3 Structures of some uncouplers.

form of non-shivering heat production associated with a naturally uncoupled state of their mitochondria. The energy normally used for ATP synthesis is released as heat in the uncoupled system. The triglyceride store of the tissue is oxidised for heat production by this means during thermal stress. The initial stage in heat production, lipolysis, is prompted by the sympathetic nervous system via adenyl cyclase which forms cyclic-AMP, leading to activation of a lipase and generation of free fatty acids. The fatty acids are activated, transferred to the mitochondria and oxidised.

There has been considerable interest in brown fat mitochondria since they exhibit a form of regulation of energy metabolism. They have a high content of respiratory enzymes, a low ATPase content and a high ionic permeability, particularly to protons. If animals are exposed to cold stress for a few days, the isolated mitochondria may be shown to have a low P/O ratio. This value returns to normal if particles are treated with purine nucleotides in the presence of bovine serum albumin which removes fatty acids (themselves uncouplers). Uncoupling appears to be, in part, non-physiological and mediated by the long-chain fatty acids which render the membrane permeable to protons. However, it now seems that uncoupling is also due to a transport system for protons. It is proposed that there is a uniport[2] for translocating protons across the inner membrane. This proton transport is inhibited by purine nucleotides (GDP, GTP, ADP, ATP) which specifically bind to a site (probably a protein, MW approx. 32 000) on the outside of the inner membrane (Heaton *et al.*, 1978). Such a system could not be found in liver mitochondria. Regulation of coupled activity is probably achieved through long-chain acyl coenzyme A esters which antagonise the inhibition of proton transport by purine nucleotides. The cytoplasmic nucleotide concentration is itself adequate to ensure coupling.

9.3   Isolation of complexes of the respiratory chain of beef heart

An extensive series of investigations carried out at the Institute for Enzyme Research at the University of Wisconsin made possible the resolution of the beef heart respiratory chain into four complexes (Fig. 9.4). The method used for the separation was based on the treatment of mitochondria with de-oxycholate followed by fractionation of the solubilised material (Fig. 9.5). The four fractions, which are lipoproteins, catalyse the following reactions.

Complex I $NADH_2$–ubiquinone reductase
Complex II   succinate–ubiquinone reductase
Complex III   ubiquinone–cytochrome *c* reductase
Complex IV   cytochrome *c* oxidase

Early studies by Hatefi *et al.* (1961) described the isolation of an $NADH_2$–cytochrome *c* reductase by deoxycholate treatment. Unlike the enzyme of Mahler *et al.* (1952) mentioned earlier (sect. 2.3(d)), this enzyme was sensitive to inhibition by both amytal and antimycin A. Subsequently, this cytochrome *c* reductase was resolved into complexes I and III.

---

[2] A simple carrier system, i.e. one not mediating an exchange of substrates or ion (antiport) nor mediating the transport of pairs of substrates or ions (symport).

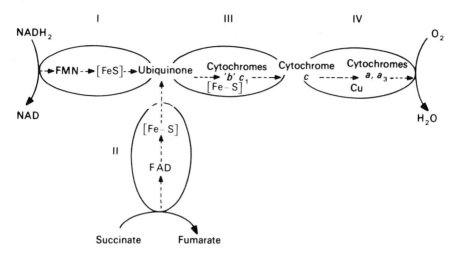

Fig. 9.4 The complexes of the respiratory chain (I, II, III and IV).

The four complexes possess between them all the components known to act in the respiratory chain (although their concentration may be rather lower) together with a substantial amount of protein and phospholipid. The isolated complexes could be recombined to give an $NADH_2$ – cytochrome c reductase (I + III), a succinate–cytochrome c reductase (II + III), an $NADH_2$/succinate– cytochrome c reductase (I + II + III), a succinate oxidase (II + III + IV), an $NADH_2$ oxidase (I + III + IV) and an $NADH_2$/succinate oxidase (I + II + III + IV) (Hatefi *et al.*, 1962a).

Apart from this analysis of beef heart mitochondria, numerous other isolations have been made both in beef heart and in other tissues and organisms. Only some of these will be considered here under the headings of the four segments of the respiratory chain shown in Fig. 9.4.

9.4 $NADH_2$ – ubiquinone reductase, NADH dehydrogenase

(a) Properties and types of enzyme

This segment of the respiratory chain is associated with a coupling site and involves both a flavoprotein with FMN as the prosthetic group and several iron–sulphur centres. It catalyses the reaction:

$$NADH_2 + ubiquinone \rightleftharpoons NAD + ubiquinol$$

$NADPH_2$ also appears to be oxidised but only very weakly. The complex is sensitive to amytal, rotenone, piericidin and rhein in mammalian mitochondria. In yeast, the activity is in many cases insensitive to rotenone. Rhein is a competitive inhibitor of NADH oxidation and presumably acts at the catalytic site. Both rotenone and piericidin probably act close to the ubiquinone reducing site. A major use of these inhibitors is to determine the functional integrity of the enzyme, as will be shown below.

$NADH_2$–ubiquinone reductase can be assayed with $NADH_2$ (or an

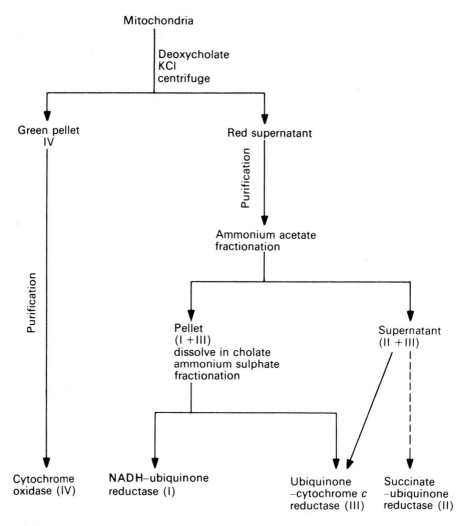

Fig. 9.5 Outline of the resolution of the respiratory chain into complexes.
N.B. Complex II is normally prepared by an independent method but could be prepared
as shown (dotted line).

$NADH_2$ analogue to give slower rates of oxidation) as substrate and a suitable
ubiquinone as electron acceptor. Since the naturally occurring quinone of
mammalian systems, Q-10, is relatively insoluble, ubiquinone-1 with a short
hydrocarbon side chain and greater water solubility has been found useful.
Ferricyanide is also widely used as an electron acceptor.

Numerous isolations of $NADH_2$–oxidising enzymes have been made.
Three forms will be considered here. The complex I from beef heart is a
lipoprotein which readily reduces ubiquinone and ferricyanide, is sensitive to
rotenone and possesses substantial amounts of phospholipid. It is not readily
kept in solution in the absence of detergent, although it can be dispersed in

sucrose solutions. There are two forms of NADH dehydrogenase which can be referred to as type I and type II. Type I dehydrogenase is a soluble high-molecular weight enzyme (MW of the order of 10^6) which possesses Fe—S centres and FMN, but not lipid and ubiquinone. The main differences between type I dehydrogenase and complex I are the lack of lipid, insensitivity to rotenone and, normally, the inability to use ubiquinone as electron acceptor. The type I enzyme was originally isolated by Ringler *et al.*, (1963) from beef heart by a technique involving lipid digestion with phospholipase A. Similar enzymes have been isolated from other sources, including the yeast *Candida utilis* (Tottmar and Ragan, 1971). A further form of NADH dehydrogenase, type II of low molecular weight (only 70 000–80 000) has been prepared by several workers. This enzyme appears to be a fragmented form of the type I enzyme. Thus treatment of the type I enzyme with elevated temperatures, organic solvents or acid pH is followed by a loss of ferricyanide-reducing activity and the gain of ability to reduce a number of electron acceptors including cytochrome *c*. The cytochrome *c* reduction is not inhibited by rotenone or antimycin. This type II enzyme hence appears to have undergone considerable changes in properties during formation.

(b) Prosthetic groups: FMN and Fe

Complex I possesses ubiquinone-10, iron and FMN, which is not covalently bound, at about 1 nmol FMN/mg protein. The Fe/FMN ratio is uncertain and may be as high as 28:1. The iron is present at an equimolar amount with acid-labile sulphur, probably mostly as [4Fe – 4S] clusters which are linked to protein (non-haem iron–sulphur protein). The exact structure of the clusters is not known, but Fig. 9.6 gives possible forms. The sulphur can be readily released as H_2S by acid treatment. If the upper figure for the Fe/FMN ratio is accepted, this suggests seven clusters, which coincides with the results obtained from some EPR studies. Although FMN is the flavin in most NADH dehydrogenases examined including that of *Candida utilis*, *Saccharomyces cerevisiae* may have FAD as the prosthetic group (Duncan and Mackler, 1966).

(c) Iron–sulphur proteins

A number of iron–sulphur proteins have been detected in the respiratory chain. They may undergo oxidation–reduction between the Fe^{3+} and Fe^{2+} states. Since progress with the study of these components has been limited by inability to isolate them in their native form, we will briefly consider some other better understood examples.

The simplest type of non-haem iron protein is represented by rubredoxin obtained from several bacteria including *Clostridium pasteurianum* and *Pseudomonas oleovorans*. In the latter organism the protein undergoes oxidation/reduction, transferring electrons between an NADH dehydrogenase and a hydroxylase. Rubredoxin of *C. pasteurianum* ($E_{m\,7.0} = -57\,mV$) has a low molecular weight (6000), is composed of 54 amino acids in a single polypeptide chain and possesses one atom of iron per molecule bound to cysteine residues (Fig. 9.6a).

Substances which contain an unpaired electron such as a semiquinone or a reduced iron will normally give rise to an electron paramagnetic resonance

(EPR) absorption. The material under investigation is placed in a variable magnetic field. The absorption of energy by the electron is measured in the microwave region of the spectrum. The g-value which is a characteristic property of the radical in a particular environment is calculated from the resonance frequency and the magnetic field strength. Rubredoxin is unusual among iron–sulphur proteins in not giving an EPR signal in the reduced state, although in the oxidised state it gives a signal at $g = 4.3$.

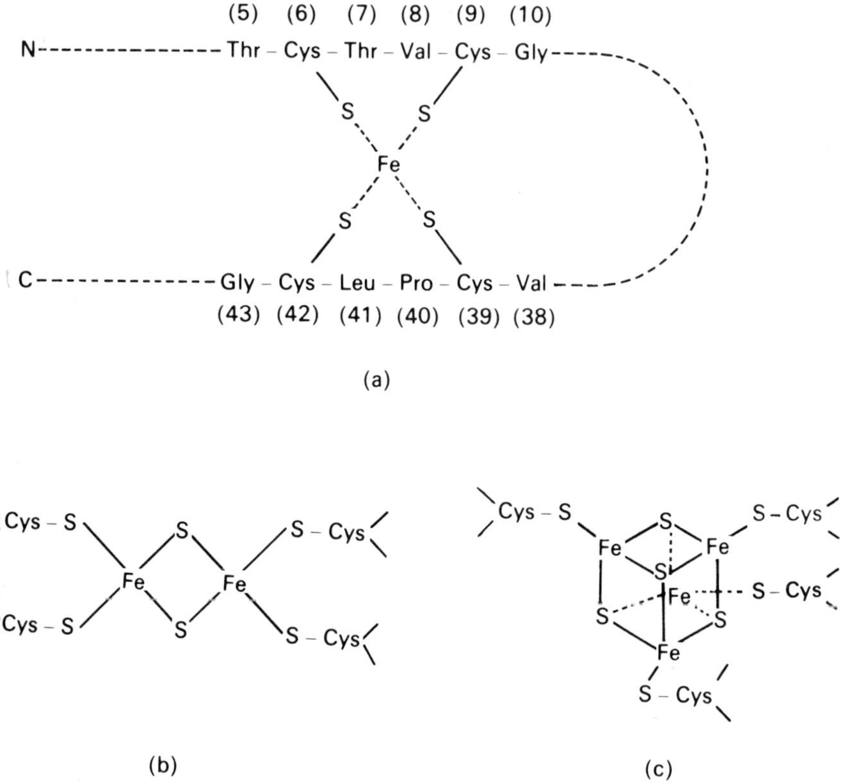

(a)

(b)

(c)

Fig. 9.6 Iron–sulphur proteins.
(a) Rubredoxin from *Clostridium pasteurianum*: primary structure showing the position of the iron. Note that this simple iron-protein has no labile sulphur. (Numbers indicate the positions of the amino acid residues in the polypeptide chain.) (b) A possible structure for the [2Fe–2S] type of iron–sulphur protein. The Fe–S cluster is linked to cysteine residues in the protein. (c) A possible structure for the [4Fe–4S] type of iron–sulphur protein.

A more complex type of iron–sulphur protein is represented by spinach ferredoxin which functions as a one-electron carrier in the chloroplast electron transport chain. This protein ($E_{m\,7.0} = -420\,\text{mV}$), which is red-brown in colour, contains two iron and two labile sulphur atoms. The latter may be released by acid treatment as H_2S. It has a molecular weight of 10 600 and consists of a single polypeptide chain of 97 residues. The arrangement of the two atoms of iron is shown in Fig. 9.6b. The cluster is linked to the protein through the sulphur atoms of the four cysteine residues. At low temperature, this iron–sulphur centre gives a strong EPR signal in the reduced state at $g = 1.94$. Other similar proteins have been characterised including a range of ferredoxins from other sources, such as adrenodoxin or adrenal ferredoxin which is involved in steroid hydroxylation.

Several [4Fe–4S] iron–sulphur proteins have been described including the HiPIP (high-potential iron protein) type from photosynthetic bacteria and other sources. HiPIPs, unlike the proteins described above, have positive mid-point potentials. They contain iron and four labile sulphur atoms in an approximately cube-like arrangement and the iron is also linked to cysteine residues (see Fig. 9.6c). The oxidised form has an EPR signal at about $g = 2$.

A related group of [4Fe–4S] proteins, found in bacteria, have negative potentials (about -300 to -400 mV). They are similar in many ways to the [8Fe–8S] bacterial iron proteins (such as *Clostridium* ferredoxin) which have two separate [4Fe–4S] clusters.

A number of enzymes are known which possess FeS centres. The best known of these is probably xanthine oxidase, which has FAD, molybdenum, iron and acid-labile sulphur in the ratio $2:2:8:8$. The FeS is present as two dissimilar pairs of [2Fe–2S] centres.

(d) Iron–sulphur centres in NADH–ubiquinone reductase

Analyses of the late 1950s showed the presence of non-haem iron in mitochondria. Beinert and coworkers in the early 1960s revealed the presence of EPR signals ($g = 1.94$) both in mitochondria and in isolated iron-flavoproteins such as xanthine oxidase and these signals were attributed to iron. Kinetic studies with the type I dehydrogenase showed that the signal appeared on reduction by NADH₂ analogues (which are oxidised more slowly than NADH₂) and disappeared on oxidation (by ferricyanide) suggesting that iron played a role in the overall oxidation–reduction process of the dehydrogenase (Beinert et al., 1965). Subsequently, beef, pigeon and yeast mitochondria have given signals attributed to iron. Beinert's group, together with Hatefi, used low temperatures to resolve the iron signal associated with complex I into four components referred to as N1, 2, 3 and 4. This analysis was taken a stage further by Ohnishi et al. (1974) who studied the EPR signal at low temperature measuring the mid-point potential (E_m) of the signals. This resulted in the resolution of the signals of N1 into 1a and 1b and the addition of signals 5 and 6 (see Table 9.1). Albracht et al. (1977) concluded these EPR signals could be interpreted adequately in terms of five centres (excluding 5 and 6 above). Centres 1a and 1b may be of the [2Fe–2S] type while the remainder are [4Fe–4S].

The function of these iron–sulphur centres in the NADH₂–ubiquinone reductase is not clear. Centre 1 may be associated with the coupling site, since addition of ATP to mitochondrial or submitochondrial particles under appropriate conditions can cause either reduction or oxidation of centre 1 coupled to oxidation or reduction of centre 2. This observation has been used to identify centre 1 as the coupling site, since the ATP effect is inhibited by uncouplers and oligomycin.

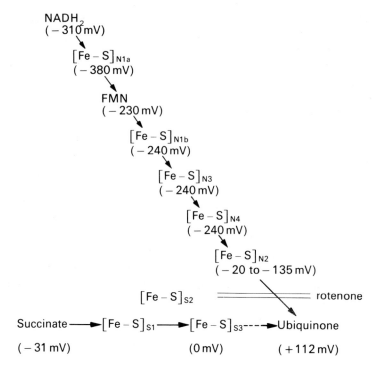

Fig. 9.7 The iron-sulphur centres of the NADH and succinate dehydrogenases (with values for E_m). The centres are arranged in a possible sequence based in part on the midpoint potentials. There is no agreement on whether $NADH_2$ reacts with an (Fe–S) centre before reduction of the flavin.

A possible approach to the function of iron–sulphur proteins is provided by growth of yeasts such as *Candida utilis* in iron-deficient or sulphur-deficient media. Mitochondria from such cultures show a loss of EPR signals due to iron–sulphur centres, a loss of piericidin A sensitivity and low P/O ratios for the oxidation of NADH. Although this direct correlation has been disputed, there appears to be a relationship between the possession of iron–sulphur centres in the NADH–ubiquinone reductase, the ability to carry out site 1 phosphorylation and sensitivity to piericidin.

The sequence of iron–sulphur centres in complex 1 is uncertain but a possible redox chain is shown in Fig. 9.7. This is based on several considerations, including the fact that only centre N2 is significantly reducible by succinate in mitochondria and submitochondrial particles, that N2 reacts with ubiquinone and that the centres might be expected to function in the order of their standard redox potentials. Oxidation of iron–sulphur centres is inhibited by rotenone. Much, however, remains to be elucidated about the structure and function of the iron–sulphur centres in the respiratory chain.

(e) Lipids

Twenty per cent of complex I is phospholipid, mainly cardiolipin, phosphatidylcholine and phosphatidylethanolamine. The lipids are probably necessary for quinone reduction since their removal by detergent (cholate) results in a loss of ability to reduce ubiquinone–1; this activity can be restored by addition of lipids under suitable conditions.

The $E_{m\,7.0}$ of the iron–sulphur centre N2 is also altered by lipid depletion, although other centres are relatively unaffected. Heron *et al.* (1977) concluded that all the lipid was necessary for full NADH₂–ubiquinone reductase activity.

(f) Resolution of complex I

Treatment with urea results in solubilisation of only 30 per cent of the complex, from which a type II dehydrogenase and denatured iron–sulphur protein can be separated. The type II dehydrogenase is an iron flavoprotein containing a small amount of iron (4Fe : 1FMN approximately). Separation of the protein subunits of complex I by Ragan gave 15 polypeptides (Table 9.1). Rather similar results have been obtained by Dooijewaard *et al.* (1978). Of these, the 75 000, 53 000 and 29 000 subunits were the major components of the iron–sulphur protein solubilised by urea treatment and the 53 000 and 26 000 subunits were associated with the type II dehydrogenase. There are probably two non-identical 53 000 subunits, one being part of an iron–sulphur protein and the other part of the type II dehydrogenase.

In summary, the segment of the mitochondrial respiratory chain that oxidises NADH₂ and reduces ubiquinone consists of about 16 polypeptides and 3 phospholipids and probably involves a single flavoprotein and up to 7 iron–sulphur clusters.

(g) Reconstitution studies

The NADH dehydrogenase complex is associated with coupling site I. The energy coupling of the complex can be demonstrated by its incorporation into vesicles. Thus, a reconstituted functional NADH₂–ubiquinone reductase has been prepared by Ragan and Hinkle (1975) using phospholipid vesicles of phosphatidylcholine and phosphatidylethanolamine into which complex I was incorporated. Oxidation of NADH by ubiquinone in these vesicles was rotenone-sensitive and associated with outward translocation of protons ($H^+/2e = 1.4$). The rate of oxidation could be stimulated with uncouplers or valinomycin + nigericin in the presence of K^+ (see sect. 7.4).

The addition of complex I to phospholipid vesicles possessing the mitochondrial ATPase (see sect. 9.1), resulted in a system in which ATP synthesis was coupled to NADH₂–ubiquinone reductase activity (Ragan and Racker, 1973). ATP synthesis was inhibited by rotenone, uncouplers and rutamycin. A P/2e ratio of 0.5 was obtained.

(h) Membrane orientation

As noted earlier, the mitochondrion is impermeable to pyridine nucleotides and in animals only oxidises internal NADH₂. Thus the catalytic site of the dehydrogenase must be on the inner face of the inner membrane. Lactoperoxidase is an enzyme which will catalyse the iodination of membrane proteins. Since the enzyme cannot diffuse through the membrane, only those proteins accessible at the surface are labelled. Using the isotope ^{125}I, it was found that all the polypeptides of complex I (see Table 9.1) became labelled except those of MW 53 000 (associated with the type II dehydrogenase), 29 000, 26 000 and 23 000. Thus one type II dehydrogenase and one Fe–S polypeptide were buried in the membrane.

Water-soluble chloroquine competitively inhibits ubiquinone reduction but only in inverted submitochondrial particles, not in systems which are 'right side out'. This suggests that ubiquinone, which is presumed free to diffuse in the membrane, is located near the inner surface of the inner mitochondrial membrane at the moment of its reduction.

9.5 Transhydrogenase

Colowick and coworkers (1952) first described a transhydrogenase in the bacterium *Pseudomonas fluorescens*. Later, an energy-dependent reduction of NADP by $NADH_2$ was described in rat liver and beef heart SMP. The energy requirement for this transhydrogenase (as with reverse electron flow) could be met either by ATP hydrolysis or by the generation of the high-energy intermediate in respiration. Oligomycin inhibited the reaction when ATP was used and the reaction was sensitive to uncouplers. The energy utilisation results in shifting the equilibrium of the reaction,

$$NADH_2 + NADP \rightleftharpoons NAD + NADPH_2,$$

strongly in the direction of NADP reduction. At low ATP concentrations, the hydrolysis of 1 ATP resulted in the reduction of 1 molecule of NADP. The transhydrogenase is bound to the inner mitochondrial membrane and translocates protons with a $H^+/2e$ ratio estimated at 2 for NADP reduction. Reversal of the transhydrogenase results in some reverse flow of protons (Moyle and Mitchell, 1973b). This has led to the view that the transhydrogenase represents a further coupling site between electron transfer and phosphorylation referred to as coupling site 0. It has also been suggested that it represents a further loop of the Mitchell type for translocating protons.

Isolated complex I normally possesses transhydrogenase activity. Reconstitution of vesicles with complex I and the ATPase results in an ATP-dependent reduction of NADP by $NADH_2$.

Recently the transhydrogenase from beef heart mitochondria has been purified to homogeneity by Höjeberg and Rydström (1977). It is composed of a single polypeptide of molecular weight about 95 000 to 100 000 and has no flavin. If this protein is incorporated into phospholipid vesicles, it will couple NAD reduction by $NADPH_2$ to ion transport, which is sensitive to uncoupling agents.

9.6 Succinate–ubiquinone reductase

(a) Properties

This segment of the respiratory chain lacks the ability to couple electron transport to phosphorylation. The complex catalyses the reaction:

$$succinate + ubiquinone \rightleftharpoons fumarate + ubiquinol$$

A flavoprotein succinate–ubiquinone reductase complex was isolated from beef heart mitochondria by Doeg *et al.* (1960). This complex II (MW 210 000) contains flavin (FAD), non-haem iron, labile sulphide, cytochrome *b*, protein and lipid. Approximately 20 per cent of the complex is lipid. The cytochrome *b* ($b_{557.5}$) does not appear to be involved in succinate–ubiquinone reductase activity and its function is obscure; it may be a contaminant, an altered form of a *b*-cytochrome from complex III. The ratio of flavin : haem : non-haem : labile sulphur is approximately 1 : 1 : 8 : 8. The complex will catalyse the direct reduction of ubiquinone by succinate, although the rate of turnover is relatively low. Like the membrane-bound enzyme but unlike the soluble enzymes described below, the activity is specifically inhibited by 2-thenoyltrifluoroacetone and also by organic mercurials.

A soluble succinate dehydrogenase was initially isolated by Singer and Kearney (1954). More recent methods for the preparation of the enzyme have depended on butanol extraction of a Keilin–Hartree heart muscle preparation (sect. 2.3) or extraction of an acetone powder of mitochondria or extraction of complex II (see Singer *et al.*, 1973, for a summary). Two types of soluble dehydrogenase have been isolated, one with a flavin : non-haem iron ratio of 1 : 4 and the other with a ratio of 1 : 8. The latter preparation is normally carried out in the presence of succinate to stabilise the enzyme. Reconstitution of a succinate oxidase system using alkali-treated Keilin–Hartree preparations (which have lost succinate oxidase and succinate dehydrogenase activity) and soluble dehydrogenase is usually successful where the dehydrogenase contains the higher level of non-haem iron (see Ohnishi *et al.*, 1976).

The flavin is covalently bound to the protein (see Fig. 2.3). The estimation of the molecular weights of the isolated dehydrogenase has been very variable, but the value is probably about 10^5 for preparations containing the greater amounts of iron. Higher values may be due to dimerisation.

(b) Resolution into subunits

Davis and Hatefi (1971), using a soluble enzyme purified from complex II with a molecular weight of 10^5 resolved the system into two subunits. One of these was a flavoprotein (MW = 70 000) with a flavin : Fe ratio of 1 : 4 and the other (MW = 37 000) was an iron–sulphur protein.

In an EPR study of the succinate–ubiquinone segment, three iron–sulphur centres have been detected by Ohnishi *et al.* (1976). S_1 ($E_{m\,7.0} = 0$ mV) and S_2 ($E_{m\,7.0} = -400$ mV in the soluble enzyme and -250 mV in the bound form) are presumed to be of the [2Fe–2S] type and S_3 ($E_{m\,7.0} = +60$ mV), a HiPIP type. S_3 which is labile towards oxygen, is regarded as the species reacting with ubiquinone, while S_1 is readily reduced by succinate; the function of S_2 is not clear (Fig. 9.7).

(c) Catalytic activity

The enzyme may be assayed with a variety of electron acceptors; phenazine methosulphate is usually preferred although ferricyanide is also used. It is possible that there are two sites for reduction of ferricyanide and phenazine methosulphate, one probably associated with the flavoprotein unit itself and the other close to the site for quinone reduction. The enzyme is sensitive to thiol reagents such as organic mercurials. The enzyme can be activated by succinate, ATP and reduced membrane-bound quinone. Oxaloacetate, which may also be tightly bound to the enzyme, inhibits activity. The role of oxaloacetate as a regulator of metabolic activity has been a subject for debate. It is probable that succinate dehydrogenase exists in active and inactive forms and that the equilibrium between the two forms is determined by the factors mentioned above. Conversion of inactive to active enzyme is linked to dissociation of oxaloacetate. Phosphate also has an influence on enzyme activity.

While less is known about the plant enzyme, it is activated by succinate and ATP. In the bacterium *Micrococcus lactyliticus*, a specific inhibitor, probably a protein, has been found.

9.7   Ubiquinol–cytochrome *c* reductase

(a) Composition

The third segment of the respiratory chain catalyses the oxidation of reduced ubiquinone and the reduction of cytochrome *c*.

$$\text{ubiquinol} + 2 \text{ cytochrome } c^{3+} \rightleftharpoons \text{ubiquinone} + 2 \text{ cytochrome } c^{2+}$$
$$+ 2H^+$$

The lipoprotein complex (complex III) has a molecular weight of about 2×10^5, of which about 15 per cent is phospholipid. Up to eight polypeptides have been separated from the complex (Table 9.1), some of which have been identified functionally (Marres and Slater, 1977).

(b) Cytochromes

The number of types of *b*-cytochrome in the complex has been disputed. Mitochondria from mammalian sources show main absorption maxima for *b*-cytochromes at 558, 562 and 566 nm. We shall assume the existence of only two cytochromes, one of which has a split absorption maximum. As noted earlier Keilin originally described the respiratory *b*-cytochrome, b_{562}, and this is also referred to as b_K in his honour. In 1970, Chance *et al.* showed the existence of a second *b*-cytochrome which differed from the first not only in absorption maxima but also in mid-point potential. The second pigment, b_T (T = transducing) or b_{566}, had maxima at 566 and 558 nm (a split peak) and an $E_{m\,7.0} = -30$ mV in uncoupled mitochondria. Treatment of coupled mitochondria with ATP alters the $E_{m\,7.0}$ of b_T to about $+240$ mV. The $E_{m\,7.0}$ of b_K is about $+40$ mV and is unaffected by ATP treatment. Similar cytochromes have also been isolated from yeast (Sato *et al.*, 1972). In plants and some micro-organisms there may be more than two *b*-cytochromes. The effect of ATP seen on mammalian cytochrome b_T has proved difficult to demonstrate with plant mitochondria. However, in mammalian mitochondria, where the cyto-chrome $b:c_1$ ratio is 2:1, it is difficult to accommodate more than two *b*-cyto-chromes.

Most isolations of b-cytochromes have shown the presence of only one polypeptide although some workers have found two. Minimum molecular weights for cytochrome b range between 22 000 and 37 000 (see Weiss, 1976). An analysis of cytochrome b from *Neurospora* gave two similar subunits of combined molecular weight 55 000. In intact mitochondria, the absorption peaks at 556 and 563 nm were found associated with $E_{m\,7.0}$ values of -40 and $+60$ mV respectively. The purified pigment had a maximum at 559 nm and $E_{m\,7.0} = -56$ mV. A recent isolation from yeast gave only one b-cytochrome, with absorption maxima at 562, 532 and 430 nm in the reduced state. The protein, which appeared to have suffered only slight modification during isolation, had a molecular weight of 28 000 (Lin and Beattie, 1978). It is possible that the two b-cytochromes represent two identical chemical entities in different environments.

Cytochrome c_1 has been isolated as a protein of molecular weight about 30 000 from various sources (see Ross and Schatz, 1976). The haem is covalently bound and has an absorption maximum at 552.5 nm. The membrane-bound cytochrome, $E_{m\,7.0} = +225$ mV, is located towards the outer surface of the inner membrane and is oxidised by cytochrome c.

The b-cytochromes are oxidised by cytochrome c_1 and this process is inhibited by antimycin A and also by 2-n-heptyl-4-hydroxyquinoline-N-oxide (HOQNO). Cytochrome b_K is readily reduced by substrates such as quinol or succinate, but cytochrome b_T is reduced only slowly in uncoupled mitochondria. However, addition of ATP to coupled mitochondria will induce the reduction of b_T. Curiously, in the presence of antimycin, the addition of an oxidant (such as oxygen) to anaerobic mitochondria causes increased reduction of cytochrome b_T. These results are difficult to interpret in terms of a simple linear sequence of electron transport and more complex mechanisms have therefore been proposed (see Rieske, 1976 for a review and also Fig. 9.9). The effect of ATP on the reaction associated with cytochrome b_T implies that the complex undergoes conformational changes under suitable conditions.

(c) Iron–sulphur protein

Rieske and coworkers (1964) isolated a non-haem iron–sulphur protein (HiPIP) from complex III where it gave an EPR signal at $g = 1.90$ when reduced. The $E_{m\,7.0}$ is about $+280$ mV. The protein possesses two iron and two labile sulphur atoms and accounts for all the non-haem iron in the complex. Its reduction by substrates is inhibited by antimycin and it is regarded as being close to cytochrome c_1. Its function in the complex is not clear.

(d) Antimycin A

Antimycin A (Fig. 9.8) binds strongly to complex III and inhibits electron transfer between b and c_1. In addition it is associated with pronounced changes in the properties of the b-cytochromes including a shift in absorption maxima and changes in redox state. It also affects the cleavage of the complex by guanidine. Antimycin A is therefore regarded as altering the conformation of the complex. The inhibitor HOQNO, which binds at the same site as antimycin, does not have the same effect on the conformational properties of the complex. It has been suggested that antimycin, which binds stoicheiometrically to complex III, attaches specifically to the b-cytochromes, the iron—sulphur protein or to a small polypeptide (MW = 12 000).

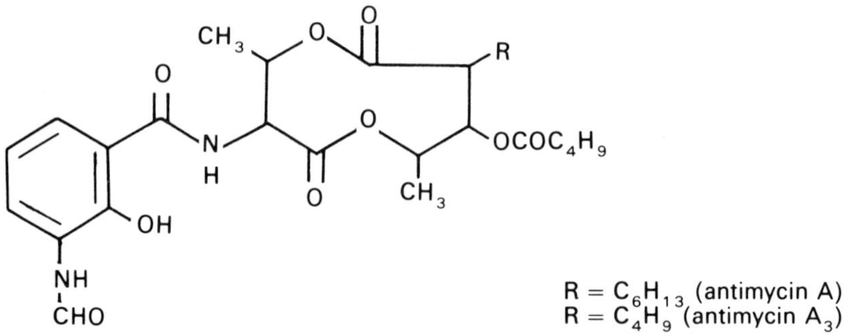

$$R = C_6H_{13} \text{ (antimycin A)}$$
$$R = C_4H_9 \text{ (antimycin A}_3\text{)}$$

Fig. 9.8　Antimycin A.

(e) Cleavage of the complex

Treatment with bile salts and ammonium sulphate cleaves the complex into a soluble fraction containing cytochrome c_1 and 'core' protein and an insoluble fraction containing the b-cytochromes and the iron—sulphur protein. Antimycin binding is lost. This cleavage, which can also be brought about by guanidine, is inhibited very effectively by antimycin or by reducing agents. It has been suggested that antimycin is bound in a crevice in the complex where it can interact with the b-cytochromes. Its presence in the crevice blocks dissociation.

(f) Reconstitution studies and ion translocation

An antimycin-sensitive succinate—cytochrome c reductase can be reconstituted from preparations of b-cytochromes, cytochrome c_1, succinate dehydrogenase, phospholipids and ubiquinone. The reconstituted complex although active in electron transport is not similar to the native enzyme system in several respects.

The incorporation of complex III into phospholipid vesicles enabled Leung and Hinkle (1975) to couple ubiquinone—cytochrome c reductase activity to outward translocation of protons across the membrane. They obtained a $H^+/2e$ ratio of 1.9. Translocation of ions was sensitive to uncouplers. In the presence of valinomycin, the oxidation reaction drove the uptake of K^+ at a ratio of $K^+/2e = 2$. Although this experiment demonstrates that complex III can translocate protons, the mechanism is not clear. The lack of any components that might be expected to function in proton transport, has led Mitchell (1975) to propose the Q cycle (Fig. 9.9). Here it is the quinone which functions as the proton carrier and involves the reduced, oxidised and semiquinone states. The b-cytochromes act as electron carriers in one arm of the loop, while the quinone serves as a proton carrier to form the other arm. Cytochrome b_T has been located on the outside of the inner membrane as required by the Q cycle.

9.8   Cytochrome oxidase

(a) General properties
Cytochrome oxidase catalyses the reaction:

$$4 \text{ cytochrome } c^{2+} + O_2 + 4H^+ \longrightarrow 4 \text{ cytochrome } c^{3+} + 2H_2O$$

The complex possesses two cytochromes (a and a_3) and two atoms of copper. Cytochrome a_3 is distinguished by its ability to react with haem ligands such as carbon monoxide and cyanide. One of the central problems in understanding this enzyme concerns the relationship of the cytochromes. According to one view, the two cytochromes are chemically identical. Reduction of the oxidase with an electron perturbs the system so that the cytochromes are no longer equivalent, now having mid-point potentials of about 220 and 350 mV; these are attributed to cytochromes a and a_3 respectively. Under these circumstances they now behave as different cytochromes. An alternative view is that there are two distinct chemical entities with different properties, a_3 being a high-potential and a a low-potential cytochrome. Here also, strong interactions between the cytochromes are assumed. Indeed a large body of experimental data shows that there is a haem–haem interaction. The absorption spectrum of one cytochrome, for example, is influenced by the state of the other. The cytochrome oxidase reaction is sensitive to several inhibitors such as carbon monoxide, azide, cyanide, sulphide and formate. Probably all these inhibitors act in the region of cytochrome a_3 as ligands; they also influence the value of E_m for cytochrome a, demonstrating the interaction between the haems of the two cytochromes.

Of the two atoms of copper, only one gives a clear EPR signal. As a consequence they are referred to as 'visible' and 'invisible'. The mid-point potentials of the coppers are similar to those of the haems and it is now widely thought that the prosthetic groups of the oxidase are two haem–copper pairs.

The absorption maxima of the reduced complex are at 603, 560, 517 and 443 nm while the oxidised complex has maxima at 830, 660, 598, 545, 515 and 418 nm. Complete reduction of the complex requires four electron equivalents. The spectral characteristics of the pigments cannot be specified with certainty although several attempts to do so have been made (see sect. 2.3). Nevertheless spectra have been used extensively in the study of this enzyme. Carbon monoxide forms a derivative of a_3 absorbing at about 590 nm (sect. 2.3). Furthermore, at temperatures below $-60°$ C, spectral changes in the 605–623 and 830–940 nm regions consequent on oxidation by oxygen amount to only half those seen at higher temperatures and have been attributed to the $Cu-a_3$ complex (Chance *et al.*, 1975).

(b) Composition
Complex IV has a molecular weight of about 2×10^5 of which about 20 per cent is lipid. The main lipids are phosphatidylcholine, phosphatidylethanol-

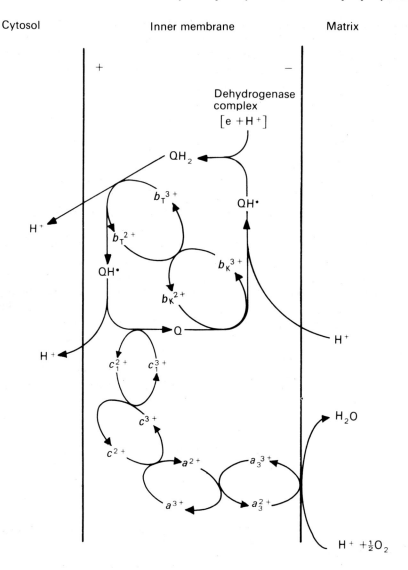

Fig. 9.9 The Q cycle as proposed by Mitchell (1975).

amine and cardiolipin, part of the last mentioned being very firmly bound. The enzyme complex has been resolved into seven polypeptides in yeast (Poyton and Schatz, 1975), *Neurospora* and beef heart (Downer *et al.*, 1976) (see Table 9.1).

The enzyme has a transmembrane orientation. Cytochrome c is located on the outside of the inner membrane, while it can be shown that the reaction with oxygen is inhibited on the inside face. If the subunits are iodinated with lactoperoxidase (which can only iodinate surface proteins of the membrane) or labelled with *p*-diazobenzenesulphonate, which does not permeate the membrane, subunits II, V and VI are labelled. With inverted submitochondrial particles, subunit III is labelled. Thus subunit III appears to be located near the inner face and subunits II, V and VI near the outer face of the inner membrane. Attempts to locate the prosthetic groups have not been so successful. There is evidence that one of the haems may be associated with subunit I (see Gutteridge *et al.*, 1977).

(c) Mechanism

Although a role in electron transfer between haems a and a_3 was earlier assigned to copper, it now seems probable that each of the copper atoms is linked to a haem, giving two haem–copper pairs. Studies of the 'visible' copper suggest that it is paired with cytochrome a. Since two electron equivalents are required for complete reduction of cytochrome a_3, it is assumed that the invisible copper is associated with this cytochrome. Kinetic studies show that the oxidation of reduced cytochrome c is coupled to reduction of cytochrome a. There is then a transfer of electrons from cytochrome a to the 'visible' EPR-detectable copper.

Oxidation of the complex involves initial binding of oxygen, which is competitively inhibited by carbon monoxide. Evidence for an oxygenated form of cytochrome a_3 with a Soret band at 428 nm was produced by several workers in the 1960s (see Lemberg and Mansley, 1966). The reaction then proceeds by the reduction of the oxygen as suggested in Fig. 9.10. However, the details of the mechanism are not agreed (see Erecinska and Wilson, 1978 for a recent discussion).

(d) Proton translocation

Cytochrome oxidase constitutes the third coupling site of the respiratory chain. Vesicles containing cytochrome oxidase can be prepared from phospholipids and the isolated complex. The addition of cytochrome c to these oxidase vesicles ensures that oxidation of the cytochrome will occur asymmetrically with respect to the membrane. Oxidation of cytochrome c by oxygen is associated with ejection of protons, which suggests that the complex is capable of acting as a proton pump (see Krab and Wikstrom, 1978).

(e) Conformational changes

Several workers have emphasised that the complex undergoes conformational changes during the oxidation of cytochrome c. This has been demonstrated using fluorescent probes and can be deduced from changes in the standard redox potentials of the haems under various conditions (see sect. 9.12(c)).

9.9 Ubiquinone

Ubiquinone is regarded as a mobile carrier between the dehydrogenase complexes and complex III. It may act either as a simple shuttle for reducing

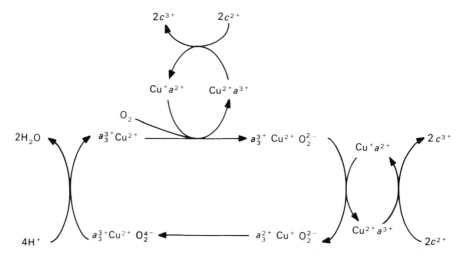

Fig. 9.10 The cytochrome oxidase reaction: a generalised scheme for the mechanism. (For more detailed models, see Erecinska and Wilson, 1978.)

equivalents between dehydrogenase complexes and b-cytochromes or in the more complex manner proposed in the Q cycle (Fig. 9.9). EPR studies have suggested that the semiquinone may be involved and that the quinone could, for example, act as a single electron carrier between the iron–sulphur proteins, themselves single electron carriers, and the cytochromes. Alternatively the Q cycle also assumes the involvement of a semiquinone.

The length of quinone side chains varies. In mammalian and higher plant systems, Q-10 with a C_{50} side chain is the effective quinone, while in microorganisms the situation is more complex. In *Saccharomyces*, Q-6 is found, in *Candida* Q-7 and Q-9, in *Mucor* Q-9 while in a variety of other moulds including *Neurospora* Q-10 is present. In the algae, *Prototheca* employs Q-7, *Cladophora* and *Euglena* Q-9 and *Ochromonas* Q-10.

9.10   The respiratory chain

(a) General

The outline given for the phosphorylating complexes of the chain (I, III and IV) in beef heart and yeast indicates that the NADH oxidase multienzyme complex consists of a limited number of polypeptides, perhaps 30–35 (Table 9.1). This number would need to be extended when substrates other than NADH were considered, such as succinate and acyl CoA esters (Fig. 2.11). Oxidative phosphorylation as a total system will require the inclusion of about a further 12 polypeptides for the ATPase. It should be noted, however, that the composition of some complexes (e.g. complex I) is not yet certain and that the functions of many of the components are far from clear.

The basic respiratory chain (NADH oxidase) has broadly similar characteristics over the range of eukaryotic organisms. Most work has been concerned with mammalian mitochondria (especially beef heart) and detailed studies of mitochondria from other sources are limited. Major differences certainly exist between respiratory systems from different sources.

(b) Plant mitochondrial respiration

The lack of detailed investigation makes it difficult to construct a clear detailed picture of the respiratory chain in higher plant mitochondria (Fig. 9.11). A membrane-bound succinate dehydrogenase oxidises succinate and two systems for NADH oxidation exist. One involves a flavoprotein dehydrogenase with its catalytic site on the inner face of the inner membrane and is associated with several iron–sulphur proteins, probably comparable with those in animal and yeast systems. The other dehydrogenase oxidises external NADH and unlike the former is insensitive to amytal, piericidin and rotenone and is not associated with a coupling site. Oxidation of both matrix and intermembrane space NADH is sensitive to antimycin A.

The quinone is ubiquinone-10 (Q-10). The study of b-cytochromes is less advanced than in mammalian systems, but the situation here could be more complex than in mammals. Cytochrome c_{549} is like c_1, while c_{547} is a salt-extractable pigment like mammalian cytochrome c.

A major difference between plant and mammalian mitochondria is the cyanide-resistant respiration found in a number of plants. While some plant mitochondria (e.g. potato) appear to possess only one terminal oxidase, many others possess both a cytochrome oxidase and a cyanide-insensitive oxidase. The latter pathway is also antimycin-insensitive, does not involve cytochromes and probably branches from the normal pathway at the quinone level. This alternative pathway is unable to couple oxidation to phosphorylation (other than in the matrix-oriented NADH dehydrogenase segment). The nature of the second plant oxidase has not been elucidated. The role of the pathway has been a matter for conjecture and satisfactory explanations have not been forthcoming.

(c) Micro-organisms

Most micro-organisms appear to possess conventional respiratory chains, although a number of minor differences are found in specific organisms. Some yeasts possess an L-lactate dehydrogenase linked directly to the respiratory chain and composed of a flavocytochrome, b_2, a complex of flavoprotein (FMN-containing) and cytochrome. This inner membrane-bound enzyme reduces cytochrome c (Gervais et al., 1977). Two genetically different forms of cyto-chrome c (with different amino acid compositions) are found in yeasts. This group of micro-organisms also shows an adaptation to low oxygen levels in which cytochrome a_1 and b_1 form the major pathway (see sect. 10.3e). *Tetrahymena* has unusual oxidases including a_1 and a_2 or a_{620} (Kilpatrick and Erecinska, 1977). Trypanosomes, in addition to cytochrome a, a_3 possess an oxidase which appears to be of the cytochrome o type (see sect. 10.3). The possibility of a cyanide-insensitive oxidase in some species also exists. The cytochrome c of these protozoa differs from that of normal eukaryotic cytochrome c (see Hill, 1976).

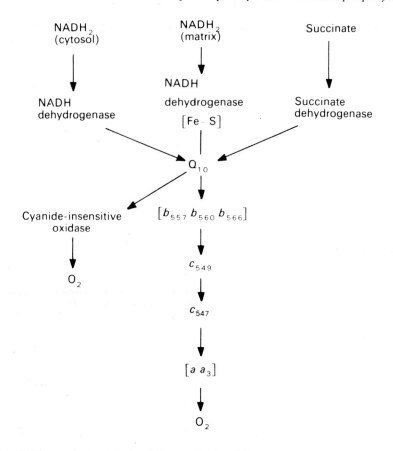

Fig. 9.11 The respiratory chain of plant mitochondria.

A number of fungi including *Neurospora*, *Aspergillus* and yeasts possess an externally oriented NADH dehydrogenase similar to the plant enzyme. An alternative oxidase is also found in several micro-organisms. Of particular interest is the *poky* strain of *Neurospora* in which the cytochrome composition is substantially altered as compared with the wild type. The *b*- and *a*-cytochromes are deficient and two-thirds of the oxidation is by an alternative pathway which, like many of the plant systems, is inhibited by salicylhydroxamic acid but not by cyanide and antimycin A. This pathway, which lacks a phosphorylation site, is a minor component of the respiratory system of the wild type.

9.11 Thermodynamic considerations

(a) Membrane potentials and ion gradients

The oxidation of succinate and NADH by oxygen through the respiratory chain is linked to an energy-conserving system in which the free energy available from oxidation is used for ATP synthesis and ion movement. Both of these processes can be measured. Ion movement may be considered as a response to translocation of protons and the accompanying membrane potential. The proton motive force is defined as shown below (see sect. 8.3)

$$\Delta p = \Delta \psi - \frac{2.303\,RT}{F} \cdot \Delta pH$$

Several attempts have been made to measure the ΔpH and the membrane potential across the mitochondrial membrane by various methods. For example, Nicholls (1974) found a value of Δp for rat liver mitochondria of 228 mV in state 4, 170 mV in state 3 and -0.6 mV in the presence of rotenone and uncoupler (using a method based on ion distribution across the membrane). In state 4, the proton gradient (ΔpH) did not contribute more than half of Δp although greater contributions of ΔpH to Δp have been obtained in other experiments. Estimation of the phosphate potential, $\left(\dfrac{[ATP]}{[ADP][Pi]} \right)$ showed a variation of Δp with the potential, although Δp was some 50–80 mV below that required to provide an adequate driving force for ATP synthesis, as suggested by Mitchell's hypothesis (assuming a stoicheiometry of two protons/ATP synthesised).

An attempt to make a direct measurement of membrane potential by Skulachev and coworkers (Drachev *et al.*, 1974) was based on the formation of reconstituted cytochrome oxidase vesicles which were fused to a membrane. The potential across the membrane ($\Delta \psi$) which was measured with a sensitive voltmeter, gave a value of 110 mV. In general, it may be assumed that the value which is obtained experimentally for Δp in state 4 (ADP-limited) is about 200 mV.

In view of the chemiosmotic theory, it is interesting to calculate the value of Δp required for ATP synthesis. At a phosphate concentration of 10 mM and an ATP/ADP ratio of 100 which is probably attained in state 4

$$\Delta G = \Delta G^{\circ} + RT \ln 10^4$$

For ATP synthesis, if $\Delta G^{\circ} = +32.2$ kJ/mol and $T = 298$ K,

then $\Delta G = +55$ kJ/mol.

Since $\Delta G = -nF\Delta E$ and assuming $n = 2$ (i.e. $H^+/P = 2$),

$\Delta E = -285$ mV

This figure would of course apply to tightly coupled mitochondria with a low proton permeability. Since the ΔG° for ATP hydrolysis varies with pH, Mg^{2+} concentration and ionic strength, the values for this parameter and the consequent values of the ΔE will be dependent on these factors (see Papa, 1976). In view of the difficulties in measurement, many workers are inclined to the view that the observed ΔpH and $\Delta \psi$ are consistent with the requirements of the chemiosmotic theory. Others have taken a contrary view. If n above is assumed to have the value of 3, then the experimentally determined value of Δp will be adequate for ATP synthesis.

(b) Redox potentials and phosphorylation

The form of the respiratory chain is such that, in state 4, the redox carriers might be expected to be in equilibrium between the sites of phosphorylation. Thus the iron–sulphur protein which reduces ubiquinone, ubiquinone itself and cytochrome b_K would be expected to have the same oxidation–reduction potential regardless of the standard potentials. Erecinska *et al.* (1974) define four such isopotential groups for which they have measured the potentials (Table 9.2).

Table 9.2 Proposed isopotential groups in the respiratory chain

Group	Components	E_h (mV)
I	NAD, NADH flavoprotein dehydrogenase iron–sulphur proteins	−300
II	Ubiquinone, cytochrome *b* iron–sulphur proteins, succinate dehydrogenase	10
III	Cytochromes c_1, *c*, *a*-Cu iron–sulphur protein of complex III	310
IV	Cytochrome a_3-Cu	620

It has been shown in several ways that coupling of ATP synthesis with electron transport is reversible, i.e. that ATP hydrolysis will drive reverse electron flow and that redox reactions of the chain will drive ATP synthesis. Consequently, if the terminal reaction with oxygen is excluded, there should be a relationship between the oxidation–reduction states of the carriers and the concentrations of ADP, ATP and P*i* in mitochondria respiring in state 4. Measurement of oxidation–reduction states of NAD and cytochrome *c* together with concentrations of ATP, ADP and P*i* in whole isolated liver cells suggests that equilibrium is approached according to the equation:

$$NADH_2 + 2 \text{ cytochrome } c^{3+} + 2 \text{ ADP} + 2 \text{ P}i = NAD$$
$$+ 2 \text{ cytochrome } c^{2+} + 2 \text{ H}^+ + 2 \text{ ATP} + 2 \text{ H}_2O$$

The difference in actual redox potentials of the nucleotide and cytochrome *c* were calculated from the standard potentials and measurement of the amount of oxidised and reduced NAD and cytochrome *c* (Wilson *et al.*, 1974). This gave a value for ΔE of about 530 mV, equivalent to about 100 kJ/mol. Estimation of the extramitochondrial[3] levels of ATP, ADP and P*i* showed that the energy requirement for ATP synthesis in these liver cells (assuming an appropriate value of ΔG° for ATP hydrolysis and P/2e = 2) would be close to 95 kJ/mol. Thus in state 4, the respiratory chain between NAD and cytochrome *c* is close to being in equilibrium with the phosphate potential.

[3] The energy available for ATP synthesis will be used both for the ATPase reaction in the matrix and for the maintenance of the lower phosphate potential of the matrix as compared with the cytoplasm (see sect. 6. 4(d)).

While this argument will hold for the earlier part of the respiratory chain, such a relationship is unlikely to be attained in the cytochrome oxidase segment where the difference in potential between cytochrome c and oxygen must be very large (see sect. 2.12). A means whereby this terminal enzyme may be regulated under physiological conditions is being sought (see Erecinska and Wilson, 1978).

(c) Phosphorylation and standard potentials

Treatment of coupled mitochondria with ATP produces apparent changes in mid-point potentials of some of the components of the respiratory chain, while others are unaffected. Cytochrome b_T, $E_{m\ 7.0} = -30$ mV, can be converted to a form having $E_{m\ 7.0} = +245$ mV. Cytochrome a_3 can be altered from $E_{m\ 7.0} = +385$ mV to $E_{m\ 7.0} = +155$ mV while cytochrome a becomes more positive. A comparable claim for an iron–sulphur protein in complex I, in which the mid-point potential is shifted by $+130$ mV, has also been made. Such changes in potential have been used as a basis for arguments that the conformations of complexes of the respiratory chain are associated with phosphate potential and phosphorylation (see sect. 8.4).

Further reading

Boyer, P. D., Chance, B., Ernster, L., Mitchell, P., Racker, E. and Slater, E. C. (1977) Oxidative phosphorylation and photophosphorylation. *Ann. Rev. Biochem.*, **46**, 955–1026.

Depierre, J. W. and Ernster, L. (1977) Enzyme topology of intracellular membranes. *Ann. Rev. Biochem.*, **46**, 201–62.

Erecinska, M. and Wilson, D. F. (1978) Cytochrome oxidase, a synopsis. *Arch. Biochem. Biophys.*, **188**, 1–14.

Lloyd, D. (1974) *Mitochondria of Micro-organisms.* Academic Press, New York.

Malmstrom, B. G. (1979) Cytochrome oxidase. Structure and catalytic activity. *Biochim. Biophys. Acta*, **549**, 281–303.

Nicholls, D. G. (1979) Brown adipose tissue mitochondria. *Biochim. Biophys. Acta*, **549**, 1–30.

Palmer, J. M. (1976) The organisation and regulation of electron transport in plant mitochondria. *Annu. Rev. Plant Physiol.*, **27**, 133–57.

Racker, E. (1970) The two faces of the inner mitochondrial membrane. *Essays in Biochemistry*, **6**, 1–22.

Ragan, C. I. (1976) NADH–ubiquinone oxidoreductase. *Biochim. Biophys. Acta*, **456**, 249–90.

Rieske, J. S. (1976) Composition, structure and function of complex III of the respiratory chain. *Biochim. Biophys. Acta*, **456**, 195–247.

Singer, T. P., Kearney, E. B. and Kenney, W. C. (1973) Succinate dehydrogenase. *Adv. Enzymol.*, **37**, 189–272.

Wikström, M. and Krab, K. (1979) Proton pumping cytochrome *c* oxidase. *Biochim. Biophys. Acta*, **549**, 177–222.

Also:

Fillingame, R. H. (1980) The proton-translocating pumps of oxidative phosphorylation. *Annu. Rev. Biochem.*, **49**, 1079–1114.

von Jagow, G. and Sebald, W. (1980) *b*-type cytochromes. *Annv. Rev. Biochem.*, **49**, 281–314.

Sweeney, W. V. and Rabinowitz, J. C. (1980) Proteins containing 4Fe–4S clusters: an overview. *Annv. Rev. Biochem.*, **49**, 139–62.

Chapter 10

Bacterial energy transformation

10.1 Bacterial energetics

Prokaryotic organisms such as bacteria do not possess mitochondria; indeed many of them are about the same size as a mitochondrion. The bioenergetic functions associated with the inner mitochondrial membrane are found in the plasma membranes of bacteria. Of particular interest are the facultatively anaerobic organisms such as *Escherichia coli*, which can grow aerobically or anaerobically. In these organisms mutations which result in a loss of respiratory or phosphorylation activity are not usually lethal.

Bacteria which are capable of aerobic growth possess a citric acid cycle, a respiratory system and a phosphorylation system similar to those found in mitochondria. Direct coupling of the respiratory system to membrane transport also occurs. The energetics of the membrane in relation to respiration, phosphorylation and transport will form the subject of this chapter.

10.2 Bacterial membranes

Bacteria possess only a simple membrane structure (see Fig. 10.1). In Gram-negative organisms like *E. coli*, an inner plasma membrane and an outer membrane are found with mucopolysaccharide wall material between the membranes. In Gram-positive organisms like *Micrococcus*, there is a single plasma membrane inside the cell wall. Other internal membranes, including the nuclear membrane and endoplasmic reticulum, are absent. However, in many organisms, the plasma membrane invaginates to form knots of internal membrane known as mesosomes. Although these bodies were once considered to be mini-mitochondria, the function of mesosomes has been a subject of controversy. Nevertheless they probably possess some respiratory activity like the plasma membrane.

The plasma membrane has a conventional double structure as seen under the electron microscope, a high protein content and a thickness varying between 6 and 10 nm. It normally accounts for about 15 to 30 per cent of the cell dry weight. Extraction of the membrane lipid does not destroy the structure. The membrane exhibits sidedness, in that a higher concentration of particles is found on the inner half as seen in freeze-fracture pictures; the outer half is more fluid. The lipid composition of the two halves of the membrane is also different. Stalked particles similar to, but smaller than, the mitochondrial ones are seen on

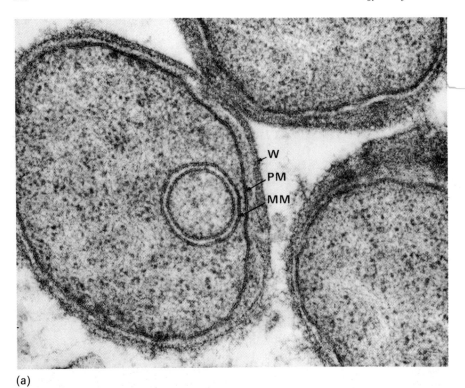

(a)

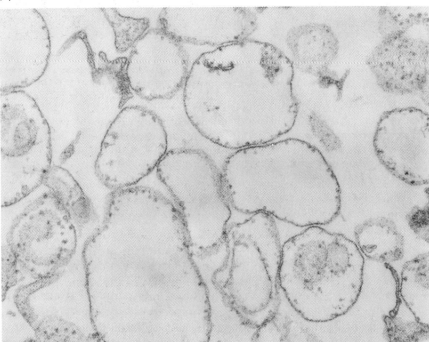

(b)

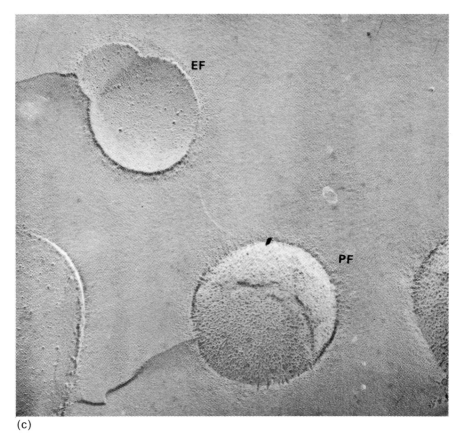

(c)

Fig. 10.1 Bacterial membranes. (a) Section through *Micrococcus luteus* showing the wall (W), plasma membrane (PM) and associated mesosome (MM), × 65 000 (b) Isolated membranes prepared by lysozyme digestion of the cell wall to form a protoplast. The protoplast is lysed in dilute solution and washed to remove cell debris, × 35 000 (c) Freeze-fracture of the bacterial plasma membrane (EF : fracture-face of the outer half; PF : fracture-face of the inner half), × 67 000.
(a) and (b) Courtesy of Dr. D. Hockley, National Institute of Biological Standards and Control, London. (c) Courtesy of Professor S. Bullivant, University of Auckland, New Zealand.

the inner surface in suitably prepared material and these are associated with the ATPase.

Both ribosomal RNA and DNA are associated with isolated membranes and it is probable that the chromosome, which is circular, is attached to the membrane at one or more points. The lipid content of membranes (Table 10.1) is variable and depends on the species, the growth phase and the conditions of culture. The main lipids are phosphatidylglycerol, diphosphatidylglycerol (cardiolipin) and, in bacilli and Gram-negative species, phosphatidylethanol-amine (and its methylated derivatives). Other nitrogen-containing phospholipids are less common. Additional membrane constituents of Gram-positive bacteria include amino acid esters of phosphatidylglycerol, lipoteichoic acids, glucosyl-

Table 10.1 Composition of bacterial membranes

Organism	Protein (%)	Lipid (%)	Carbohydrate (%)	Nucleic acid (%)	Phospholipid (% total phospholipid)	
Gram-positive						
Micrococcus luteus	42	39	15	5	diphosphatidylglycerol	(66)
					phosphatidylglycerol	(27)
					phosphatidylinositol	(6)
Staphylococcus aureus	65	24	—	—	phosphatidylglycerol	(54)
	—	—	—	—	diphosphatidylglycerol	(40)
					diglucosyldiglyceride	(5)
Paracoccus denitrificans					phosphatidylcholine	(31)
					phosphatidylglycerol	(52)
					phosphatidylethanolamine	(6)
					diphosphatidylglycerol	(3)
Bacillus subtilis	60	15			diphosphatidylglycerol	(49)
					phosphatidylglycerol	(11)
					phosphatidylethanolamine	(34)
					lysophosphatidylglycerol	(6)
Gram-negative						
Escherichia coli	—	—	—	—	phosphatidylethanolamine	(75)
					phosphatidyldiglycerol	(10)
					diphosphatidylglycerol	(10)
Proteus mirabilis	56	38	—	—	phosphatidylethanolamine	(58)
					diphosphatidylglycerol	(30)
					phosphatidylglycerol	
Pseudomonas stutzeri					phosphatidylethanolamine	(72)
					phosphatidylglycerol	(24)

glycerides and lipomannans. Except in *Mycoplasma* and related organisms, steroids are absent. Quinones (menaquinones and benzoquinones) are widely distributed and polyisoprenoids such as squalene and carotenoids are also found.

10.3   Bacterial respiratory chain

(a)   Components of the chain

In principle, the respiratory chain of bacteria is similar to that of the mitochondrion, in that it is composed of a sequence of flavoprotein dehydrogenases, quinones and cytochromes which may be in the order *b*, *c*, *a*. Many bacteria also possess a transhydrogenase system which is membrane-bound and which catalyses an energy-dependent reduction of NADP by $NADH_2$. When considered in detail, the respiratory systems of bacteria are complex. They exhibit a greater range of membrane-bound dehydrogenases; for example, *E. coli* strains possess such enzymes for succinate, $NADH_2$, L-lactate, D-lactate, hydroxybutyrate, formate and α-glycerophosphate, etc. Two types of quinone are found, either a benzoquinone (ubiquinone) or a naphthoquinone (menaquinone, vitamin K_2) or both. The cytochrome chain may not include a *c*-type cytochrome but it may have more than one type of oxidase. Under good aerobic conditions, *Paracoccus denitrificans* has a similar respiratory chain to that found in mitochondria while in *E. coli* and *Bacillus megaterium*, the chain is very different.

P. denitrificans

$$NADPH_2 \longleftrightarrow NADH_2 \rightarrow [FP \rightarrow FeS] \rightarrow Q_{10} \rightarrow cyt\ b_{558,}^{562}$$
$$\rightarrow cyt\ c_{550}^{546}, \rightarrow cyt\ a, a_3 \rightarrow O_2$$

E. coli

$$NADPH_2 \longleftarrow\dashrightarrow NADH_2 \rightarrow [Fp \rightarrow FeS] \rightarrow Q(MK)$$
$$\rightarrow cyt\ b_{556,}^{(562)}, \rightarrow cyt\ o \rightarrow O_2$$

B. megaterium

$$NADH_2 \rightarrow [Fp \rightarrow FeS] \rightarrow MK \rightarrow cyt\ b \rightarrow cyt\ a, a_3$$

In some facultative anaerobes, fumarate, nitrate, sulphate or carbonate may act as terminal electron acceptors in place of oxygen. Further, the composition of the chain may be altered by environmental conditions. In *E. coli*, for example, transhydrogenase activity which is lower than in many bacteria, varies with the conditions of growth. Sensitivity to inhibitors is frequently different to that in the mitochondrial respiratory chain. Thus most bacterial respiratory chains are insensitive to rotenone and antimycin A and sometimes show low sensitivity to cyanide. *n*-Heptyl-hydroxyquinoline-*N*-oxide, however, usually acts as a useful inhibitor in the *b*-cytochrome region.

(b)   Dehydrogenases

The flavoprotein dehydrogenases of some bacteria are shown in Table 10.2. In general these are membrane-bound and linked to the cytochrome system in a manner similar to that in the mitochondrion. Such dehydrogenases

Table 10.2 Bacterial flavoprotein dehydrogenases

Dehydrogenase substrate	Prosthetic group	Reaction	Organism
$NADH_2$	FAD	$NADH_2 \rightarrow NAD$	*Escherichia coli* *Bacillus megaterium*
	FMN	$NADH_2 \rightarrow NAD$	*Azotobacter vinelandii*
*Succinate	FAD	succinate $\rightleftharpoons$ fumarate	*Micrococcus lactylyticus*
*(Fumarate reductase)	FAD	succinate $\rightleftharpoons$ fumarate	*Vibrio succinogenes*
α-Glycerophosphate	FAD	α-glycerophosphate $\rightleftharpoons$ dihydroxyacetone phosphate	*Streptococcus faecalis*
Malate	FAD	malate $\rightleftharpoons$ oxaloacetate	*Mycobacterium phlei*
D-Lactate	FAD	D-lactate $\rightleftharpoons$ pyruvate	*Escherichia coli*
L-Lactate	—	L-lactate $\rightleftharpoons$ pyruvate	*Escherichia coli*
Glutamate	—		*Azotobacter vinelandii*
Formate	Mo (molybdoprotein)	formate $\rightleftharpoons CO_2 + H_2$	*Vibrio succinogenes*
D-Alanine		alanine $\rightleftharpoons$ pyruvate	*Pseudomonas aeruginosa*

are usually composed of flavoproteins with FMN or FAD as prosthetic groups and also have iron–sulphur centres. In some instances, the dehydrogenase complex appears to act directly with the cytochrome *b* complex without an intermediary quinone, for example the succinate dehydrogenase of *Mycobacterium phlei* (Kurup and Brodie, 1966). In several cases NAD-linked dehydrogenases catalysing similar reactions to the membrane-bound enzymes may also be found in the cytoplasm of the same organism. Many bacteria possess both a membrane-bound NAD-independent malate dehydrogenase and an NAD-dependent enzyme in the soluble fraction.

Because of the lack of compartmentalisation in bacterial cells, the $NADH_2$ formed in catabolism, will be directly available for both oxidation and biosynthesis. It is therefore not surprising to find that the $NADH_2$ dehydrogenase appears to be subject to regulation by the $NAD/NADH_2$ ratio (Dancey and Shapiro, 1976).

(c) Quinones

The quinones and their structures are shown in Fig. 10.2 and Table 10.3. Either ubiquinones or menaquinones may function in the respiratory chain between the dehydrogenases and the cytochromes. However, the standard redox potentials are different, ubiquinone being more positive (about $+112$ mV) than menaquinone (about -74 mV). Some enterobacteria possess both quinones. In *Escherichia coli*, examination of the role of quinones suggests that ubiquinone acts in the normal respiratory chain with oxygen as terminal electron acceptor while menaquinone functions primarily with fumarate as electron acceptor or in the oxidation of dihydro-orotate, a step in pyrimidine synthesis. This division of function probably reflects the difference in redox potential.

(d) Cytochromes

A range of cytochromes recorded for bacteria is shown in Table 10.3. The most notable point is that while the *b*-cytochromes are ubiquitous, the *c*-type cytochromes are missing in a number of bacteria. Four types of oxidase are found, the *a*, a_3 system, cytochromes *o*, a_1 and *d*, the latter also being known as a_2. Many organisms possess more than one type of oxidase giving rise to branched chains. Structurally the cytochromes appear to differ to a greater or lesser extent from those found in higher organisms, although relatively few detailed studies have been carried out.

The *b*-cytochromes possess, as prosthetic group, protohaem which is non-covalently bound. Reference to Table 10.3 suggests that there are essentially two groups of *b*-type cytochromes in many organisms. One has an α-band at 556–560 nm and has been equated with cytochrome b_1 found originally in yeast; the second has an α-band between 562 and 566 nm. However, *E. coli* has three *b*-cytochromes with maxima at 556, 558 and 562 nm. Cytochrome b_{556} has been purified from aerobically grown *E. coli*. It is an oligomer of identical subunits (MW = 17 500) and has an $E_m = -45$ mV. It can be reduced by a lactate dehydrogenase system in the presence of menadione (Kita *et al.*, 1978). Cytochrome b_{562} has also been isolated. It has a molecular weight of about 12 000 and an $E_m = +113$ mV (Itagaki and Hager, 1966). At the present time

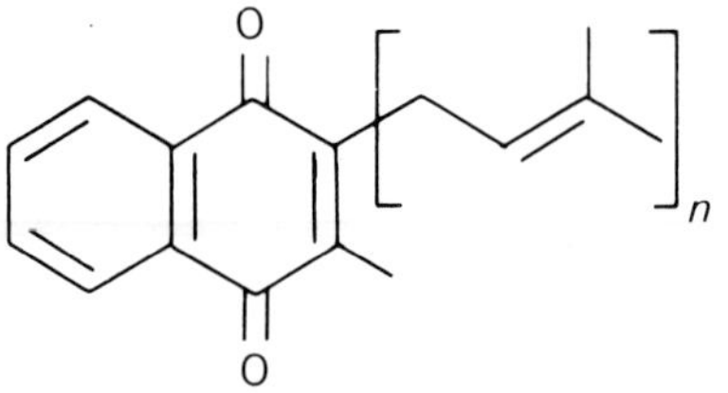

Menaquinone, vitamin k_2 (MK) ($n = 4, 6, 7, 8$ or 9)

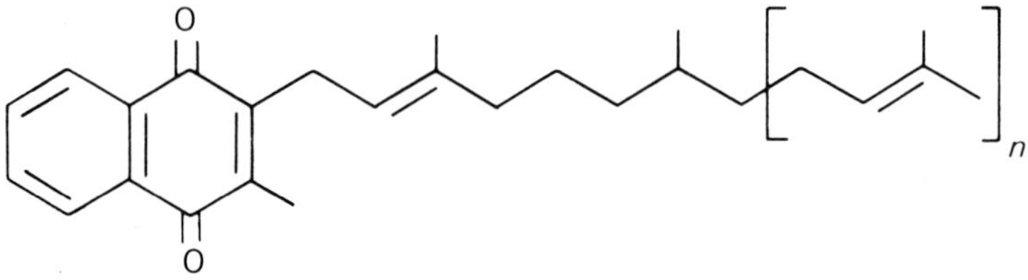

Dihydromenaquinone (MKH$_2$) ($n = 7$)

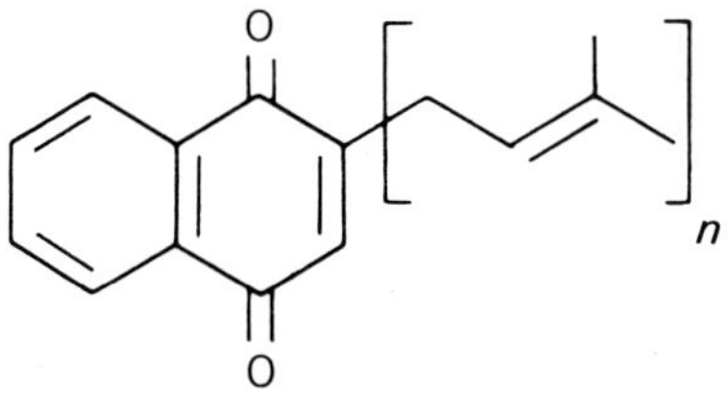

2-Demethylmenaquinone (DMK)   Ubiquinone (Q) ($n = 7, 8, 9$ or 10)
($n = 5, 6, 7, 8$ or 9)

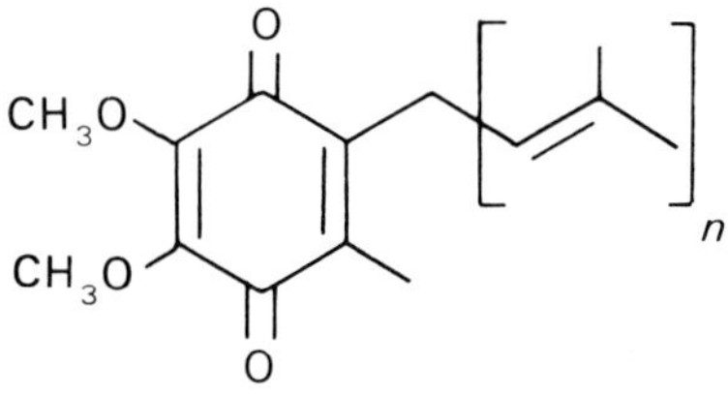

Fig. 10.2 Some bacterial quinones. Note that all these quinones are capable of undergoing oxidation–reduction as shown in Fig. 2.4.

there is no agreement on either the number of b-cytochromes participating in respiration or on the sequence in which they act. While b_1 is an accepted component, the role of b_{562} is not clear.

Cytochrome c_{551} of *Pseudomonas* has been isolated and found to be closely related to c_{550} of *Paracoccus*, c_2 of the purple non-sulphur bacteria and mitochondrial cytochrome c (Almassy and Dickerson, 1978). Where bacterial respiratory chains possess c-type cytochromes, spectrographic data show that two are usually involved. *Escherichia coli* does possess c-type cytochromes but these do not appear to function in respiration and are not membrane-bound.

(e) Cytochrome oxidases
Several autoxidisable cytochromes have been found in bacteria and all of these react with CO. They belong to the a, b and d types but have not been fully investigated. Cytochromes a are characterised by the non-covalent binding of a

Table 10.3 Quinones, transhydrogenases and cytochromes in bacteria

Organism	Quinones	Trans-hydrogenase	b-Cytochromes	c-Cytochromes	Oxidases
Escherichia coli	Q8 (MK8)	(+)	b_{556} b_{562}		o
Escherichia coli O$_2$ limited	Q8 MK 8	(+)	b_{556} b_{558}		o (a_1) d
Klebsiella pneumoniae	Q8	+	b_{559} b_{563}		o
Azotobacter vinelandii	Q8		b_{560}	c_{551}* c_{555}*	o a_1 d
Haemophilus parainfluenzae	DMK		b_{557} b_{562}	(c_{550} c_{552})	o a_1 d
Haemophilus parainfluenzae O$_2$ limited	DMK		b	c	d
Beneckea natriegens	Q		b_{557} b_{562}	c_{547} c_{550} c_{554}	o (a_1) (d)
Pseudomonas ovalis	Q	+	b	c	o
Pseudomonas aeruginosa				c_{551}	cd_1
Micrococcus luteus	MK8		b_{557} b_{562}	c_{549} c_{552}	a, a_3
Paracoccus denitrificans	Q10	+	b_{556} b_{562}	c_{546} c_{550}	(o) a, a_3
Paracoccus denitrificans O$_2$ limited	Q10	+	b	c	(o) (a, a_3) (cd_1)
Bacillus megaterium	MK 7		b		(o) a, a_3
Bacillus licheniformis	MK		b		(o) a, a_3
Bacillus subtilis	MK 7				
Mycobacterium phlei	MK 9.2H	+	b_{559} b_{563}	c_{548} c_{554}	(o) a, a_3

( ) activity low

* The *Azotobacter* c-cytochromes were originally described as c_4 (c_{551}) and c_5 (c_{555})

haem with a formyl side chain and by their spectral properties. Two groups of *a*-type oxidases have been found. The first has an α-band at 600–605 nm and a γ-band at 440–445 nm in the reduced state. The α-band shifts to about 590 nm with CO treatment and these spectral characteristics are attributed to the cytochrome a, a_3 complex which is similar to the mammalian oxidase. However, mitochondrial cytochrome *c* does not act as a good electron donor for this oxidase in most cases. The second group known as a_1 and originally found in brewer's yeast and *Acetobacter* in the early 1930s by several groups of workers, has an α-band at 585–595 and a γ-band at 435–445 nm. Treatment with CO gives an α-band at 592 nm. Cytochrome a_1 serves as the terminal oxidase in certain bacteria such as *Acetobacter*. It is also functional in *Azotobacter*, but in *E. coli*, where it is also found, it appears to be inactive.

The *b*-type oxidase, discovered by Chance *et al.* (1953) and known as cytochrome *o*, has about the same reactivity as the a, a_3 system and occurs in many organisms with a, a_3. The reduced oxidase has absorption bands at 560 and 435 nm. The cytochrome has a molecular weight of 27 000 and consists of two chemically indistinguishable subunits and two haems of the protohaem type having mid-point potentials of $+118$ and -112 mV. The cytochrome binds both CO and cyanide. It is usually detected by reaction with CO, which gives a reduced-CO-minus-reduced difference spectrum with bands at 415–420, 532–537 and 557–563 nm (Tyree and Webster, 1978; Yang and Jurtshuk, 1978).

The *d*-type cytochromes were discovered in *E. coli* and *Shigella dysenteriae* by Yaoi and Tamiya (1928) and named cytochrome a_2 by Keilin. Subsequently it was realised that the prosthetic group of this cytochrome (Fig. 10.3) was not of the *a*-type and it was renamed cytochrome *d*, although the original designation is still used. The α-band of the reduced pigment is around 620–630 nm and, on oxidation, a band at 645 nm appears. Treatment with cyanide results in the disappearance of the oxidised band and no reduced band, but the activity is less sensitive to cyanide than other oxidases. The cytochrome is found in *E. coli* only under certain environmental conditions.

A related form of cytochrome *d* is found in some *Pseudomonas* species (*Pseudomonas aeruginosa*) and has been designated cd_1. The absorption maxima of the pigment in the reduced state are at 625, 554, 549, 521 and 418 nm corresponding to both *d* and *c* cytochromes. The pigment with a molecular weight of 120 000 possesses two haem *c* and two haem d_1. Haem *c* is linked covalently to the protein and haem d_1 is bound non-covalently. The protein is composed of two subunits, each of molecular weight 63 000. Oxidation is inhibited by CO and cyanide and the oxidase appears to catalyse nitrite reduction as well as oxygen reduction (see Hill and Wharton, 1978).

(f) Environmental effects

In many organisms, changes in the environment are known to alter the electron transport pathway. In *E. coli* grown under highly aerobic conditions with a non-fermentable carbon source such as succinate, the pathway for oxidation of $NADH_2$ is:

$$NADH_2 \longrightarrow F_p \longrightarrow Q \longrightarrow \text{cyt } b_{556}, \text{cyt } b_{562} \longrightarrow \text{cyt } o \longrightarrow O_2$$

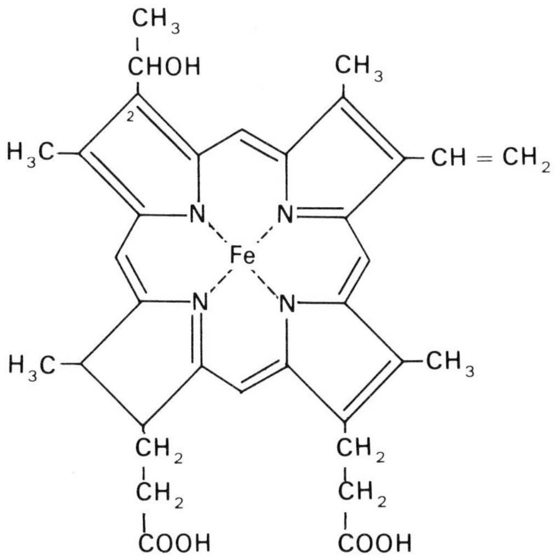

Fig. 10.3 Prosthetic group of cytochrome d: probable structure for haem d. There is uncertainty about the nature of the substituent group on position 2.

Under a variety of other conditions such as aerobic growth with glucose, there is a tendency to replace cytochrome o by cytochrome d giving rise to a branched chain pathway with separate b cytochromes.

$$NADH_2 \longrightarrow F_p \longrightarrow Q \begin{array}{l} \longrightarrow \text{cyt } b_{556}, \text{cyt } b_{562} \longrightarrow \text{cyt } o \longrightarrow O_2 \\ \searrow \text{cyt } b_{558} \longrightarrow \text{cyt } d \longrightarrow O_2 \end{array}$$

In *Paracoccus denitrificans* or in *Alcaligenes eutrophus*, reduction of oxygen concentration results in a replacement of cytochromes a, a_3 by cytochrome o.

(g) Anaerobic electron transport

Some organisms are able to grow anaerobically by using a modified electron transport chain and either fumarate or nitrate as alternative electron acceptors to oxygen. In *Escherichia coli*, either nitrate reductase or fumarate reductase may be induced under appropriate conditions (Fig. 10.4). Fumarate reductase is a different protein to succinate dehydrogenase. Any one of several dehydrogenases may function in these anaerobic electron transport chains. The system proposed for the reduction of fumarate in *E. coli* when the organism is grown in the presence of fumarate is shown in Fig. 10.4b. When grown under anaerobic conditions in the presence of nitrate (and Mo), the nitrate reductase and a specific b-cytochrome, $b_{558}^{NO_3}$ (b_{NR}), are synthesised. Nitrate serves as an alternative electron acceptor with a variety of substrates (Fig. 10.4a). The nitrate reductase itself contains molybdenum and an iron–sulphur centre. In *Paracoccus*, anaerobic systems for both nitrate and nitrite reduction are found (Fig. 10.4c).

10.4   Energy conservation

Some important differences exist between bacterial and mitochondrial oxidative phosphorylations. In the bacterium, the adenine nucleotides

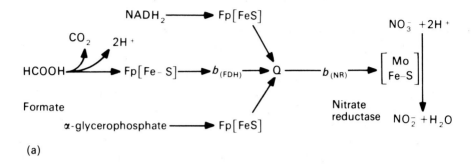

(a)

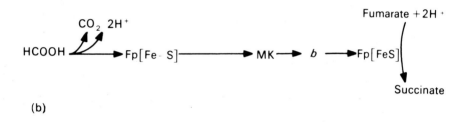

(b)

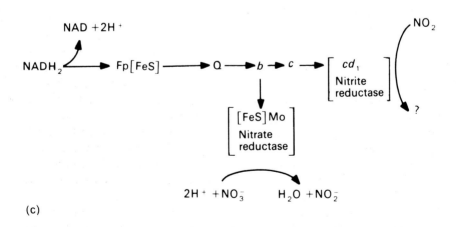

(c)

Fig. 10.4 Anaerobic respiratory chains: (a) Nitrate reductase in *Escherichia coli*; (b) Fumarate reductase in *Escherichia coli*; (c) *Paracoccus denitrificans*.

do not cross the plasma membrane, whereas in the mitochondrion these metabolites are transported across the inner membrane. As a consequence P/O ratios cannot readily be measured with whole bacteria. Furthermore, bacterial particles which can carry out oxidative phosphorylation rarely show respiratory control. A number of approaches have been made to the problem of oxidative phosphorylation in bacteria, but none has proved entirely satisfactory.

Bacterial vesicles which will phosphorylate, can be prepared by softening or digesting the cell wall with lysozyme and then subjecting the resultant sphaeroplast (the osmotically sensitive structure prepared from Gram-negative cells) or protoplast (the structure obtained from Gram-positive cells after removal of the wall) to an osmotic shock by very rapid dilution in buffer of low osmotic strength. In many cases ultrasonic treatment also gives active particles. Measurement of the P/O ratio for such vesicles gives low values, rarely better than 1.0 to 1.5 and often lower.

Alternatively, phosphorylation can be measured in relation to segments of the respiratory chain. Thus phosphorylation associated with NADH dehydrogenase and with the cytochrome *b* complex have been detected in respiratory vesicles. A third site has been detected in several organisms possessing *c*-type cytochromes, although assay of phosphorylation of this site is hampered by the lack of satisfactory electron donors for the reduction of endogenous cytochrome *c*; cytochrome *c* from horse heart or yeast does not react with the bacterial system in most organisms.

Reverse electron flow can also be used to demonstrate coupling sites. Thus reduction of NAD by succinate may be driven by ATP hydrolysis demonstrating coupling site 1. Similar ATP-dependent reduction of NADP by $NADH_2$ can also be shown.

A further approach has depended on the assumption that two protons are ejected for each pair of electrons passing through a coupling site. (Note however that the stoicheiometry of the respiratory proton system in mitochondria has recently been a matter of debate; cf. Chapter 7.) H^+/O ratios have been measured for a variety of substrates under various conditions. In general, organisms with a *c*-type cytochrome active in the respiratory chain and an active transhydrogenase can show an H^+/O ratio of about 8, while those with weak transhydrogenase activity give a value close to 6 and organisms such as *Escherichia coli* and *Bacillus* spp., which lack a *c*-type cytochrome and significant transhydrogenase activity, have values approximating to 4 (Fig. 10.5a). Thus P/O ratios of 2 for *E. coli*, 3 for *Micrococcus luteus* and up to 4 for *Pseudomonas* might be predicted from these results.

A rather simpler type of respiratory chain is obtained with anaerobic nitrate reduction. Here with α-glycerophosphate, succinate or lactate as substrate, $H^+/2e$ ratios of about 2 have been obtained with starved cells (Fig. 10.5b). A higher value (4) is obtained with malate as substrate, which probably reflects a further proton-translocating loop with this substrate.

The difficulties of measuring P/O ratios have led to the exploration of a different approach (see Stouthamer and Bettenhaussen, 1973). This is based on

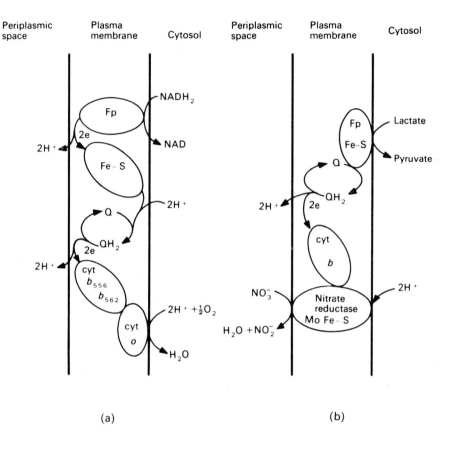

Fig. 10.5 A scheme for proton translocation by respiratory chains in *E. coli*: (a) Aerobic respiration; (b) Anaerobic nitrate reduction with lactate as substrate.

the assumption that, when the energy source for bacterial growth is rate-limiting, there will be a direct relationship between the ATP synthesised in oxidative phosphorylation (or in substrate level phosphorylation) and the amount of growth measured as cell mass. The energy source in these experiments is normally glucose. Thus the $Y_{glucose}$ (g cells per mole glucose metabolised) can be measured. If in the case of anaerobic growth, the moles of ATP synthesised per mole of glucose metabolised can be predicted from a knowledge of the fermentation pathway, a Y_{ATP} (g cells per mole ATP) can be calculated. However, it is also necessary to take into account the ATP or glucose required for maintenance of the cells, the $m_{glucose}$ (mol glucose h^{-1} per g cells). The true molar growth yield, $Y_{glucose}^{max}$ is therefore related to the observed value by the equation:

$$\frac{1}{Y_{glucose}} = \frac{1}{Y_{glucose}^{max}} + \frac{m_{glucose}}{\mu}$$

where μ is the specific growth rate (h^{-1}). The true molar growth yield for ATP, the Y_{ATP}^{max}, can be calculated as above from a knowledge of anaerobic growth and substrate level phosphorylation using a substrate with a known fermentation pathway:

$$\frac{1}{Y_{ATP}} = \frac{1}{Y_{ATP}^{max}} + \frac{m_{ATP}}{\mu}$$

With aerobic growth it is more convenient to measure oxygen uptake when the relationship is:

$$\frac{1}{Y_{O_2}} = \frac{1}{Y_{O_2}^{max}} + \frac{m_{O_2}}{\mu}$$

$Y_{O_2}^{max}$ is the true molar growth yield (g cells per mole O_2). The relationship between oxygen uptake and ATP synthesis can then be expressed as:

$$Y_{O_2}^{max} = N \cdot Y_{ATP}^{max}$$

where N is the overall efficiency of aerobic energy conversion expressed as moles ATP (or its equivalent) per mole O_2 consumed. N approximates to twice the P/O ratio, since the contribution of substrate-level phosphorylation is small in relation to oxidative phosphorylation. If Y_{ATP}^{max} is known for anaerobic growth, the value of N and hence the P/O ratio can be determined from measurements of growth parameters of aerobic cultures.

For enterobacteria, such as *E. coli* and *Klebsiella aerogenes*, the value of N approximates to 4 (most values lie between a little over 3 and a little under 5). The P/O ratio for such organisms would therefore be 2, a value which agrees with results obtained by other means. It is difficult to apply this method to obligately aerobic bacteria since Y_{ATP}^{max} cannot be obtained for the same organisms grown anaerobically. However, use of Y_{ATP}^{max} values obtained for growth of organisms such as *E. coli* or *Klebsiella* with measurements of $Y_{O_2}^{max}$ for aerobes such as *Bacillus* spp. and *Paracoccus* for example, give N values of about 4 and 6 corresponding to P/O ratios of 2 and 3 respectively. This is in general agreement with values obtained by other methods although such a procedure is open to criticism (see Stouthamer, 1977). The interest of these estimates of oxidative phosphorylation lies in the fact that they are made on living systems.

It may therefore be concluded that organisms such as the enterobacteria (*E. coli* and *Klebsiella*) and the bacilli and others which do not have a *c*-type cytochrome in the respiratory chain, have one coupling site associated with the NADH dehydrogenase complex and the other with the *b*-cytochrome segment of the chain. Organisms such as *Micrococcus*, *Mycobacterium* and *Paracoccus* which possess functional *c*-type cytochromes have three coupling sites.

Before leaving the subject of energy conservation, we will note a problem that arises from the application of the chemiosmotic theory to bacterial systems. In mitochondria, part of the energy conserved in respiration is used to maintain, in the matrix, a phosphate potential favourable to ATP synthesis. This is achieved, in the case of inorganic phosphate, by a net movement of one proton inwards for each phosphate transported into the mitochondrion (see sect. 7.5(b)). Thus not all the protons transported across the membrane are associated directly with ATP synthesis *per se*. In bacteria, with no intracellular compartments, almost all the proton gradient will be available for ATP synthesis (relatively little being dissipated in transport systems). Such an argument might lead one to anticipate stoicheiometries in bacteria rather different from those found in mitochondria. This problem remains to be resolved.

10.5 Bacterial ATPases

The bacterial ATPase which is involved as an ATP synthase in oxidative phosphorylation is located on the inside of the plasma membrane and, like the mitochondrial enzyme, can be seen as a stalked particle (diameter 9 nm) attached to membranes after negative staining. The ATPase activity (ATP + H_2O → ADP + Pi) is dependent on activation by a divalent cation, Mg^{2+} and/or Ca^{2+} but is insensitive to oligomycin. However, carbodiimides such as *N,N*-dicyclohexylcarbodiimide (DCCD) do inhibit the membrane-bound ATPase. A DCCD-binding proteolipid has been isolated. Activity is cold-labile and is stimulated by treatment with trypsin. The enzyme from *Escherichia coli* (MW = 345 000) possesses five subunits of molecular weight 56 800 (α), 51 800 (β), 32 000 (γ), 20 700 (δ) and 13 200 (ε). The stoicheiometry of the subunits is uncertain but appears to be similar to that of the mitochondrial F_1-ATPase. Enzyme activity is still retained in particles which possess α, β and γ subunits only. The δ subunit, which is not required for ATP hydrolysis, serves to bind the enzyme to the membrane. Trypsin, which stimulates activity, destroys either the δ or the ε subunit and one or both of these subunits are involved in masking the ATPase activity. The enzyme normally possesses bound adenine nucleotides.

The soluble enzyme may be prepared from respiratory particles (formed by French press treatment of a cell suspension) which are subjected to dialysis at low ionic strength. Reattachment of the ATPase to the stripped respiratory particles can be demonstrated by measuring the ATP-dependent transhydrogenase activity (that is reduction of NADP by NADH associated with ATP hydrolysis).

In association with ATP hydrolysis, the membrane-bound enzyme has been shown to translocate protons (West and Mitchell, 1974) outwards at a stoicheiometry of two protons/ATP hydrolysed.

In the thermophilic bacterium PS3, which has a very stable ATPase, the F_0 part of the complex has been analysed and found to possess only three subunits (MW = 19 000, 13 500 and 5400) with a stoicheiometry of 1 : 2 : 5. F_0 confers proton permeability on artificial phospholipid vesicles in which it is incorporated. This permeability can be reduced by treatment of the preparation with DCCD or bacterial F_1 (Okamoto *et al.*, 1977). Thus the properties of the bacterial ATPase are very similar to those of the mitochondrial enzyme.

10.6 Chemolithotrophic bacteria

A number of bacteria obtain their energy by the oxidation of inorganic compounds such as ammonia, nitrite, thiosulphate, hydrogen, etc; oxygen acts as the terminal electron acceptor. Most of these bacteria synthesise their cellular components from carbon dioxide by a form of the Calvin reductive pentose phosphate cycle (see sect. 11.1). The source of energy for the cycle is ATP, formed by oxidative phosphorylation, and in a few cases may involve substrate-level phosphorylation in addition. In several cases the redox potentials of the inorganic substrates are not sufficiently electronegative to give efficient reduction of the pyridine nucleotides needed for carbon fixation and their reduction requires an additional source of energy.

In nitrite oxidation by the soil organism *Nitrobacter*, nitrite reduces cytochrome a_1 which is oxidised by a simple electron transport chain (Fig. 10.6). That this respiratory chain possesses a single coupling site is inferred from the fact that two protons are transferred across the membrane for each oxygen atom reduced. The pyridine nucleotides required for carbon assimilation are reduced by nitrite through an energy-dependent electron transport system. (see Fig. 10.6).

A more complex oxidation is seen in the ammonia-oxidising organism *Nitrosomonas*. Three oxidations are involved in the conversion of ammonia to nitrite, of which two are of the oxygenase type while the third involves a respiratory chain coupled to ATP synthesis. Since the NH_2OH/NOH couple has about the same potential as the succinate $\sim$ fumarate couple, pyridine nucleotide reduction is achieved by an energy-dependent electron transport up a potential gradient.

In *Thiobacillus*, oxidation of sulphur compounds involves both oxidative and substrate-level phosphorylation by pathways as yet not fully understood.

10.7 Bacterial transport

Bacteria are capable of growth on a very wide range of substrates. For example, *Escherichia coli* may grow on any of the following carbohydrates: glucose, fructose, galactose, lactose, maltose, arabinose, xylose, rhamnose, mannitol, sucrose, raffinose, dulcitol. Since all these substrates are metabolised intracellularly, all must be transported across the selectively permeable plasma membrane. Two mechanisms are known for this transfer – the permease system and the phosphotransferase system.

The phosphotransferase system discovered by Roseman and colleagues (Kundig *et al.*, 1964) is not directly linked to the respiratory chain and will be mentioned here only briefly. This system is found in anaerobic and facultatively anaerobic bacteria and transports certain sugars across the membrane, releasing them internally as the sugar phosphates. The source of the phosphate is phosphoenolpyruvate (PEP) and the system involves several proteins.

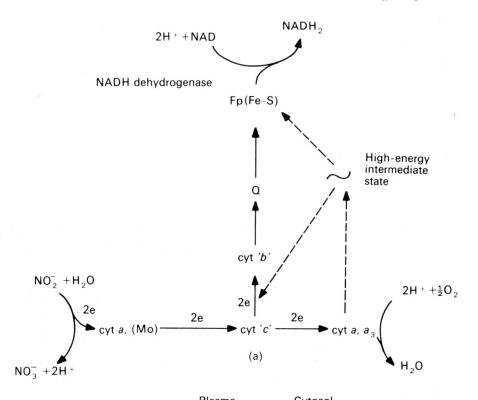

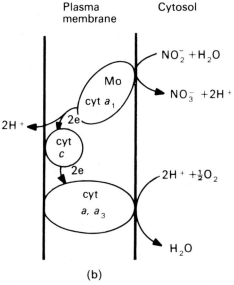

Fig. 10.6 Nitrite oxidation in *Nitrobacter*.
(a) Outline of the electron transport chain showing the oxidation of nitrite, reduction of NAD and reduction of oxygen. The energy from the oxidative pathway is used for driving reverse electron flow (see also Fig. 2.14). (b) Outline scheme for proton translocation coupled to nitrite oxidation.

$$\text{PEP} + \text{HPr} \xrightarrow{\text{Enzyme I}} \text{Pr--P} + \text{pyruvate}$$

or

$$\text{Pr--P} + \text{sugar} \xrightarrow{\text{Enzyme IIa/IIB}} \text{HPr} + \text{sugar phosphate}$$

$$\text{Pr--P} + \text{sugar} \xrightarrow{\text{Enzyme III/IIB}} \text{HPr} + \text{sugar phosphate}$$

The phosphorylated–dephosphorylated protein (Pr–P, HPr) and enzymes I and III are cytoplasmic while enzymes IIA and IIB are membrane-bound. Enzyme I and HPr are non-specific and either constitutive or inducible. In *Staphylococcus aureus*, where a variety of sugars are transported by a phosphotransferase system, a III/IIb system transports lactose into the cell as lactose phosphate. In *Escherichia coli*, where lactose is transported by a permease, a IIA/IIB system transports fructose, α-methylglucoside and *N*-acetylmannosamine, a different IIA protein being required for each. *Escherichia coli* also possesses a III/IIB system specific for glucose. It should be noted that it is the extracellular sugar which is translocated inwards and phosphorylated. Experiments showed that glucose inside vesicles is not readily phosphorylated whereas glucose outside vesicles is accumulated internally as glucose-6-phosphate.

The permease system is responsible for the transport of most of the substrates across bacterial membranes. The classical permease is that for lactose (galactosides) in *E. coli* which was first described by Monod and his coworkers (Monod, 1956) during their study of the induction of cultures to growth on lactose. The permease was shown to be a stereospecific galactoside-transporting protein (MW = 30 000), the product of the *y* gene (in the *lac* operon), capable of accumulating very large amounts of galactoside when a non-metabolisable sugar (a thiogalactoside) is used as substrate; under appropriate conditions, a 1000-fold concentration of galactoside can occur across the membrane. Attempts by Fox and Kennedy to isolate the protein were only partially successful (see Fox *et al.*, 1967). Subsequently, a number of permeases have been detected. In the case of amino acids, a variety of permeases have been identified on kinetic evidence. A single amino acid may be translocated by more than one type of permease and a single permease may transport a range of structurally related amino acids. The permeases can be divided into those whose activity is lost as a result of osmotic shock and those that are resistant to this treatment. As a result of osmotic shock, substrate-binding proteins can be isolated and these are presumed to be part of a transport system; it seems likely, however, that they do not themselves act as membrane carriers.

The energetics of transport systems is complex. In the case of the lactose permease, lactose crosses the membrane with one or two protons, depending on the pH of the medium. However, there is also evidence to suggest that the carrier in non-energised conditions behaves in a symmetrical way across the membrane. In the presence of a membrane potential or pH gradient, it undergoes conformational changes which result in a high affinity site on the outside and a low affinity site on the inside (cytosol side). In either case, the movement of lactose is a response to the proton motive force set up by the respiratory system (or by the ATPase). The uptake of a number of other sugars also appears to be coupled to proton transport.

A dicarboxylate system which transports fumarate, succinate and malate appears to consist of two membrane proteins. Uptake of the dicarboxylates takes place in an electroneutral manner with the uptake of two protons.

In the case of amino acids there is also evidence that their energy-dependent uptake is driven by the proton motive force. Thus it has been suggested that neutral amino acids are transported with a proton, that acidic amino acids, like the dicarboxylates above, are taken up in an uncharged form and therefore translocate protons, while basic amino acids are taken up in an electrogenic manner in response to the membrane potential. In the case of proline, for example, uptake appears to be dependent on the membrane potential. A proline-binding protein isolated from *E. coli* membranes and incorporated into vesicles will transport the amino acid in response to a membrane potential set up by valinomycin–potassium treatment.

A rather different mechanism applies to some amino acids such as glutamate in *E. coli* (and also *Halobacterium* – see sect. 15.9). Here uptake of the amino acid can be shown to be dependent on a sodium ion gradient. Sodium is co-transported with the amino acid. The sodium gradient itself is maintained by a respiration-dependent Na^+/H^+ exchange system (MacDonald *et al.*, 1977).

The foregoing interpretation of the energetics of membrane transport in bacteria is consistent with the chemiosmotic approach to cell energetics. However, an alternative view based on a conformational system has been proposed by Kaback. This interpretation has arisen from the fact that, in isolated vesicles obtained from *E. coli* plasma membranes, transport of lactose and several amino acids appeared to be linked to the membrane D-lactate dehydrogenase. In mutants lacking this dehydrogenase, the NADH dehydrogenase and succinate dehydrogenases were as effective as the lactate dehydrogenase, although in the wild type neither dehydrogenase matched the lactate enzyme. It has therefore been suggested that there are specific sites in the membrane responsible for transport which couple to the lactate dehydrogenase normally, but which, in the absence of this enzyme, will couple to other dehydrogenases (Kaback, 1972).

10.8 Evolution of the mitochondrion from prokaryotes

There are a number of similarities between mitochondria and bacteria. For example, the DNA and the system for protein synthesis in the mitochondrion are in many ways more closely related to that in the bacterium than to the nuclear DNA and the nuclear cytoplasmic system of protein synthesis in eukaryotes. Bioenergetic structures and properties of the mitochondrial inner membrane are similar to those of the bacterial plasma membrane, but not to the plasma membrane of eukaryotic cells. The lipid composition of the mitochondrial inner membrane, where steroids are probably absent and cardiolipin is present, also more closely resembles that of the bacterial membrane than most other eukaryotic membranes. These similarities between mitochondrial and bacterial structure have led to the endosymbiotic theory of mitochondrial evolution (see Fig. 10.7). Eukaryotic cells first appear in the fossil records in pre-Cambrian time about 1.2 to 1.4 billion years ago. This is rather later than the probable first appearance of oxygen, which began to accumulate in the atmosphere as a result of photosynthetic activity. Prior to this, anaerobic prokaryotes were the only form of life; heterotrophic organisms would have relied on fermentative pathways such as glycolysis for energy. It has been suggested that the earliest form of oxidative metabolism was of the peroxisome type and did not involve oxidative phosphorylation (see De Duve, 1973). However, it is thought that, early in the evolution of the eukaryotic cell, an amoeboid cell took up a free-living prokaryote capable of respiratory activity coupled to oxidative phosphorylation and resembling aerobic bacteria found today. In the newly evolved eukaryote, the plasma membrane of the bacterium gave rise to the inner membrane of the mitochondrion which invaginated to give the cristae. It is assumed that, hitherto, the amoeboid cell had had a fermentative metabolism possibly giving rise to lactate and this was used by the symbiotic bacterium for growth.

More recently, it has been suggested that *Paracoccus denitrificans*, which resembles the mitochondrion more closely than most bacteria, may be closely related to the original bacterium of the endosymbiotic theory (John and Whatley, 1975). Interesting features of this organism which suggest its relationship to the mitochondrion are as follows:

1. Phosphatidylcholine is a major constituent of the plasma membrane (see Table 10.1) and straight-chain saturated and unsaturated fatty acids make up the bulk of the membrane fatty acid.
2. Ubiquinone-10 (see Table 10.3) is the functional quinone.
3. The cytochrome content of aerobically grown cells closely resembles that found in mitochondria (see Table 10.3)
4. The cytochrome c (c_{550}), unlike most bacterial cytochromes, is interchangeable with mitochondrial cytochrome c as a substrate for both the mitochondrial cytochrome oxidase and the *Paracoccus* oxidases, a, a_3 and cd_1. Cytochrome c_{550} is soluble and resembles the mitochondrial protein in physical and structural properties.
5. The respiratory chain is sensitive to low concentrations of antimycin A and rotenone.
6. Respiratory control and the estimated P/O ratio are similar to those found in the mitochondrion.

Figure 10.7 shows the major changes required in the transition of a protoeukaryote with its symbiotic bacterium to the eukaryote. The products of fermentation, possibly lactate or ethanol, are replaced by pyruvate as a substrate for the mitochondrion. Whereas the bacterial membrane was impermeable to adenine nucleotides, the promitochondrion developed an adenine nucleotide translocase. The bacterial membrane is permeable to dicarboxylic acids and the exchange mechanisms of the mitochondrion may have developed from the uptake processes of the bacterial plasma membrane. The other major change required is the transfer of much of the genetic information for the proteins of the endosymbiotic bacterium, coded for in the bacterial chromosome, to the eukaryotic nuclear chromosome.

An alternative to this view has been proposed by several workers including

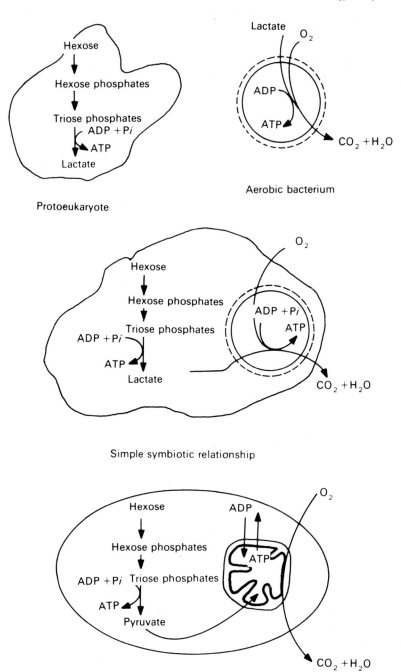

Fig. 10.7 Evolution of the mitochondrion: the endosymbiotic theory (John and Whatley, 1975)

Raff and Mahler (1972). They view the origin of a eukaryotic cell as an enlarged prokaryote which has developed a respiratory phosphorylation system in the plasma membrane (see Fig. 10.8). The enlargement of the cell requires an increase

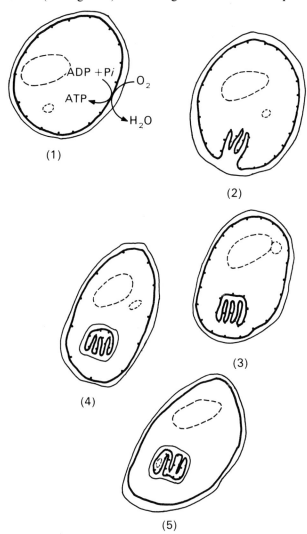

Fig. 10.8 Evolution of the mitochondrion: non-symbiotic theory (Raff and Mahler, 1972).
1. Protoeukaryotic cell with chromosome and plasmid (dotted).
2. Invagination of the plasma membrane.
3. Separation of a membrane vesicle from the plasma membrane.
4. Outer membrane developed around the vesicle.
5. Incorporation of the plasmid into the mitochondrion.
Note that before the incorporation of the plasmid, the respiratory vesicles themselves would require constant replacement since several proteins would be unable to pass from their site of synthesis in the cytoplasm into the vesicle. The acquisition of an appropriate plasmid would enable the mitochondrion to synthesise such proteins.

in the amount of plasma membrane if the ratio of cell volume to plasma membrane area is to be kept constant. Such an increase could occur by invagination in a manner rather similar to that of the bacterial mesosome. This would result in a vesicle with respiratory elements which are separated from the rest of the cell thus introducing a means of regulating metabolism and giving a selective advantage to this new eukaryote. Since the membrane would be impermeable, further advantage would be conferred if the vesicle incorporated a suitable plasmid (extrachromosomal double-stranded DNA composed of a number of genes, usually circular and commonly found in bacteria). It would be necessary for the plasmid to code for the membrane proteins and ribonucleic acids now synthesised in the mitochondrion. Extrachromosomal DNA of several sorts can be readily shown to exchange parts of its genome with that of the chromosomal DNA. Thus a plasmid might readily acquire the properties normally associated with the mitochondrial genome.

At the present time there is no sound rational basis for selecting one or other of the two hypotheses outlined here. However, the attraction of the endosymbiotic theory is obvious, although it is difficult to account for the transfer of genetic information from the symbiotic bacterium to the eukaryotic nuclear chromosome. This problem does not arise with the second theory which explains the genetic aspects relatively well, but does not provide so convincing a view of the evolution of the inner mitochondrial membrane.

Further reading

Baird, B. A. and Hammes, G. G. (1979) Structure of oxidative and photophosphorylation coupling factor complexes. *Biochim. Biophys. Acta*, **549**, 31–53.

Haddock, B. A. and Jones, C. W. (1977) Bacterial respiration. *Bacteriol. Rev.*, **41**, 47–99.

Konnings, W. N. (1977) Active transport of solutes in bacterial membrane vesicles. *Adv. Microbiol. Physiol.*, **15**, 175–251.

Wilson, D. B. (1978) Cellular transport mechanisms, *Annu. Rev. Biochem.*, **47**, 933–65.

Chapter 11

Photosynthesis: the fixation of carbon dioxide

11.1   The Calvin pentose phosphate pathway

(a) Introduction

Early in the present century, it became clear from experiments of Blackman, which were later interpreted by Warburg, that the photosynthetic process proceeded by light and dark reactions. This interpretation was supported by experiments of Emerson and Arnold (1932a) with flashing light (see sect. 1.1(c)). On the basis of this work, carbon dioxide fixation into organic carbon compounds such as sucrose or starch, the end products of photosynthetic carbon assimilation, was regarded as a 'dark reaction' which did not directly involve a photochemical event. The discovery of CO_2 fixation as part of the fermentation process in propionic acid bacteria (see sect. 3.5) underlined the general belief that carbon dioxide incorporation was a simple metabolic process to be understood in the same way as other metabolic reactions. The central problem then became the identification of the carboxylation reaction.

When isotopes of carbon became available, a new approach to the problem was possible. Early experiments with the radioactive ^{11}C, which has a very short half-life (20 min), were not successful. Equally difficult for photosynthetic experiments was the heavy non-radioactive isotope, ^{13}C, requiring a mass spectrograph for its analysis. However, in 1945, a radioactive isotope ^{14}C with a long half-life (5000 years) became available as a product of nuclear reactors. It was the use of this isotope that enabled Melvin Calvin and his coworkers to pioneer the study of photosynthetic carbon fixation and to make clear the metabolic pathways involved.

When finally elucidated, the Calvin reductive pentose phosphate cycle involved few reactions not already known to biochemists in other systems. The major new reaction, the carboxylation reaction, involves the carboxylation of a pentose phosphate to form two molecules of 3-phosphoglyceric acid (PGA). The PGA is metabolised by enzymes of the glycolytic pathway to hexose. The fate of the hexose is either regeneration of the pentose phosphate which includes several reactions of the alternative or pentose phosphate pathway of carbohydrate metabolism, or the synthesis of starch. The enzymes for the cycle are located in the chloroplast in plants.

The experimental approach adopted by Calvin to study the incorporation

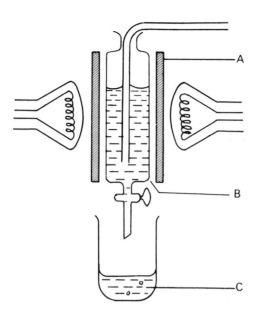

Fig. 11.1 Apparatus for the fixation of $^{14}CO_2$ by suspension of unicellular algae such as *Chlorella* and *Scenedesmus*. A 'lollipop' (B) with an inlet tube for gassing is mounted over boiling ethanol (C). The lollipop containing the photosynthesising suspension is illuminated with light from two lamps; heat filters (A) are interposed between lamp and preparation. At the end of the incubation time, the contents of the lollipop are plunged into boiling ethanol.

of carbon dioxide was as follows. An illuminated photosynthesising preparation such as a suspension of unicellular algae or later, chloroplasts, was incubated with radioactive carbon dioxide or bicarbonate for a short time (Fig. 11.1). This labelled the intermediates of carbon assimilation. The incubation mixture was then plunged into boiling ethanol to terminate all metabolic activity. The radioactive compounds in the ethanol extract were separated by two-dimensional paper chromatography. When the chromatogram was placed against X-ray film for some time, the radiations from those spots carrying ^{14}C produced a blackened area on the film as seen after development (Fig. 11.2). Thus the X-ray film gave a print of the chromatogram in terms of labelled compounds. The chromatogram itself could be treated with reagents which showed up the intermediates as coloured spots. When the chromatogram was run with known standards, a tentative identification of the intermediates could be made. Confirmation of such identifications could be achieved by several procedures including the elution of the radioactive spots from chromatograms, mixing with the suspected non-radioactive compound and re-chromatography to confirm the identity.

(b) Role of phosphoglyceric acid

Calvin found that incubation of the algae *Chlorella* or *Scenedesmus* with $^{14}CO_2$, followed by chromatography of the ethanol extract, gave twenty or more radioactive compounds. Radioactive CO_2 is very rapidly converted into a

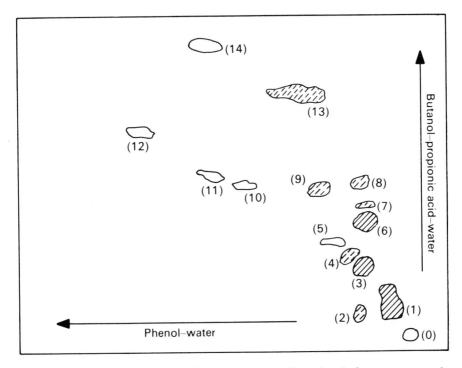

Fig. 11.2 Diagram of the autoradiography of two-dimensional chromatograms obtained with extracts from algae allowed to fix $^{14}CO_2$ for about 1 min. The radioactive spots are as follows: (0) origin, (1) hexose diphosphate and ribulose diphosphate, (2) UDP–glucose, (3) glucose phosphate and sedoheptulose phosphate, (4) mannose phosphate and fructose phosphate, (5) pentose phosphate, (6) PGA, (7) phospho-glycollate, (8) phosphoenolpyruvate, (9) aspartate, (10) serine, (11) glycine, (12) alanine, (13) malate, (14) glycollate. (Strongly labelled spots are heavily shaded)

number of metabolic intermediates including sugar phosphates, sugar diphosphates, phosphoglyceric acid, triose phosphates and small amounts of amino acids and carboxylic acids (Benson *et al.*, 1950). By shortening the incubation time to 10 seconds, it could be shown that radioactivity was preferentially being incorporated into 3-phosphoglyceric acid, although several other substances were also labelled. Generally, incubation periods of a few seconds using either algal or leaf preparations showed PGA to be the main compound labelled. Longer periods of incubation resulted in the labelling of an increasingly large number of compounds.

Confirmation of PGA as the initial labelled compound was provided by an analysis of the activity of individual carbon atoms. After only a few seconds, 95 per cent of the radioactivity in PGA could be found in C-1. When the incubation time was increased, the proportion of the activity in C-1 was found to fall while that in C-2 and C-3 rose. Analysis of the distribution of label in hexose, extracted simultaneously with PGA, showed the labelling pattern given in Fig. 11.3. The arrangement of this label prompted Calvin to suggest that hexose is synthesised from PGA by a route similar to the glycolytic one (Fig. 11.3).

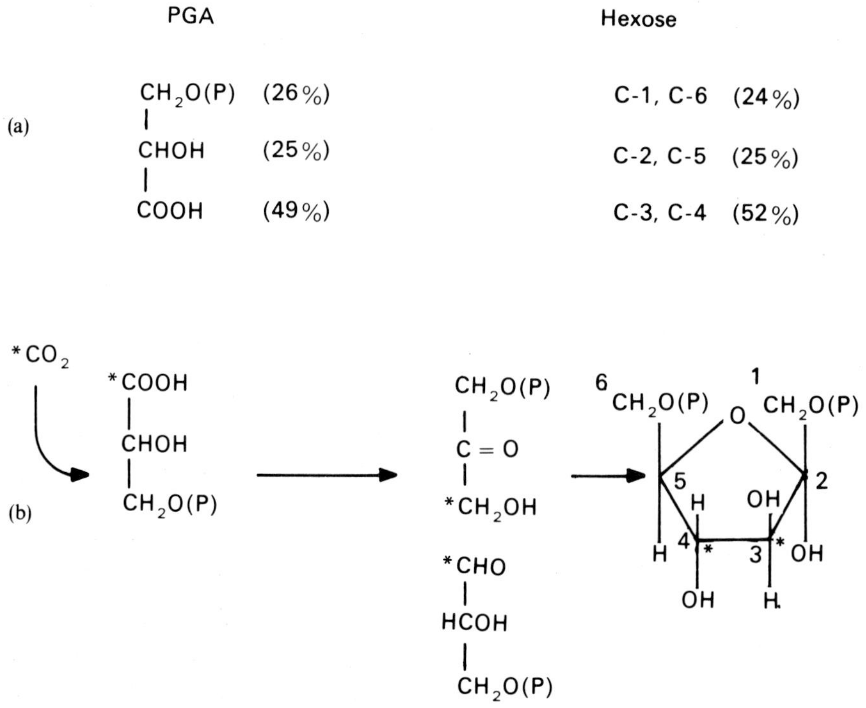

PGA   Hexose

(a)

$CH_2O(P)$ (26%)   C-1, C-6   (24%)

CHOH   (25%)   C-2, C-5   (25%)

COOH   (49%)   C-3, C-4   (52%)

$*CO_2$

(b)

$*COOH$
$\mid$
CHOH
$\mid$
$CH_2O(P)$

$\longrightarrow$

$CH_2O(P)$
$\mid$
$C = O$
$\mid$
$*CH_2OH$

$*CHO$
$\mid$
HCOH
$\mid$
$CH_2O(P)$

$\longrightarrow$

$6\,CH_2O(P)$ O $CH_2O(P)\,1$
5   H   OH   2
H   4 *   3 * OH
OH   H

Fig. 11.3 Labelling of 3-phosphoglyceric acid (PGA) and hexose with $^{14}CO_2$. (a) Percentage activity in the individual carbon atoms in PGA and in pairs of carbon atoms in hexose. (b) Pathway for hexose synthesis showing how $^{14}CO_2$ labels primarily C-3 and C-4 of hexose.

The nature of the precursor of PGA remained to be elucidated. PGA could be formed by carboxylation of either a C_2 compound or a C_5 compound, followed in the latter case by cleavage to form two molecules of PGA. Evidence that ribulose diphosphate was the CO_2 acceptor came from an experiment in which the levels of intermediates were followed in a photosynthesising preparation initially provided with a high concentration of CO_2 (1 per cent) followed by an abrupt change to 0.0003 per cent CO_2. The sudden drop in CO_2 supply is associated with a very rapid fall in the PGA concentration and a rapid rise in the ribulose diphosphate followed by ribulose monophosphate. Such results suggest that ribulose diphosphate is normally carboxylated to form PGA (Fig. 11.4). This same experiment also showed a relationship between the levels of triose phosphates (products of PGA reduction) and the pentose phosphates, suggesting a cyclical system (see Wilson and Calvin, 1955).

(c) The Calvin cycle

As already suggested, ribulose diphosphate is regenerated from the products of PGA metabolism. Calvin proposed a scheme for the conversion of triose phosphates to pentose phosphate. The overall pathway for carbon dioxide fixation, the Calvin cycle, was proposed by Calvin and coworkers in 1954, but

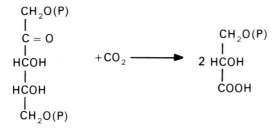

Fig. 11.4 Carboxylation of ribulose diphosphate

subsequently revised to include xylulose-5-phosphate, as shown in Fig. 11.5 and Table 11.1. (see Bassham *et al.*, 1954; Calvin, 1956).

Essentially the cycle consists of:

1. A carboxylation phase. The carboxylation of ribulose diphosphate to give two molecules of PGA, catalysed by the enzyme ribulose diphosphate carboxylase (carboxydismutase).

2. Synthesis of hexose phosphate. A series of reactions which in essence are the reverse of the glycolytic sequence from PGA to triose phosphates, fructose diphosphate and fructose-6-phosphate.

3. Regeneration of ribulose diphosphate. A series of reactions from fructose-6-phosphate and triose phosphate to pentose phosphates, the pathway being similar to that seen for the alternative pathway of carbohydrate metabolism.

4. Synthesis of sucrose and starch. Although this process is not part of the Calvin cycle itself, it is necessary to see the cycle as a biosynthetic pathway for the formation of cellular constituents, for example sucrose and starch. Such carbohydrates can be synthesised from hexose phosphate. Starch synthesis can occur inside the chloroplast while sucrose is synthesised in the cytosol. As will be discussed in more detail, it is the triose phosphates that are exported from the chloroplast and used for extraplastidic metabolism.

The criteria on which the cycle was based were essentially threefold. Firstly, radioactive carbon from $^{14}CO_2$ should appear in specific intermediates of the cycle in an appropriate sequence in time. Secondly, radioactivity should appear in specific carbon atoms of the cycle intermediates in a manner predicted from the cycle. Thirdly, there should be an adequate demonstration of the enzymic activities required for the function of the cycle. In the twenty or so years since the cycle was first proposed, a debate ensued on the adequacy of enzymic activity to account for CO_2 fixation. In almost all cases the weak activity of cycle enzymes in isolated chloroplast preparations has been accounted for. Low activities were due to a number of factors, including the loss of enzymes from chloroplasts during preparation and the failure to realise the need for activating agents.

The labelling of specific carbon atoms was of particular importance in demonstrating the pathway for the regeneration of ribulose diphosphate from PGA. A number of sugar phosphates became labelled during Calvin's experiments including ribose, ribulose, the C_7 sugar sedoheptulose and hexose

Table 11.1 Enzymes of the reductive pentose phosphate cycle (Calvin cycle)

Enzyme	Reaction	EC number
Ribulose diphosphate carboxylase	ribulose diphosphate + CO_2 + H_2O → 3-phosphoglyceric acid	4.1.1.39
Phosphoglycerate kinase	3-phosphoglyceric acid + ATP = 1,3-diphosphoglyceric acid + ADP	2.7.2.3
Glyceraldehyde-phosphate dehydrogenase (NADP)	1,3-diphosphoglyceric acid + $NADPH_2$ = glyceraldehyde-3-phosphate + Pi + NADP	1.2.1.13
Triose-phosphate isomerase	3-phosphoglyceraldehyde = dihydroxyacetone phosphate	5.3.1.1
Fructose-diphosphate aldolase	3-phosphoglyceraldehyde + dihydroxyacetone phosphate = fructose-1,6-diphosphate	4.1.2.13
Hexose diphosphatase (Fructose diphosphatase)	fructose-1,6-diphosphate + H_2O → fructose-6-phosphate + Pi	3.1.3.11
Transketolase	fructose-6-phosphate + 3-phosphoglyceraldehyde = erythrose-4-phosphate + xylulose-5-phosphate	
	sedoheptulose-7-phosphate + 3-phosphoglyceraldehyde = ribose-5-phosphate + xylulose-5-phosphate	2.2.1.1
'Aldolase' (probably identical with fructose-diphosphate aldolase)	erythrose-4-phosphate + dihydroxyacetone phosphate = sedoheptulose-1,7-diphosphate	
Sedoheptulose diphosphatase	sedoheptulose-1,7-diphosphate + H_2O → sedoheptulose-7-phosphate + Pi	3.1.3.37
Ribosephosphate isomerase	ribose-5-phosphate = ribulose-5-phosphate	5.3.1.6
Ribulosephosphate-4-epimerase	xylulose-5-phosphate = ribulose-5-phosphate	5.1.3.4
Phosphoribulokinase	ribulose-5-phosphate + ATP → ribulose diphosphate + ADP	2.7.1.19

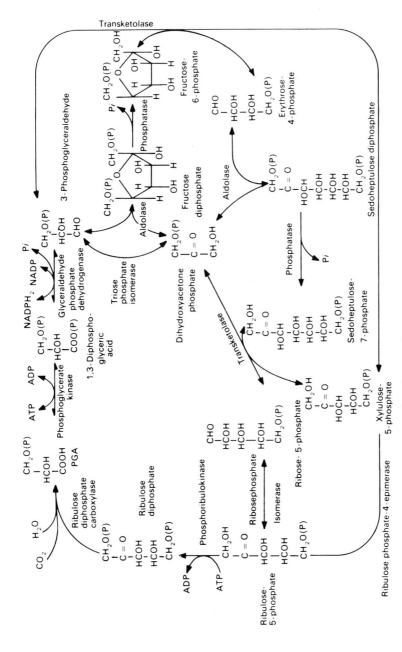

Fig. 11.5 The Calvin reductive pentose phosphate cycle.
Note: the enzymes and reactions are tabulated in Table 11.1.

THAMES POLYTECHNIC LIBRARY

phosphates. An indication of the intensity of labelling of the carbon atoms after incubation with $^{14}CO_2$ is shown in Fig. 11.6. It may be seen that the labelling pattern of ribulose did not correspond to that of any five carbons in hexose or sedoheptulose. Hence ribulose diphosphate could not have been derived simply from either of these sugars and it became necessary to consider more than one source of pentose phosphate, if the relatively heavier labelling in C-3 was to be explained. Similar problems also applied to sedoheptulose, where C-3, C-4 and C-5 were labelled. The solution here lay in the different origins of C-1, C-2 and C-3 (triose phosphate) and C-4 – C-7 (C-3–C-6 of fructose-6-phosphate). The labelling patterns were best accounted for by the pathway shown. It may been seen that of the three molecules of pentose formed, one was labelled in C-1, C-2 and C-3 while the other two were labelled in C-3 only.

(d) Ribulose diphosphate carboxylase

Calvin and coworkers obtained a cell-free preparation from both algae and higher plants which, when incubated with ribulose diphosphate and labelled bicarbonate, gave rise to labelled PGA. The enzyme ribulose diphosphate carboxylase, was subsequently isolated and later purified in several laboratories.

In 1947, the name 'Fraction 1 protein' was used to describe a protein fraction obtained in large quantities from spinach leaves. Later the protein was found to have ribulose diphosphate carboxylase activity and shown to be a major component of the soluble chloroplast proteins. The carboxylase activity proved to be inseparable from Fraction 1 protein. Both the protein and the carboxylase were found exclusively in the chloroplast.

The enzyme from spinach has a molecular weight of 560 000 and is composed of two types of subunit (MW = 54 000 and 16 000). The complex is composed of eight large and eight small subunits, catalytic activity being associated with the former. Contrary to earlier suggestions, the enzyme does not contain copper. The mechanism of the carboxylase reaction, which is irreversible, is shown in Fig. 11.7. Ribulose diphosphate is split between C-2 and C-3. CO_2 rather than bicarbonate is the source of the carboxyl group. Oxygen acts as a competitive inhibitor with CO_2 giving rise to one molecule of PGA and one molecule of phosphoglycollate. Both reactions require Mg^{2+}.

In earlier work, measurement of the K_m for CO_2 presented something of a problem. The isolated enzyme gave a value of the order of 500 μM although this was very variable. In the intact chloroplast, the apparent K_m was about 10 μM. Further, the concentration of CO_2 in water in equilibrium with air at 1 atm (0.03% CO_2) is 10 μM at 25° C. However, it is now known that the enzyme requires activation. Preincubation (in the absence of ribulose diphosphate) in the presence of CO_2 and Mg^{2+} converts inactive or partially activated enzyme to the activated form with a K_m comparable to that found in chloroplasts. Although the CO_2 activation site is independent of the catalytic site it is probably also associated with the large subunit.

(e) Photorespiration

This form of respiration, which is light-dependent and independent of normal respiration, has been known for some time. Interest in the process has

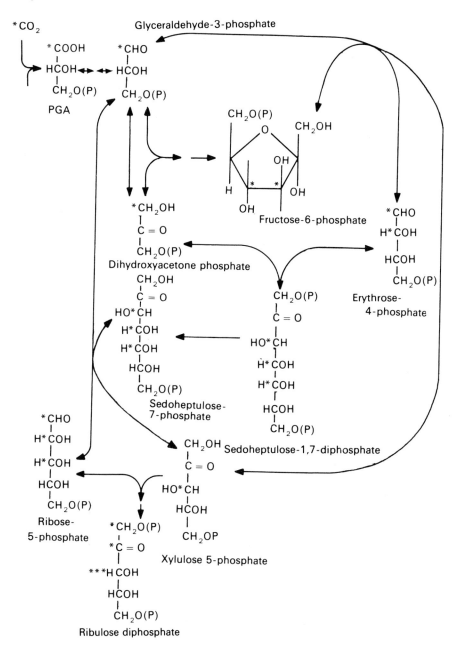

Fig. 11.6 Labelling of specific carbon atoms in intermediates of the Calvin pentose phosphate cycle. Experimentally, activity is found in C-3 and C-4 of hexose (see Fig. 11.3), C-3, C-4 and C-5 of sedoheptulose and C-1, C-2 and C-3 of ribulose diphosphate, but C-3 of this pentose is much more heavily labelled than C-1 and C-2. It should be noted that the ribulose diphosphate is derived from ribose phosphate labelled in C-1, C-2 and C-3 and two xylulose phosphates both labelled in C-3.

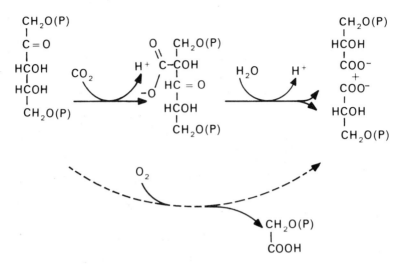

Fig. 11.7 The ribulose diphosphate carboxylase and oxygenase reactions (see Lorimer, 1978).

been stimulated by the fact that in some species, it is responsible for a significant loss of fixed carbon dioxide. A greater photosynthetic yield might be expected if a satisfactory means of controlling photorespiration were found.

Photorespiration is now attributed to reactions initiated by the ribulose diphosphate carboxylase acting as an oxygenase, with phosphoglycollate as the product. A possible route for the metabolism of phosphoglycollate is shown in Fig. 11.8. Phosphoglycollate can be converted in the chloroplast to glycollate which can then be oxidised mainly in the leaf peroxisomes. Serine is probably formed in the mitochondria. The pathway contributes to amino acid synthesis; it is nevertheless an apparently wasteful process and the reason for its existence is not understood.

A source of glycollate canvassed by some workers is the transketolase reaction. This enzyme possessed thiamin pyrophosphate as a prosthetic group which forms a glycolaldehyde derivative as an intermediate in the reaction (Fig. 11.9). The oxidation of this intermediate has been suggested as an additional possible source of glycollate.

(f) Factors affecting Calvin cycle activity

Reference to Fig. 11.5 shows that the requirements for cycle activity are two equivalents of $NADPH_2$ and three equivalents of ATP for each equivalent of CO_2 assimilated. The provision of this cofactor requirement is a property of the light reaction which can conveniently be summarised:

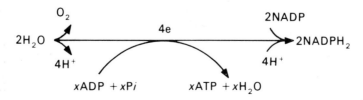

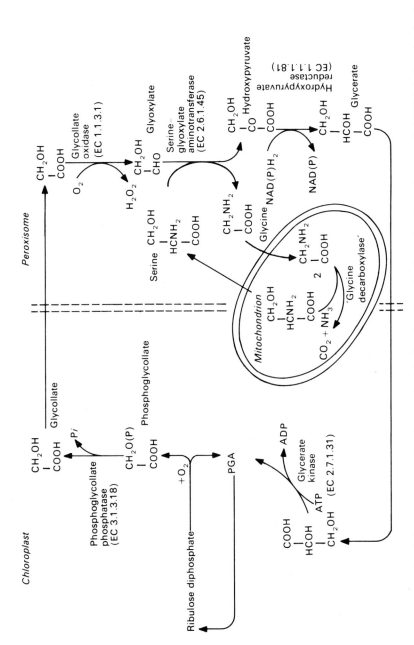

Fig. 11.8 Photorespiration in leaves: proposed pathway. The 'glycine decarboxylase' reaction (see Moore *et al.*, 1977) is believed to be the major source of CO_2, although decarboxylation of glyoxylate to formate may also contribute to CO_2 production. In some algae, glycollate oxidation occurs in the mitochondria and is catalysed by a pyridine-nucleotide-dependent glycollate dehydrogenase. The fate of the glycerate is most likely to be phosphorylation to PGA in the chloroplast.

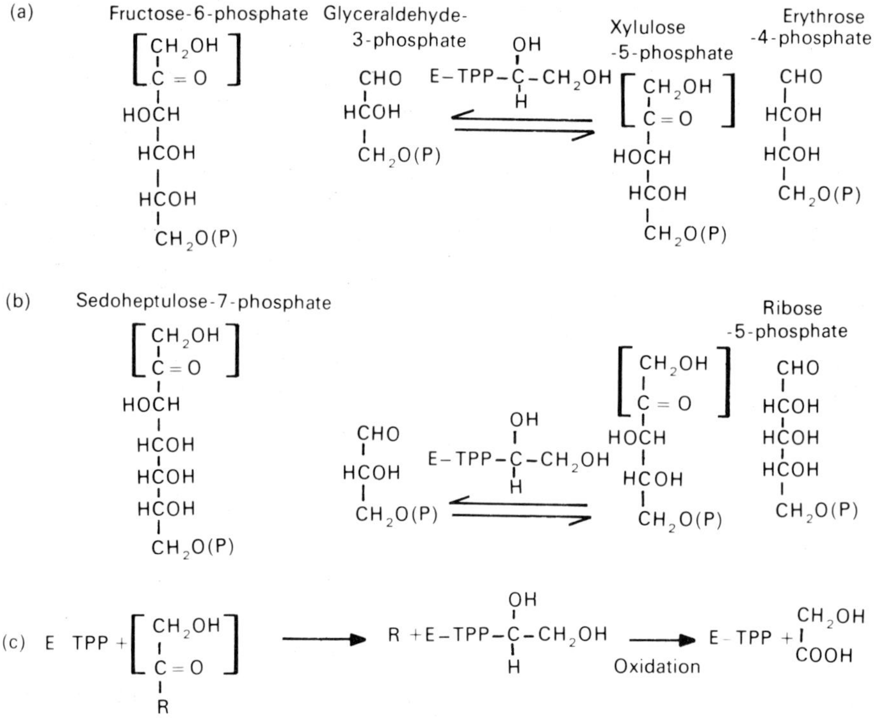

(a)

Fructose-6-phosphate Glyceraldehyde-3-phosphate Xylulose-5-phosphate Erythrose-4-phosphate

$$\begin{bmatrix} CH_2OH \\ | \\ C=O \\ | \\ HOCH \\ | \\ HCOH \\ | \\ HCOH \\ | \\ CH_2O(P) \end{bmatrix}$$

$$\begin{array}{c} CHO \\ | \\ HCOH \\ | \\ CH_2O(P) \end{array}$$

$$E-TPP-\overset{\overset{\displaystyle OH}{|}}{\underset{\underset{\displaystyle H}{|}}{C}}-CH_2OH$$

$$\rightleftharpoons$$

$$\begin{bmatrix} CH_2OH \\ | \\ C=O \\ | \\ HOCH \\ | \\ HCOH \\ | \\ CH_2O(P) \end{bmatrix}$$

$$\begin{array}{c} CHO \\ | \\ HCOH \\ | \\ HCOH \\ | \\ CH_2O(P) \end{array}$$

(b)

Sedoheptulose-7-phosphate   Ribose-5-phosphate

$$\begin{bmatrix} CH_2OH \\ | \\ C=O \\ | \\ HOCH \\ | \\ HCOH \\ | \\ HCOH \\ | \\ HCOH \\ | \\ CH_2O(P) \end{bmatrix}$$

$$\begin{array}{c} CHO \\ | \\ HCOH \\ | \\ CH_2O(P) \end{array}$$

$$E-TPP-\overset{\overset{\displaystyle OH}{|}}{\underset{\underset{\displaystyle H}{|}}{C}}-CH_2OH$$

$$\rightleftharpoons$$

$$\begin{bmatrix} CH_2OH \\ | \\ C=O \\ | \\ HOCH \\ | \\ HCOH \\ | \\ CH_2O(P) \end{bmatrix}$$

$$\begin{array}{c} CHO \\ | \\ HCOH \\ | \\ HCOH \\ | \\ HCOH \\ | \\ CH_2O(P) \end{array}$$

(c) $E\ TPP +$
$$\begin{bmatrix} CH_2OH \\ | \\ C=O \\ | \\ R \end{bmatrix}$$
$\longrightarrow$ $R + E-TPP-\overset{\overset{\displaystyle OH}{|}}{\underset{\underset{\displaystyle H}{|}}{C}}-CH_2OH$ $\xrightarrow{\text{Oxidation}}$ $E-TPP+\begin{array}{c} CH_2OH \\ COOH \end{array}$

Fig. 11.9 The transketolase reaction: (a) and (b) reactions of the Calvin cycle, showing the role of thiamin pyrophosphate, a prosthetic group of the enzyme (see also sect. 3.2); (c) a proposed mechanism for glycollate formation.

Although light might be expected to regulate cycle activity through the supply of cofactors, control of the cycle appears to be rather more complex. A study of the levels of intermediates found before and after transition from light to dark suggests that the cycle is regulated by light at two points. One of these is the ribulose diphosphate carboxylase, while the other is the diphosphatase (see Jensen and Bassham, 1968).

On illumination of chloroplasts there is a considerable lag in CO_2 fixation. This is not due to low levels of ribulose diphosphate, but is attributed to the need to activate the carboxylase. The mechanism for the activation probably involves the level of Mg^{2+} and the pH of the stroma where the enzyme is located. Illumination of the chloroplast results in pumping of protons out of the stroma and an associated counter-flow of Mg^{2+} into the stroma. Thus illumination causes both a rise in Mg^{2+} concentration and in pH of the stroma (see Portis and Heldt, 1976). As noted earlier, activation of the carboxylase can be achieved with Mg^{2+} in the presence of CO_2 and this process is also favoured by a higher pH. Furthermore, the higher pH favours the activity of the activated enzyme, which has a pH optimum at about 8.3–8.6. A number of metabolites have been shown to influence the activity of the carboxylase. For example, $NADPH_2$ and phosphogluconate increase activity while several pentose phosphates act as competitive inhibitors. However, the physiological significance of these latter effects is not clear.

The fructose-1,6-diphosphatase has a molecular weight of 160 000 and is composed of four subunits. Dithiothreitol, Mg^{2+} and fructose-1,6-diphosphate activate the enzyme which has a high pH optimum. The activating effect of light on the enzyme might be accounted for in terms of an increase in pH and Mg^{2+} concentration in the stroma, as discussed in relation to the carboxylase. The phosphatase reaction has a requirement for Mg^{2+} but this cation also has an effect on enzyme activity (Portis *et al.*, 1977). Moreover, a protein, thioredoxin, has been described which is reduced in the light by ferredoxin (Fd), a member of the chloroplast electron transport chain (see sect. 14.2). Thioredoxin, whether reduced by ferredoxin or dithiothreitol, will activate the diphosphatase (see Schurmann and Wolosiuk, 1978).

$$H_2O + 2Fd_{ox} \longrightarrow 2Fd_{red} + 2H^+ + \tfrac{1}{2}O_2$$

$$2Fd_{red} + thioredoxin_{ox} + 2H^+ \longrightarrow thioredoxin_{red} + Fd_{ox}$$

$$\text{Fructose-1,6-diphosphatase}_{inactive} \xrightarrow{\text{thioredoxin (red)}} \text{fructose-1,6-diphosphatase}_{active}$$

oxidised glutathione

It is not clear whether sedoheptulose diphosphatase is a distinct enzyme from fructose diphosphatase in chloroplasts. Nevertheless, similar principles of regulation appear to apply to this enzyme. Regulation of several other enzymes has been described but its significance is not clear. For example, adenine nucleotides appear to regulate 3-phosphoglycerokinase.

11.2 Starch and sucrose synthesis

As was realised by the physiologist Sachs at the end of the nineteenth century, starch is also a product of photosynthesis and may be deposited inside the chloroplast. The metabolic pathway for starch synthesis and degradation is shown in Fig. 11.10. ADP-glucose is formed from glucose-1-phosphate by the glucose-1-phosphate adenyl transferase (ADP-glucose pyrophosphorylase). The starch synthetase extends $\alpha 1-4$ linked chains by the addition of glucose residues to the non-reducing ends. Plants possess branching enzymes which introduce branches into the structure forming $1-6$ linkages.

Control of starch synthesis appears to be exercised through the adenyl transferase (ADP-glucose pyrophosphorylase). This enzyme is activated by PGA and inhibited by inorganic phosphate, but the inhibition may be overcome by PGA. Fructose-6-phosphate is also an activator of the enzyme, less effective than PGA. When photosynthetic systems are illuminated, the level of inorganic phosphate falls because of photophosphorylation, and the level of fructose-6-phosphate rises; the synthesis of starch will hence be promoted in the light and depressed in the dark. The steady-state levels of PGA are about the same in the light as in the dark. Activation of the enzyme by PGA is, however, more effective

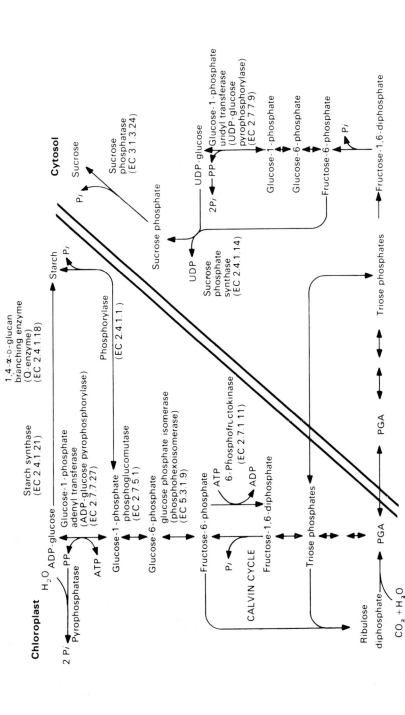

Fig. 11.10 Synthesis of starch and sucrose. The pathway for starch synthesis and the probable pathway for breakdown are shown. Sucrose is synthesised outside the chloroplast and may be translocated to other parts of the plant. (ADP-glucose = adenosine diphosphate glucose, UDP-glucose = uridine diphosphate glucose).

at the higher pH of the stroma achieved in the light. Carbohydrate may be exported from the chloroplast as triose phosphate. Starch may be converted to triose phosphate with the additional enzymes phosphorylase and phospho-fructokinase.

Sucrose is a major product of photosynthetic activity. This sugar is translocated through the phloem to other parts of higher plants. Sucrose synthesis takes place outside the chloroplast in the cytoplasm (Fig. 11.10).

11.3   The dicarboxylic acid pathway (C_4 pathway)[1]

(a) Outline of the pathway

The Calvin pentose phosphate cycle is found in a wide range of plants and in algae as the primary means of photosynthetic CO_2 fixation. However, most detailed studies of the pathway have been confined to spinach and pea chloroplasts, which fix CO_2 primarily by this means. In maize and sugar cane and a number of other higher plants, a different pattern of fixation is found in association with the Calvin cycle and recently an understanding of this pathway as an integrated metabolic system has emerged.

Not long after Calvin proposed the pentose phosphate cycle, research workers in Hawaii and later in Australia examining $^{14}CO_2$ fixation by sugar cane found a pattern of labelled intermediates which did not correspond with that obtained from spinach preparations. In particular, a large proportion of label was rapidly incorporated into malate and aspartate. This work led to the formulation of a scheme for photosynthetic carbon assimilation by the Australian workers M. D. Hatch and C. R. Slack in 1966. The kinetics of labelling in C_4 dicarboxylic acids, including oxaloacetic acid, showed them to be formed as products of a primary carboxylation reaction. A little later, the main carboxylating enzyme was shown to be the phosphoenolpyruvate carboxylase which forms oxaloacetate. In addition to incorporation of label by carboxylation into C_4 acids, a small amount of activity (5–15 %) was incorporated very early into PGA. Activity in PGA built up more slowly than in the dicarboxylic acids, suggesting that this CO_2 fixation could be, at least in part, secondary to the primary fixation into the C_4 acids. Initially the formation of PGA was seen as involving a transcarboxylation, the carboxyl group of PGA being derived from a carboxyl group on malate. It is now clear that the dicarboxylic acids lose a carboxyl group which is released as CO_2 and refixed by Calvin cycle activity. The logic and the advantages of this more complex system will be seen later.

Understanding of the processes involved in CO_2 fixation in plants such as sugar cane and maize has been developed partly from consideration of the leaf tissues involved. Chloroplasts from the mesophyll tissues (Fig. 11.11) differ enzymically as well as structurally from those of the large parenchyma cells surrounding the conducting tissues of the leaf, the bundle sheath. The cells are well connected by protoplasmic strands, plasmodesmata. The bundle sheath

---

[1] Plants which fix CO_2 primarily by the Calvin cycle are said to possess the C_3 pathway and are occasionally referred to as C_3 plants. Plants possessing the dicarboxylate pathway as an adjunct to the Calvin cycle are said to fix CO_2 by the C_4 pathway and may be called C_4 plants.

The fixation of carbon dioxide

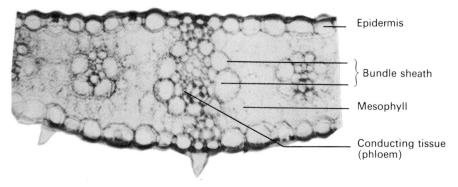

Fig. 11.11 Section of a leaf of *Zea mays* (maize) showing the relationship of the bundle sheath cells to the conducting tissue and mesophyll cells.

chloroplasts are larger than those in the mesophyll cells and, in some grasses, lack grana; they also accumulate starch soon after illumination, whereas those in the mesophyll of rapidly growing plants do not (Fig. 12.1b). It has been shown in maize leaves that the bundle sheath chloroplasts contain the malic enzyme and ribulose diphosphate carboxylase activities while the mesophyll chloroplasts possess a loosely bound phosphoenolpyruvate carboxylase and an NADP-linked malate dehydrogenase. Thus the main carboxylating enzyme for the Calvin pentose phosphate pathway is found in one tissue (bundle sheath) in association with a decarboxylating enzyme, while the adjacent tissue (mesophyll) possesses the carboxylating enzyme for oxaloacetate formation and the enzyme for reduction of oxaloacetate to malate (Fig. 11.12). This enzyme distribution is consistent with a pathway for C_4 photosynthesis in which there is an initial CO_2 fixation in mesophyll cells and release and refixation of CO_2 in bundle sheath cells. Carbon dioxide fixation occurs in the cytoplasm of the mesophyll to form oxaloacetate which is either transaminated to aspartate in the cytoplasm or reduced to malate in the chloroplast. Evidence suggests that carbon initially fixed in malate or aspartate in the mesophyll is transferred to the bundle sheath and becomes the carboxyl carbon of PGA. Malate or aspartate migrate from the mesophyll to the bundle sheath. In the bundle sheath cells, malate is decarboxylated to pyruvate which migrates back to the mesophyll chloroplasts where it is converted to phosphoenolpyruvate by the pyruvate orthophosphate dikinase in the chloroplast. The carbon dioxide is metabolised by the ribulose diphosphate carboxylase. In the case of aspartate, the amino acid is transaminated and decarboxylated to release CO_2 for conversion to PGA. The remaining three carbons return to the mesophyll as alanine.

Three types of C_4 pathway have been proposed as shown in Fig. 11.13 (Hatch and Kagawa, 1976). The first type, exemplified by *Zea mays* (maize), involves migration of malate to the bundle sheath and its decarboxylation to oxaloacetate by an NADP-dependent malic enzyme found in the chloroplast. The second type, found in the grasses *Chloris gayana* and *Panicum maximum*, involves migration of aspartate to the bundle sheath where it is transaminated in the cytoplasm and the product, oxaloacetate, decarboxylated in the chloroplast

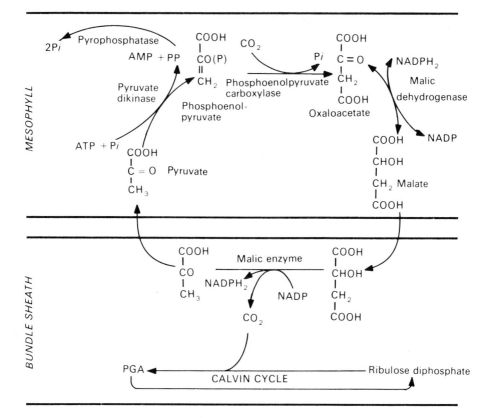

Fig. 11.12 Simple outline of the C_4 pathway of photosynthesis.

to phosphoenolpyruvate by the phosphoenolpyruvate carboxykinase. Phospho-enolpyruvate is converted to pyruvate and transaminated in the cytoplasm. In the third type, shown by *Atriplex spongiosa* and the grass *Panicum miliaceum*, the aspartate is transaminated in the bundle sheath mitochondria to oxaloacetate which is reduced to malate. The mitochondrial NAD-linked malic enzyme decarboxylates the malate to pyruvate, which is transaminated to alanine in the cytoplasm for transfer to the mesophyll.

The NADP-linked malic enzyme in plants fixing carbon by the C_4 pathway, is located in the bundle sheath chloroplasts. The NAD-linked enzyme is mitochondrial and of very high activity in C_4 species lacking significant amounts of the other decarboxylating enzymes. The carboxykinase is located in chloroplasts of bundle sheath cells and present in high activity in *Panicum maximum* (Rathnam and Edwards, 1975).

The cofactor requirements of the pathway are substantially higher than those of the Calvin cycle. Whereas the latter system requires only three equivalents of ATP and two of $NADPH_2$ for each equivalent of CO_2 fixed, the C_4 pathway requires five equivalents of ATP and two of $NADPH_2$. It should be noted that when malate exchanges for pyruvate (in *Zea mays*, for example), two

Table 11.2 Enzymes of the C_4 pathway of photosynthesis

Enzyme	Reaction	EC number
Phosphoenolpyruvate carboxylase	phosphoenolpyruvate + $H_2CO_3 \rightarrow$ oxaloacetate + Pi	4.1.1.31
Phosphoenolpyruvate carboxy-kinase (ATP)	oxaloacetate + ATP = phosphoenolpyruvate + ADP + CO_2	4.1.1.49
Malic enzyme (NADP-linked)	L-malate + NADP = pyruvate + $NADPH_2$ + CO_2	1.1.1.40
Malic enzyme (NAD-linked)	L-malate + NAD = pyruvate + $NADH_2$ + CO_2	1.1.1.39
Pyruvate orthophosphate dikinase	pyruvate + ATP + Pi $\rightarrow$ phosphoenolpyruvate + AMP + PP	2.7.9.1
Pyrophosphatase	pyrophosphate + $H_2O \rightarrow 2Pi$	3.6.1.1
Malate dehydrogenase	Malate + NAD = oxaloacetate + $NADH_2$	1.1.1.37
Aspartate aminotransferase (transaminase)	aspartate + oxoglutarate = oxaloacetate + glutamate	2.6.1.1
Alanine aminotransferase (transaminase)	alanine + oxoglutarate = pyruvate + glutamate	2.6.1.12

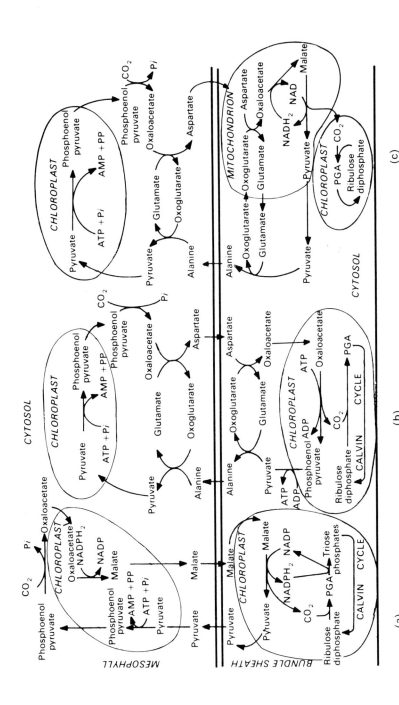

Fig. 11.13 Pathways for C_4 photosynthesis: (a) the NADP-dependent malic enzyme type; (b) the phosphoenolpyruvate carboxykinase type; and (c) the NAD-dependent malic enzyme type. The enzymes for these pathways are listed in Table 11.2.

reducing equivalents are transferred to the bundle sheath. The function of the pathway is not properly understood, but it appears to be a means of supplying high concentrations of CO_2 to the ribulose diphosphate carboxylase, that is as a CO_2 pump to the bundle sheath. This may be of significance in leaves where the rate of CO_2 diffusion is limited resulting in low concentrations of mesophyll CO_2.

The pathway outlined shows all the CO_2 to pass through dicarboxylic acids before incorporation into PGA. However, there is evidence to suggest that up to 15 per cent of the CO_2 fixed in PGA arises by direct diffusion to the bundle sheath cells.

(b) Regulation

The regulation of the pathway is not yet understood. However, the pyruvate orthophosphate dikinase of mesophyll chloroplasts can be extracted in an inactive form from darkened leaves and an active form from illuminated leaves. Activation of the inactive enzyme can be achieved by reducing agents in the presence of inorganic phosphate. *In vivo* activation appears to be associated with the activity of the chloroplast electron transport chain.

The NADP-linked malic dehydrogenase of the mesophyll chloroplasts shows similar properties to the dikinase. A number of effectors are known for the enzymes of the C_4 pathway, but their significance is not clear.

11.4   Other carboxylation pathways

(a) Malate formation in C3 plants

Dicarboxylic acids are known to be formed as a minor pathway for the fixation of CO_2 in photosynthetic systems of algae and higher plants. PGA may be converted to phosphoenolpyruvate by the normal glycolytic route. Carboxylation of the phosphoenolpyruvate then gives rise to the dicarboxylic acids.

(b) The CAM pathway

Plants of the order Crassulaceae with fleshy photosynthetic tissue have an unusual metabolism associated with photosynthesis referred to as Crassulacean Acid Metabolism (CAM). This type of metabolism is, in fact, found in a wide range of plants of various orders growing in an arid environment. However, in comparison with C_4 and C_3 photosynthesis, it is of minor importance. The CAM pathways are shown in Fig. 11.14. In the dark, phosphoenolpyruvate formed from starch and dextrans is carboxylated to form oxaloacetate by the cytoplasmic phosphoenolpyruvate carboxylase. The oxaloacetate is reduced to malate and stored in the cell vacuole. This dark CO_2 fixation can be demonstrated with $^{14}CO_2$; the label is found in C-4 malate. Contrary to earlier findings, other carbons of malate receive almost no label. In the light, the malate is decarboxylated by one of two routes. Some plants appear to use primarily the NADP-dependent malic enzyme while in others the major route involves oxidation to oxaloacetate and decarboxylation by the phosphoenolpyruvate carboxykinase. The CO_2 released is fixed into PGA by the ribulose diphosphate carboxylase in the chloroplast, which possesses normal Calvin cycle activity. The

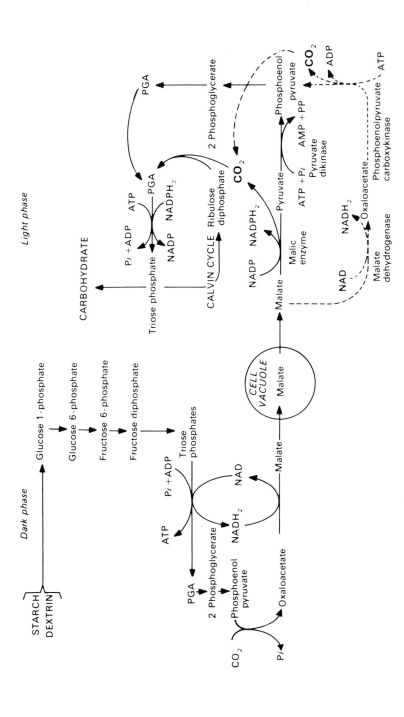

Fig. 11.14 Crassulacean acid metabolism. Carbon dioxide is fixed in the dark phase into malate. In the light phase CO_2 may be derived from malate by the malic enzyme or from atmospheric CO_2. It appears that in some plants CO_2 from malate is obtained by an alternative route in which the major decarboxylating enzyme is the phosphoenolpyruvate carboxykinase (shown ---). The cellular location of many of the enzymes is not known with certainty.

C_3 product of decarboxylation is believed to be used for carbohydrate synthesis.

In principle, the CAM pathway resembles the C_4 pathway in that CO_2 is initially fixed in dicarboxylic acids and subsequently released for metabolism by Calvin's reductive pentose phosphate pathway. The pathways differ in that the CAM pathway takes place within a single cell whereas the C_4 pathway operates across two tissues.

The CAM pathway probably represents an adaptation to growth in arid conditions. Although some CO_2 is fixed directly into PGA during the day, there is major CO_2 fixation at night when the plant is not under water stress. During at least part of the day, the stomatal pores in the leaves are probably closed to reduce water loss, but this also reduces the entry of CO_2 into the leaf tissue.

Further reading

Calvin, M. and Bassham, J. A. (1962) *The Photosynthesis of Carbon Compounds*. Benjamin, New York.

Halliwell, B. (1976) Photorespiration. *FEBS Lett.*, **64**, 266–70.

Hatch, M. D. (1978) Regulation of enzymes in C_4 photosynthesis. *Current Topics in Cell Regulation*, **14**, 1–28.

Jensen, R. G. and Bahr, J. T. (1977) Ribulose 1,5-bisphosphate carboxylase-oxygenase. *Annu. Rev. Plant Physiol.*, **28**, 379–400.

Kelly, J. G., Latzko, E. and Gibbs, M. (1976) Regulatory aspects of photosynthetic carbon metabolism. *Annu. Rev. Plant Physiol.*, **27**, 181–205.

Osmund, C. B. (1978) Crassulacean acid metabolism: a curiosity in context. *Ann. Rev. Plant Physiol.*, **29**, 379–414.

Walker, D. A. (1976) Regulatory mechanisms in photosynthetic carbon metabolism. *Current Topics in Cell Regulation*, **11**, 204–41.

Also:

Buchanan, B. B. (1980) Role of light in the regulation of chloroplast enzymes. *Annu. Rev. Plant Physiol.*, **31**, 341–74.

Chapter 12

The chloroplast: structure, properties and biogenesis

12.1 Structure

The chloroplast in higher plants is typically a lens-shaped structure with a diameter of the order of 10 μm and thus substantially larger than the mitochondrion. The particle is surrounded by an envelope consisting of a double membrane (Fig. 12.1). Internally there are numerous flattened sacs referred to as *thylakoids* (see Menke, 1962) which are shown diagrammatically in two dimensions in Fig. 12.2a. Regions of tight stacking of thylakoids (like a pile of coins) are known as *grana* (granum sing.) and these are interspersed by regions of loosely arrayed thylakoids. The regions of the chloroplast outside the grana are known as the *stroma*. The grana are shown as loosely connected by stroma lamellae (membranes). A number of attempts have been made to develop this two-dimensional view of the chloroplast in three dimensions (see Park and Sane, 1971) and one such model is shown in Fig. 12.2b. A common feature of these models is that the internal space of each granum thylakoid is connected through stroma thylakoids to the internal spaces of thylakoids in other grana and even to thylakoids in the same granum. Thus a view of the chloroplast structure emerges in which a number of thylakoids give rise to grana and stroma lamellae so that two membrane-bound spaces are defined, one within the thylakoids (the loculus) and a second extrathylakoid space (matrix) bounded by the chloroplast envelope.

Evidence from several sources suggests that the enzymes for carbon fixation (Calvin cycle) are found outside the thylakoids in the soluble part of the stroma, the matrix. The photosynthetic pigments are associated with the membranes of both the grana and the stroma. Electron micrographs of chloroplasts frequently show the presence of large starch grains in the stroma. Suitable preparations also show inclusions such as DNA (several DNA bodies per chloroplast), ribosomes and osmiophilic granules which are composed mainly of chloroplast quinones.

The stacking of thylakoids in discrete grana, which is so characteristic of most higher plant chloroplasts, may not be essential to photosynthetic function. Indeed the reason for the formation of grana has puzzled many workers. In a low salt medium, chloroplasts lose their granal stacks and have parallel sheets of membranes, but are still able to evolve oxygen photosynthetically. Restacking can be achieved by the addition of salts. These effects of electrolytes (essentially monovalent and divalent cations) suggest the involvement of surface charges on the membranes which separate by electrostatic repulsion but which reaggregate when electrolytes weaken the surface charge. Hydrophobic interactions between the membranes may also be involved in the stacking. The formation of grana appears to require the presence of photosystem II (see below).

Not all chloroplasts in higher plants have granal stacks. In particular, the bundle-sheath cells of C_4 plants such as maize lack grana although the mesophyll cell chloroplasts of the same plant possess them (Fig. 12.1b). In the algae, chloroplasts are variable. Red algae (Rhodophyta) have a simple chloroplast with no grana and the thylakoids distributed through the stroma (Fig. 12.1c). In the Cryptophyta, the thylakoids may stack in pairs while in brown algae they are found in threes. This adhesion of thylakoids in small numbers is found in a variety of algae including some Chlorophyta (green algae). However, in *Acetabularia* (also a member of the Chlorophyta) something approaching the granal arrangement of higher plants is found. Many algal chloroplasts also contain pyrenoids, dense granules associated with ribulose diphosphate carbo-xylase activity.

A special case is provided by the blue-green algae (blue-green bacteria, cyanobacteria) which are prokaryotes and do not possess chloroplasts. Nevertheless, they carry out oxygen-evolving photosynthesis as opposed to other prokaryotes (photosynthetic bacteria), which are unable to evolve oxygen. In blue-green algae, the photosynthetic apparatus is bound in thylakoids which ramify through the outer regions of the cell (see Fig. 15.1).

12.2   Chemical composition

(a) Envelope

The envelope consists of two membranes. Gentle swelling of the chloroplast detaches the envelope which can be separated by sucrose-gradient centrifugation. The membranes have a lipid composition distinct from that of the rest of the chloroplast (see Table 12.1). There is no chlorophyll but a significant amount of carotenoid, particularly violaxanthin. The phospholipid composition also differs from that in the lamellar membrane. The chloroplast sterols appear to be located in the envelope, a distribution reminiscent of that in the mitochondrion, where sterols are in the outer membrane. The envelope proteins are distinct from those of the thylakoid membranes.

(b) Thylakoid membranes

Spinach chloroplast thylakoid membranes are composed of about 52 per cent lipid and 48 per cent protein with small amounts of Fe, Mn and Cu. The lipid composition is given in Table 12.1 and shows that 21 per cent of the lipid fraction is chlorophyll while about 40 per cent is galactolipid. There is a high concentration of unsaturated fatty acid; more than 90 per cent of the fatty acids of the galactolipid have three double bonds. The most prominent polypeptides associated with these membranes are derived from the ATPase, the ribulose

Table 12.1 Percentage lipid composition of chloroplasts from higher plant (spinach)

	Envelope	*Lamellae*
Monogalactosyldiglyceride	33	22
Digalactosyldiglyceride	34	15
Trigalactosyltriglyceride	1.5	0
Phosphatidylglycerol	2.9	8
Sulpholipid	0.8	7
Phosphatidylcholine	9	2.9
Phosphatidylinositol	1.1	1.3
Phosphatidylethanolamine	1.4	0
Sterols, sterol esters and sterol glycosides	5.6	0
Cerebroside	2.7	0
Chlorophyll	0	21
Carotenoid	+	+
Quinone	0	3

Based on Poincelot (1973).
See also Lichtenthaler and Park (1963).

diphosphate carboxylase and the light-gathering pigment–protein complexes (Henriques and Park, 1976).

12.3 Thylakoid membrane structure

In the mid-1960s, several workers observed 9–10 μm particles on the surface of negatively stained lamellae comparable to those seen in mitochondrial preparations (see Oleszeko and Moudriánakis, 1974). These particles may be removed from membranes by washing with dilute EDTA solutions (binding being mediated by divalent cations) and the washings possess a latent Ca^{2+}-dependent ATPase which can be activated by trypsin. The enzyme can be shown to have a role in photophosphorylation. The particle itself has been isolated in Racker's laboratory (Vambutas and Racker, 1965) and is designated CF_1, chloroplast factor 1. If stroma-depleted chloroplasts (with ruptured envelopes) are treated with antibodies to CF_1, enzymic activity is inhibited, suggesting that the ATPase is situated on the outer surface of the thylakoid – a location confirmed by other studies. It is estimated that there is one CF_1 particle for every 500–800 chlorophyll molecules. Some of the ribulose diphosphate carboxylase is also bound to the outer surface of the thylakoid membrane and this may be seen as 12 μm cuboid particles in negatively stained preparations. The enzyme can be removed with dilute sodium pyrophosphate.

In addition to these extrinsic thylakoid membrane proteins, members of the photosynthetic electron transport chain are also found on the outer surface. Ferredoxin, an iron–sulphur protein, and the enzyme ferredoxin–NADP reductase form stoicheiometric complexes bound into the surface but may be removed by washing. In general there are more extrinsic protein particles on the outside surface of the thylakoid membrane than on the internal surface, indicating membrane asymmetry.

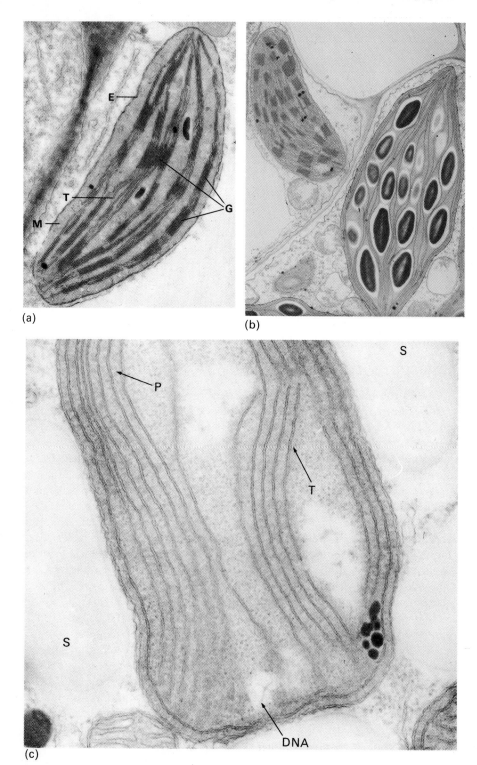

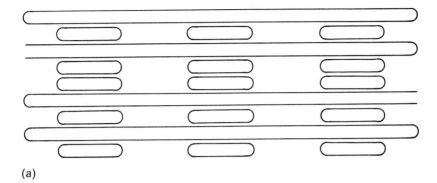

(a)

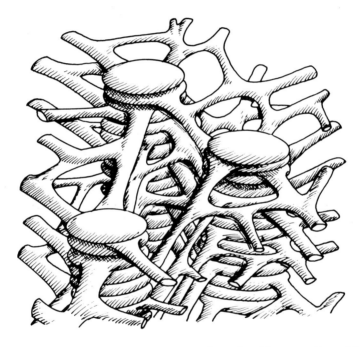

(b)

Fig. 12.2 Models of the chloroplast. (a) Two-dimensional interpretation of thylakoid structure (b) A three-dimensional interpretation of thylakoid structure according to Weier *et al.* (1963)

Fig. 12.1 Electron micrographs of thin sections through a chloroplast.
(a) Chloroplast from *Phaseolus vulgaris* × 14 500.
    E = envelope, G = granum, M = matrix, T = thylakoid.
    (Courtesy of Dr. J. Whatley, University of Oxford).
(b) Chloroplasts from *Zea mays*. To the left is a mesophyll chloroplast with grana while to the right is a bundle sheath chloroplast lacking grana but possessing starch grains, × 7000. (Courtesy of Dr. J. Whatley, University of Oxford).
(c) Chloroplast of a red alga, *Corallina officinalis*, × 56 500.
    DNA = deoxyribonucleic acid, P = phycobilisome (see sect. 13.3(a)), S = starch grain, T = thylakoid.
    (Courtesy of Dr. M. C. Peel, Queen Elizabeth College, London).

The technique of freeze-fracturing and freeze-etching (sect. 4.2(c)) has been applied to the thylakoid membrane by various workers including Branton and Park (1967). The fracture-faces and surface of the granum thylakoids are shown in Fig. 12.3 and a model based on the observations in Fig. 12.4. The outer

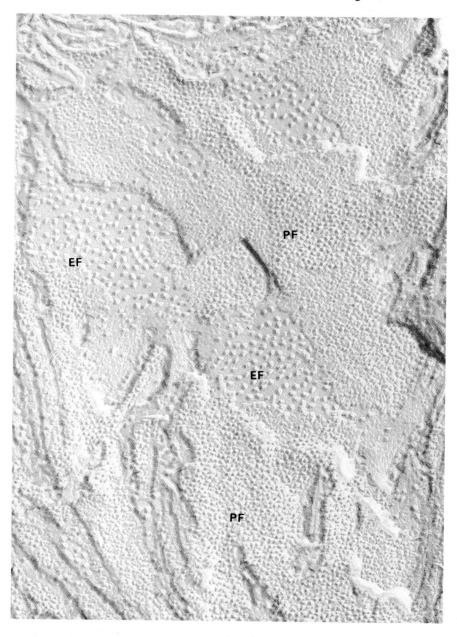

Fig. 12.3 Freeze-fracture faces of chloroplast thylakoids × 150 000. The fracture-faces are identified on Fig. 12.4. (Courtesy of Dr Pfeifhofer).

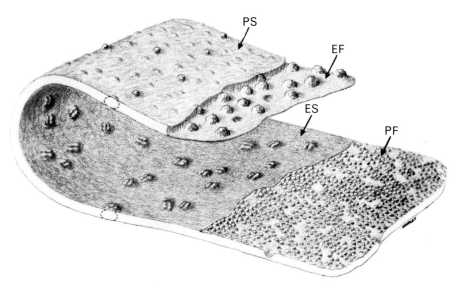

Fig. 12.4 Model of chloroplast thylakoid membrane showing surfaces (inner = ES, outer = PS) and fracture-faces (outer half of membrane = PF, inner half = EF). (After Park and Pfeifhofer, 1969.)

surface (PS) shows the presence of a variety of particles in unwashed preparations, while the inner surface (ES) shows a number of particles, 18.5×15.5 μm, only slightly raised above the surface but showing the existence of four subunits; otherwise the inner surface is smooth. The particles of the ES face appear to protrude through the inner half of the membrane so that they are seen in fracture-face (EF) as a slightly smaller particle. Fracture-face PF shows a large number of smaller particles (12 μm diameter) which are embedded in the outer half of the membrane.

Studies such as these have been extended to include stroma lamellae, which are found to possess the smaller intrinsic particles and a few of the larger ones. However, the large particle of the stroma lamella is smaller than that of the granum (Armond and Arntzen, 1977). It is widely believed that the large particles are associated with photosystem II activity and the smaller ones with photosystem I activity.

12.4 Chloroplast fractionation

The photosynthetic process involves 'light reactions' associated with pigments together with an electron transport system and a 'dark reaction', which is primarily the metabolic pathway for CO_2 fixation. There are two light reactions in photosynthesis (Fig. 12.5) referred to as photosystem I (or pigment system I), PS I, and photosystem II (or pigment system II), PS II. Both are necessary for the flow of electrons from water to NADP but PS I is capable of an independent cyclic electron flow coupled to ATP synthesis. Fractionation of chloroplasts has contributed to our understanding of the nature of the two photosystems.

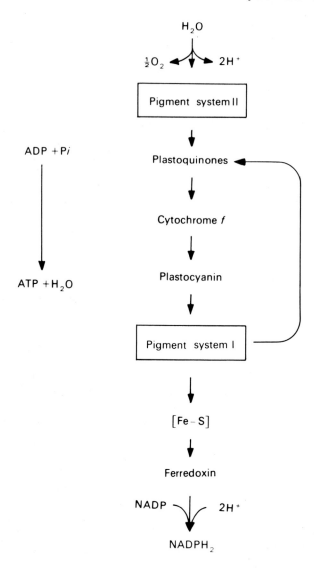

Fig. 12.5 Outline of the chloroplast electron transport chain. ATP synthesis is coupled to the oxidation–reduction reactions.

Early experiments by Trebst *et al.* (1958) in Arnon's laboratory showed that the photochemical reactions of the membranous portions of the chloroplasts (primarily the grana) could be separated from the soluble fraction, which was capable of fixing CO_2 when supplied with cofactors. The idea that chlorophyll and thus the essential photochemical reactions of photosynthesis were specifically localised in grana persisted with the observation that chlorophyll fluorescence observed with the light microscope appeared to originate from the grana.

The separation of grana and stroma lamellae by mechanical means (French press) was achieved in the late 1960s and convincingly demonstrated by Sane *et al.* (1970). These workers confirmed the identity of the fractions by freeze-fracturing studies and showed that grana lamellae possess PS I and PS II activities, whereas stroma lamellae have mainly PS I.

Partial physical separation of PS I and PS II was initially achieved by Boardman and Anderson in 1964 using digitonin-treated chloroplasts, which were subjected to differential centrifugation. A small particle fraction with significant PS I activity was separated from a large particle fraction enriched in chlorophyll *b* and showing PS II and some PS I activity. Comparable results were also obtained using French press treatment (which involves forcing a suspension of chloroplasts through a small orifice under pressure, thus subjecting the particles to considerable sheering forces) followed by density gradient centrifugation. By the early 1970s it became clear that the PS I particles arose primarily from stroma lamellae (though probably including some PS I particles from grana) and the PS II were derived from grana lamellae. A refined approach to separation of particles used by Arntzen *et al.* in 1972 involved initial separation of grana and stroma lamellae by French press treatment and differential centrifugation. Treatment of the grana with digitonin was followed by sucrose density gradient centrifugation to separate low density PS I and high density PS II fractions. PS I and PS II from grana may be reconstituted to give a system which will evolve oxygen and reduce NADP. However, PS I derived from the stroma lamellae would not reconstitute with PS II from the grana. Examination of the ultrastructure of the fractions showed that while PS I consisted of small particles, PS II was composed of large particles on membranous sheets suggesting that PS II fractions were composed of fragments of membrane carrying the PS II particles, but having lost PS I. This supports the interpretation given to freeze-fracture studies of the chloroplast discussed above, in that the large particles may be PS II and the small, PS I. Such an interpretation has not gone unchallenged since some chloroplasts do not appear to possess large particles yet have PS II activity. It has been suggested that rather than representing photosystems, the particles are antenna pigment–protein complexes (see sect. 13.3(c)). However, Armond and Arntzen (1977) have shown a correlation between PS II activity and the presence of large particles in stroma lamellae.

A note on the quantasome. In the 1950s, some studies of the fine structure of chloroplast membranes revealed the presence of a repeating unit in the inner face of the thylakoid membrane. Subsequently Park and coworkers applied the term quantasome to such a regular repeating subunit. The term was intended to imply that these subunits were a morphological expression of the fundamental unit of photosynthesis originally proposed by Emerson and Arnold, and possessing about 2500 chlorophyll molecules. The two-dimensional lattice of particles described by Park and coworkers (see Park and Biggins, 1964) had dimensions of 18.5×15.5 μm. It is now accepted that these particles did not represent the basic photosynthetic unit, but were probably identical with the large subunit of the inner membrane face of the thylakoid, which may be PS II.

12.5   Permeability

(a) Permeability of chloroplast membranes

A study of isolated chloroplast permeability comparable to the studies with mitochondria and using tritiated water (^3H_2O) and [^{14}C] sucrose, leads to the conclusion that a limited volume of the particle is permeable to sucrose while a much greater volume is impermeable to sucrose, but permeable to metabolites such as 3-phosphoglycerate, for which a specific carrier system is known to exist (Heldt and Sauer, 1971). The correlation of data from permeability studies with electron micrographs of chloroplasts suggests that the outer membrane of the chloroplast envelope is permeable to low molecular weight substances, whereas the inner membrane is selectively permeable – impermeable to sucrose but permeable to phosphoglycerate. From this it may be deduced that carrier systems must be located in the inner membrane.

There are few data on the ionic composition of chloroplasts since often ions are readily lost during preparation of the organelle. The major monovalent cation in plants is potassium which constitutes about 20 per cent of the ash. In leaves, some 40–50 per cent of this potassium together with about half of the sodium (3 % of the ash) is found in chloroplasts if they are isolated in non-aqueous media. Calcium is the major divalent ion (13–20 % of the ash) and about 60 per cent of this is found in the chloroplasts while about 70 per cent of the magnesium (3–4 % of the ash) is in these particles.

As in the mitochondrion, the major cation movement occurring with chloroplast activity is that of the proton. Following the publication of Mitchell's chemiosmotic hypothesis, Jagendorf and coworkers (see Neumann and Jagendorf, 1964) demonstrated a light-dependent proton pump. Using type II chloroplasts with broken envelopes (thylakoid preparation), protons were found to be taken up from the medium when the preparation was activated by light. Uncouplers abolished the uptake and the pH change decayed in a subsequent dark period. Use of chloroplasts with intact envelopes gives only a slow uptake from the medium, since the intact envelope impedes equilibrium between the matrix and the external medium. Evidence from several sources shows that the protons are free within the thylakoid space (loculus).

Thus the thylakoid membrane like the inner mitochondrial membrane possesses a proton pump which is active in the energised system (when illuminated) and which pumps protons from the matrix into the loculus of the thylakoids. However, the membrane has reverse polarity with respect to that of the inner mitochondrial membrane. The pH difference across the membrane may be as high as 2.5 to 4.0 units. (The external pH is important in these measurements and the optimum is near 8.0.) The reverse polarity is consistent with the location of the ATPase (CF_1) on the outside of the thylakoid. The importance of proton translocation in relation to photophosphorylation will be discussed later.

Other ion movements have been observed in association with proton movements. Chloride ions will move into the thylakoid with protons while K^+

and Mg^{2+} move outwards in exchange for H^+. Although the existence of cation carriers might be suspected, little is known about this aspect of chloroplast biochemistry.

(b) Specific membrane translocators

Sucrose and starch, major products of photosynthetic activity, cannot permeate the chloroplast envelope. How then are the products of photosynthesis made available to the rest of the plant? If $^{14}CO_2$ is fed to leaves or to the brown alga, *Fucus*, within 3 minutes a steady state of distribution is reached in which only 45 per cent (spinach) or 15 per cent (*Fucus*) is in the plastid. After very short incubations (10 s) significant amounts of labelled PGA, fructose-6-phosphate, glucose-6-phosphate and UDP-glucose are found in the cytoplasm and label accumulates very rapidly in extraplastidic sucrose. Labelling also appears relatively early in malate, aspartate, glycine, serine and alanine. Ribulose and sedoheptulose diphosphates are apparently confined to the chloroplast and are unable to penetrate the envelope; similarly sugar monophosphates also fail to penetrate the envelope (see Heldt and Rapley, 1970). The membrane is permeable to inorganic phosphate, 3-phosphoglycerate, aspartate, malate, glutamate and succinate. A dicarboxylate translocator(s) specific for malate, succinate, oxoglutarate, fumarate, aspartate, glutamate and oxaloacetate (but not malonate, citrate or monocarboxylic acids) has been described (Heldt and Rapley, 1970; Lehner and Heldt, 1978). The carrier(s) catalyse(s) an exchange of dicarboxylates and possibly a relatively slow unidirectional transport. It is inadequate for the export of significant amounts of photosynthetic products in C_3 plants, although it could mediate the transfer of reducing equivalents across the chloroplast envelope (Fig. 12.6c). However, probably the major translocator proposed by Heldt's group is a phosphate translocator specific for dihydroxy-acetone phosphate, 3-phosphoglycerate, glyceraldehyde phosphate and inorganic phosphate, but not pyrophosphate or 2-phosphoglycerate (see Fliege *et al.*, 1978).

The most probable view of chloroplast biochemistry is that triose phosphate is the major product of photosynthesis to be exported; it is translocated in exchange for inorganic phosphate by a translocator situated in the inner membrane of the envelope (see Fig. 12.6a). Most other labelled compounds arising in the cytoplasm may be regarded as products of cytoplasmic metabolism from triose phosphate. It should be noted that the export of this triose would rapidly deplete the matrix of phosphate if it were not exchanged for extraplastidic phosphate. In light, although not in the dark, triose phosphates are exported much more rapidly than PGA. In addition to the export of products of CO_2 fixation, this translocator may also function in a phosphoglycerate–phosphoglyceraldehyde shuttle in the export of reducing equivalents and possibly in cytoplasmic ATP synthesis (see Fig. 12.6b). The phosphate translocator is inhibited by *p*-chloromercuriphenylsulphonate, pyridoxal-5-phosphate and trinitrobenzene sulphonate. Using inhibitor binding as a means of labelling the translocator, a protein (MW $= 29\,000$) has been isolated.

In the mesophyll chloroplasts of C_4 plants, Huber and Edwards (1977) have obtained evidence for a carrier-mediated transport of pyruvate which was inhibited by phenylpyruvate, oxoisovalerate, oxoisocaproate and cyano-hydroxycinnamate. These workers also showed that phosphate exchanged specifically for phosphoenolpyruvate. Thus in C_4 mesophyll chloroplasts, translocators exist to exchange pyruvate and phosphate for phosphoenol-pyruvate as required by the schemes of Hatch and coworkers (Fig. 12.6d).

Although the mitochondrion is able to export ATP to the cytoplasm, it seems unlikely that the chloroplast has the same ability, though a carrier able to import ATP in the dark has been described. This carrier, with relatively weak activity, may function to provide ATP for chloroplast metabolism in the dark. It has also been suggested that triose phosphate may be imported in the dark and converted to PGA in association with a substrate-level phosphorylation (Fig. 12.6b).

The envelope is impermeable to pyridine nucleotides, NAD and NADP. However, both NAD- and NADP-dependent malate dehydrogenase activity is found in the chloroplast and an NAD-dependent enzyme exists in the cytoplasm, making possible a malate–oxaloacetate shuttle for transferring reducing equivalents across the envelope membranes (Fig. 12.6c).

12.6   Chloroplast enzymes

The enzyme composition of chloroplasts is not as well documented as that of mitochondria. A number of metabolic systems have, however, been identified as active in the chloroplast, but in most cases a detailed examination of the enzymes involved has not been undertaken. The main metabolic systems are set out as follows.

(a) Systems involved in carbon fixation
These are well documented and have been set out elsewhere (see Tables 11.1, 11.2 and Figs. 11.5 and 11.13).

(b) Carbohydrate metabolism and galactolipid synthesis
The pathways for the synthesis of sucrose and starch have also been outlined previously (Fig. 11.10). An oxidative system for glucose-6-phosphate metabolism provides a means of reducing NADP in the dark and converting glucose-6-phosphate to pentose phosphates, enabling chloroplasts to carry out the alternative pentose phosphate pathway described in mammalian and microbial tissues.

Glucose-6-phosphate dehydrogenase (EC.1.1.1.49)

Glucose-6-phosphate + NADP $\longrightarrow$ 6-phosphogluconolactone
$$+ NADPH_2$$

6-phosphogluconolactone + H_2O $\longrightarrow$ 6-phosphogluconate

6-Phosphogluconate dehydrogenase (EC.1.1.1.44)

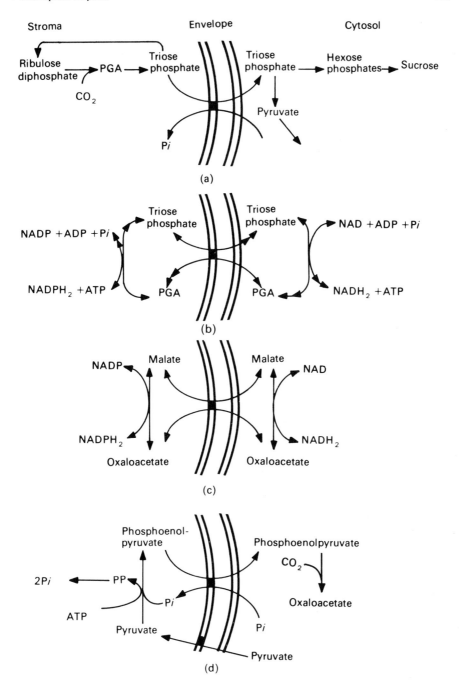

Fig. 12.6 Some proposed roles for chloroplast translocators. (a) Export of the products of CO_2 fixation in the light. (b) Export (in the light) or import (in the dark) of reducing equivalents and ATP. (c) Transfer of reducing equivalents across the cell envelope by the dicarboxylate carrier. (d) Photosynthetic oxaloacetate synthesis in C_4 plants (see Fig. 11.12).

$$6\text{-phosphogluconate} + NADP \longrightarrow \text{ribulose-5-phosphate}$$
$$+ CO_2 + NADPH_2$$

As noted earlier, galactose in the form of galactolipids is a major constituent of chloroplasts. The pathway involves the incorporation of UDP-galactose (formed from UDP-glucose by the UDP-glucose 4-epimerase) into lipid. Synthesis of galactolipids is associated with the chloroplast envelope and the following enzymes have been proposed (van Beseouw and Wintermans, 1978).

UDP-galactose-diglyceride galactosyl transferase (EC 2.4.1.−)

diglyceride + UDP-galactose $\longrightarrow$ monogalactosyl diglyceride + UDP

Galactolipid-galactolipid galactosyl transferase (EC 2.4.1.−)

2 monogalactosyl diglyceride $\longrightarrow$ digalactosyl
diglyceride + diglyceride

monogalactosyl diglyceride + digalactosyl diglyceride $\longrightarrow$
trigalactosyl diglyceride + diglyceride

The synthesis of the diglyceride is also associated primarily with the envelope. A glycerol phosphate acyltransferase is present in the stroma, but an acylglycerolphosphate acyltransferase, the acyl CoA synthetase, and the phosphatidate phosphatase are bound to the envelope which is able to synthesise diglycerides from glycerol-3-phosphate in the presence of a soluble chloroplast extract (Joyard and Douce, 1977).

(c) Fatty acid synthesis

Following the initial studies of Smirnov (1960), a fatty acid synthesising system in the chloroplast has been characterised. Broken chloroplasts will incorporate acetate or acetyl CoA into fatty acids, if the necessary cofactors (including ATP, HCO_3^- and $NADPH_2$) are added. The bicarbonate and ATP requirement can be replaced by malonyl CoA when acetyl CoA is the substrate. The probable pathway, similar to that in animals and micro-organisms, is shown in Fig. 12.7 and the major product is probably palmitate. The enzymes for fatty acid synthesis (other than those involved in malonyl CoA synthesis) are present as a multienzyme complex similar to that found in bacteria and composed of several enzymes. In isolated chloroplasts, light stimulates synthesis even in the presence of ATP suggesting that the $NADPH_2$ requirement is met by photosynthetic reduction of pyridine nucleotides

$$H_2O + NADP \xrightarrow{\text{light}} NADPH_2 + \tfrac{1}{2}O_2$$

In fact, the main cofactor requirement for lipid synthesis is similar to that of the Calvin cycle; ATP and $NADPH_2$. Unlike the mammalian fatty acid synthetase but like the bacterial system, the chloroplast enzyme complex appears to be in a readily dissociable form.

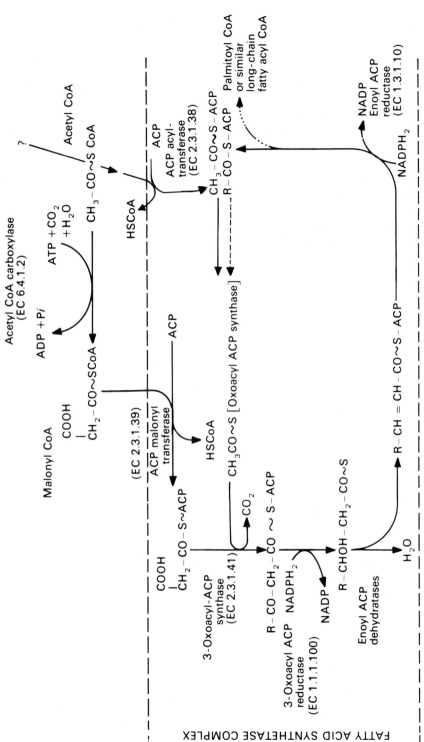

Fig. 12.7 Outline of fatty acid synthesis. Note that the main cycle of reactions of the synthase are repeated to extend the chain length by two carbons per cycle. The $NADPH_2$ and ATP requirements are met by the photosynthetic light reaction.

The chloroplasts also possess a desaturation system for formation of unsaturated fatty acids.

Although it is clear from experiments using acetyl CoA (or acetate) that chloroplasts are capable of acetate utilisation (the stroma contains an acetyl CoA synthetase), the source of the acetyl group *in vivo* is uncertain. Convincing evidence for a pyruvate dehydrogenase system or a citrate lyase system is lacking and there is no firm evidence that the acetyl group is obtained from outside the chloroplast but it has been suggested that glyoxylate could be a source of acetate. However, in developing plastids from plants grown in the dark, the envelope does appear to be permeable to acetate although this permeation ceases shortly after illumination.

(d) Isoprenoid synthesis

Chloroplasts contain a number of compounds of the isoprenoid type including carotenoids, the phytyl side chain of chlorophyll and the side chains of phylloquinones, plastoquinones and tocopherol. The conventional pathway for carotenoid synthesis is outlined in Fig. 12.8. The pathway from mevalonate to carotenes is readily demonstrable in chloroplasts although much of our detailed knowledge of the system comes from other sources, such as mammalian cholesterol synthesis and other plant and microbial systems. The origin of mevalonic acid in chloroplasts has been a problem, although $^{14}CO_2$ can readily be incorporated into carotenoids in the light. It is improbable that the mevalonate is synthesised outside the plastid, since the chloroplast envelope is impermeable to mevalonate although etioplasts are permeable. The difficulty in identifying the source of acetate in chloroplasts has been discussed above. It is possible however, that the branched-chain amino acid leucine could provide a substrate for mevalonate synthesis. The chloroplast sterols are synthesised outside the plastid from acetyl CoA through mevalonate and imported into the organelle as sterols.

(e) Porphyrin synthesis

The synthesis of chlorophyll is a property of chloroplasts. The details of the pathway involved are still uncertain, but a proposed route based in part on porphyrin synthesis in animal tissues is shown in Fig. 12.9. Although animal tissues synthesise γ-aminolaevulinate (ALA) from succinyl CoA and glycine, it is not clear whether this is the route followed in the chloroplast. Either glutamate or oxoglutarate (without succinyl CoA as an intermediate) may serve as the main source of the carbon atoms. The ALA synthetase has not been satisfactorily demonstrated in chloroplasts, but an ALA transaminase using γ, δ-dioxyvalerate has been found.

The conversion of ALA to porphobilinogen (by porphobilinogen synthase), uroporphyrinogen III and protoporphyrin IX is a property of chloroplasts. Protoporphyrin IX is then converted to chlorophyll by a pathway involving chelation of Mg and addition of phytyl pyrophosphate formed by the isoprenoid route in Fig. 12.8. Developing plastids also possess a ferrochelatase enabling them to synthesise protohaem and cytochrome prosthetic groups. While some higher plants and the majority of algae can form chlorophyll in the

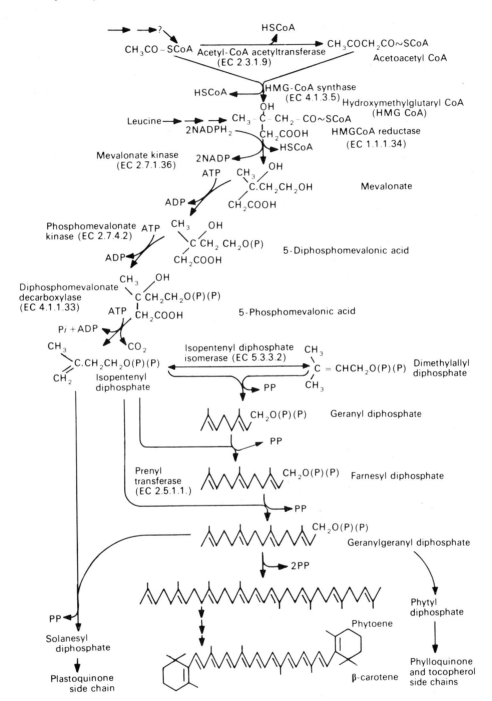

Fig. 12.8 Outline of the isoprenoid pathway. Note that this pathway is found both inside and outside the chloroplast. Chloroplast steroids are synthesised by an extraplastidic system.

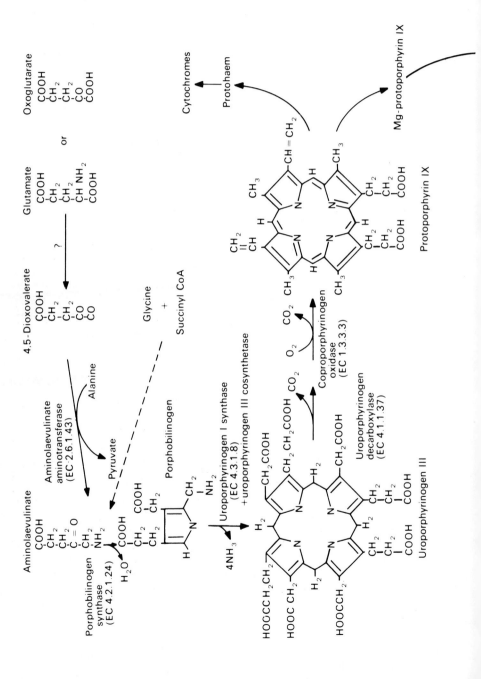

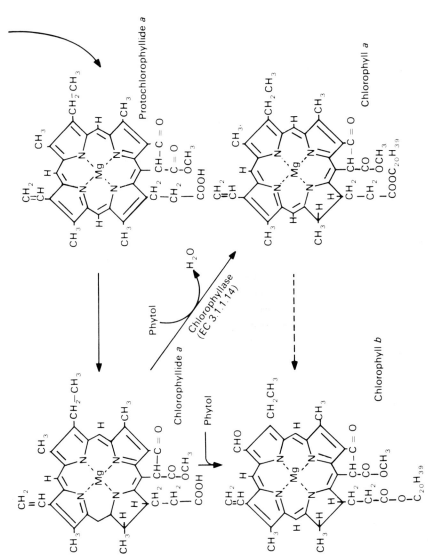

Fig. 12.9 Proposed pathway for the synthesis of chlorophylls in plastids.

dark, flowering plants do not form chlorophyll in leaves without light. Plants grown in the dark are described as etiolated and are yellow in colour; they contain low levels of carotenoid and protochlorophyllide. On illumination, the protochlorophyllide is converted first to chlorophyllide *a* and then to chlorophyll *a*. Chlorophyll *b* may be derived from chlorophyll *a*.

(f) Other reactions

A variety of claims for the existence of enzyme systems in the chloroplast have been made. At present a detailed understanding of many aspects of chloroplast metabolism is lacking. Nevertheless, the plastid, in addition to possessing the enzyme systems already described, is the site for some nitrogen metabolism including a nitrate-reducing system and glutamine synthetase, sulphate reduction and sulpholipid synthesis. An organelle-specific superoxide dismutase has also been found in the chloroplast.

12.7   Chloroplast biogenesis

(a) A system for protein synthesis

The chloroplast, like the mitochondrion, possesses its own genetic machinery for synthesising proteins. In the late 1950s and early 1960s, cytological studies suggested the presence of DNA in chloroplasts. Such a view was supported from genetic studies of plastids. In 1963, Sagar and Ishida succeeded in isolating DNA from a chloroplast preparation of the unicellular alga, *Chlamydomonas*.

The DNA from higher plant chloroplasts (ctDNA) which is attached to the membranes, consists of double-stranded molecules of which there are several copies in each plastid (of the order of 20). The molecular weight is in the range 85×10^6 to 110×10^6 and the contour length is about 40–45 μm, equivalent to approx. 1.5×10^5 base pairs. Such a molecule could code for about 125 polypeptides of MW 50 000. In higher plants the DNA is not readily distinguishable from other DNA in the cell as its buoyant density (approx. 1.697 g/cm^3) is very close to that of the nuclear DNA (1.695) and mitochondrial DNA (1.706) (Kolodner and Tewari, 1975). In algae, where the nuclear DNA is of higher density, ctDNA is distinguishable. The chloroplast possesses its own DNA polymerase, which appears to be distinct from the enzymes found in other cell fractions. Replication is probably initiated by the formation of two D-loops on opposite strands (see sect. 5.2 (a)).

Photosynthetic cells can be shown to possess three types of ribosome, the cytoplasmic (80S), mitochondrial (60S in higher plants and 70–80S in lower organisms) and those in the chloroplast (70S with 50S and 30S subunits). Nearly half of the leaf ribosomes are found in the chloroplast and a negligible number in the mitochondria, the remainder being cytoplasmic. The ribosomal RNAs can also be distinguished by their sedimentation characteristics; cytoplasmic RNA 25S, 18S and 5S; mtRNA 21S and 16S and chloroplast rRNA 23S, 16S, 5S and 4.5S. The 4.5S RNA is probably missing from algal chloroplasts. Plastid ribosomes have two subunits but are smaller than the cytoplasmic ones and contain fewer proteins (about 55 as opposed to 75–100 in the cytoplasmic ribosomes).

Chloroplasts possess their own aminoacyl tRNA synthetases and tRNAs which are distinct from the cytoplasmic and mitochondrial ones. Although the plastids of mature leaves synthesise little protein (or RNA), particles from young expanding leaves can be shown to synthesise several of their own proteins. In general the protein-synthesising system shows similarities with the mitochondrial and prokaryotic rather than the cytoplasmic system.

Some ribosomes are bound to thylakoid membranes, while others are free in the stroma. There is some evidence to suggest that these two types are functionally different, the former being involved in the synthesis of membrane proteins and the latter in soluble proteins such as Fraction 1 protein (ribulose diphosphate carboxylase).

(b) Gene products

A DNA-dependent RNA polymerase which, unlike the nuclear enzyme, is rifampicin-sensitive, is found in the chloroplast. Isolated chloroplasts have been shown to incorporate ribonucleotide triphosphates into their RNA. Both chloroplast rRNA and tRNA have been shown to hybridise with chloroplast DNA. As with other types of rRNA synthesis, both the rRNAs are derived from a common precursor. Thus there is now convincing evidence that the chloroplast synthesises its own RNA on the basis of information coded in the plastid DNA.

Protein synthesis is light-dependent and in intact chloroplasts it is sensitive to inhibitors of photophosphorylation. Light energy can be replaced in part by ATP. Chloramphenicol and lincomycin, inhibitors of prokaryotic protein synthesis, inhibit synthesis. With lysed chloroplasts, protein synthesis can be demonstrated if ATP and GTP are present. Protein synthesis is initiated by *N*-formylmethionine (as opposed to cytoplasmic synthesis which is initiated with unformylated methionine) and involves chloroplast-specific elongation factors.

Using labelled substrates for chloroplast protein synthesis, label can be found incorporated into polypeptides in the stroma, thylakoid membranes and envelope. In the stroma, the large subunit of Fraction 1 protein (ribulose diphosphate carboxylase) is synthesised in the plastid, but the small subunit is of nuclear–cytoplasmic origin. The membrane-bound polypeptides that are synthesised by chloroplasts themselves (at least five in the thylakoids and two in the envelope) are being identified but include some subunits of the chloroplast ATPase, chlorophyll–protein polypeptides and possibly cytochrome *f*. Indeed some experimental approaches suggest that chloroplasts may be able to synthesise a very much larger number of polypeptides. However, some subunits of the ATPase, the RNA polymerase, the tRNA synthetases (most or all), the chlorophyll *a/b* binding proteins (see sect. 13.3 (c)), ferredoxin, ribosomal proteins and the enzymes for some isoprenoid biosynthesis are products of extraplastidic protein synthesis. In the case of the ribulose diphosphate carboxylase, where the large subunit is synthesised within the chloroplast and the small subunit in the cytoplasm, it has been shown that the cytoplasmic precursor of the small subunit is significantly larger than the mature polypeptide. The precursor is transported into the chloroplast where it is modified and combined with the large subunit to form the carboxylase (Chua and Schmidt, 1978). As with the mitochondrial system, most of the proteins found in chloroplasts are synthesised outside the particle. Such a situation poses problems of regulation of synthesis, which were outlined for the mitochondrion and also apply here.

The regulation of synthesis of a functional chloroplast can be studied in etiolated cells which are allowed to green. The process is not fully understood, but the various stages in greening are being documented. For example, the synthesis of chlorophyll, which is light-triggered, is not tightly coupled to the synthesis of polypeptides in PS I (pigment system I). Under appropriate conditions, polypeptides associated with the pigment system are synthesised early, while chlorophyll synthesis continues over a much longer period and photochemical activity of the system continues to rise well after polypeptide synthesis is complete.

(c) Mapping the chloroplast DNA

As with the mitochondrial system, considerable progress has been made in mapping the ctDNA. Studies with restriction endonucleases *Eco*Rl and *Bam*HI have enabled basic maps of the genome to be drawn (see Malnoe and Rochaix, 1978). Two copies of the genes for rRNA have been found by hybridisation studies. Each group appears to consist of the sequence: 16S-(2100 base-pair spacer)-23S-5S. In addition more than 20 genes coding for tRNAs have also been located. As noted above, the large subunit of the ribulose diphosphate carboxylase is a product of chloroplast protein synthesis. Studies on the transcription *in vitro* and translation of restriction endonuclease fragments of ctDNA are leading to the identification of the gene coding for this protein. In maize chloroplasts, a membrane-bound polypeptide (MW = 32 000) whose synthesis is light-induced, has been shown to be a product of a specific restriction endonuclease fragment (Bedbrook *et al.*,1978).

12.8   Evolution of chloroplasts

The arguments for the origin of chloroplasts from more primitive plastid-less systems are comparable to those already advanced for the origin of mitochondria. Thus chloroplasts may be regarded as being derived from endosymbiotic prokaryotes. Obvious candidates for such prokaryotes are blue-green algae (cyanobacteria) which, unlike the mechanistically simpler photo-synthetic bacteria, are able to evolve oxygen photosynthetically. The simpler members of this group are unicellular and do not possess chloroplasts. However, they do have simple thylakoids which appear singly in the cell but do not aggregate in twos or threes (as in the more advanced red algae) and show no stacking (see sect.15.1). The structure of the DNA, the pathways of carbon fixation, photosynthetic electron transport and protein synthesis in these organisms are generally similar to those found in the higher plant chloroplast. A study of base sequences of the 16S RNA from chloroplasts of the unicellular eukaryote *Euglena* and from the red alga *Porphyridium* and comparison with the corresponding RNA from *Anaystis* (a blue-green alga) shows sufficient homology to support this view of the phylogenetic origin of the chloroplast. Homology between the 18S cytoplasmic ribosome and the 16S chloroplast RNA is not seen to any great extent. Hence this hypothesis of the origin of the chloroplast suggests that primitive blue-green algae (related to those found today) are the forerunners of the endosymbiotic organelle. Supporters of this theory also draw attention to the existence of algal symbionts which can be found, particularly in simpler invertebrates.

The alternative view (see Klein, 1970) is that eukaryotes are derived directly from prokaryotes with a later partitioning of the genome between nucleus and chloroplast. The progressive modification of the blue-green algal cell and its thylakoids may be seen as the source of the chloroplast-containing eukaryotic cell.

Further Reading

Anderson, J. M. (1975) The molecular organisation of chloroplast thylakoids. *Biochim. Biophys. Acta*, **416**, 191–235.

Arntzen, C. J. (1978) Dynamic structural features of chloroplast lamellae. *Current Topics in Bioenergetics*, **8**, 111–60.

Bedbrook, J. R. and Kolodner, R. (1979) The structure of chloroplast DNA. *Annu. Rev. Plant Physiol.*, **30**, 593–620.

Ellis, R. J. (1977) Protein synthesis in isolated chloroplasts. *Biochim. Biophys. Acta*, **463**, 185–215.

Gillham, N. W., Boynton, J. E. and Chua, N. H. (1978) Genetic control of chloroplast proteins. *Current Topics in Bioenergetics*, **8**, 211–60.

Heber, U. (1974) Metabolite exchange between chloroplasts and cytoplasm. *Annu. Rev. Plant Physiol.*, **25**, 393–421.

Kung, S. (1977) Expression of chloroplast genomes in higher plants. *Annu. Rev. Plant Physiol.*, **28**, 401–37.

Chapter 13

Chloroplast photochemistry

13.1 Light energy

Fundamentally, photosynthesis is the conversion of light energy into chemical energy. Light is absorbed by the pigment molecules and the energy is utilised in the synthesis of ATP and the reduction of NADP. We turn first to a simple consideration of light and its absorption by pigment molecules.

Light may be described as an electromagnetic wave or as a stream of particles (photons). For our purpose it is the latter which will be most useful since molecules absorb light in packets (quanta). The energy in a quantum of light (E) may be represented by

$$E = hv$$

where h is Planck's constant and v the frequency. The frequency is related to the wavelength by

$$\lambda \cdot v = c$$

where λ is the wavelength and c is the velocity of light in a vacuum.[1] We can therefore write

$$E = \frac{h \cdot c}{\lambda}$$

Thus the energy of the quantum is inversely proportional to wavelength, excitation of a molecule by blue light (shorter wavelengths, e.g. 450 nm) involving more energy than that by red light (e.g. 650 nm). Since it is useful to measure the energy of light in chemical terms, we define the energy of one Einstein (one 'mole' of quanta) as the energy in one quantum multiplied by Avogadro's number ($N = 6 \cdot 10^{23}$) equal to $\dfrac{hc}{\lambda} \cdot N$.

Since Planck's constant is $6.6 \cdot 10^{-34}$ J s and the velocity of light is 3.10^{10} cm s^{-1}, the energy in one Einstein of light can be given by

$$\frac{6.6 \cdot 10^{-34} \cdot 3.10^{10} \cdot 6.10^{23}}{\lambda \, 10^{-7}} \, J = \frac{1.18 \cdot 10^5}{\lambda} \, kJ$$

---

[1] The velocity of light c is ideally measured in a vacuum, but the value in air is approximately the same.

where λ is the wavelength in nm. Thus one Einstein of 700 nm light is equivalent to 169 kJ.

13.2   Excitation of organic molecules

(a) Excited singlet states

The absorption spectrum of an organic molecule represents its ability to absorb light quanta of different wavelengths. Thus chlorophylls *a* and *b* have a spectrum with two major peaks of absorption, one at the red end and the other at the blue end of the visible spectrum (Fig. 13.1). Molecules without activation are said to be in the ground state. Normally this is a singlet state where electrons can be grouped in oppositely directed or antiparallel pairs (spin magnetic quantum numbers for each pair are $m_s = +\frac{1}{2}$ and $-\frac{1}{2}$ thus cancelling each other). The absorption of a quantum of red light by chlorophyll raises the energy level of the molecule to that of the first excited singlet state. This excited state is normally unstable and energy equivalent to that of the absorbed quantum may be emitted as the molecule returns to the ground state. Because of the complexity of the electronic states of organic molecules, there are, in fact, many sub-states in both the ground and the excited states with a consequent broadening of the absorption bands (see Fig. 13.2). If one considers the case of chlorophyll *a* with an absorption band at the red end of the spectrum (about 660 nm in organic solvents such as petroleum), the first excited state will have an energy level of approximately 179 kJ/mol above the ground state, that is the energy derived from one quantum of 660 nm light. Such excited molecules have a short life (about 10^{-9} s) and may lose their energy in one of several ways.

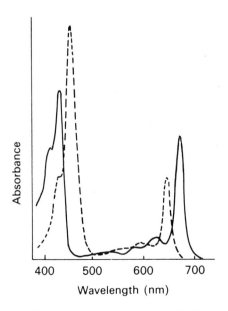

Fig. 13.1 Absorption spectra of chlorophyll *a* (———) and chlorophyll *b* (- - - -).

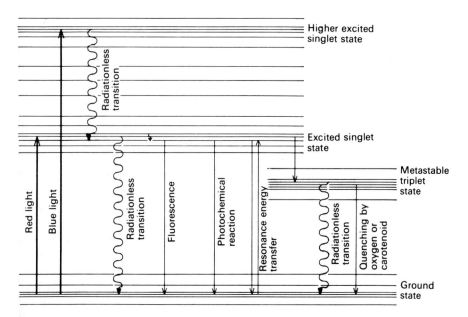

Fig. 13.2 Excitation and de-excitation of chlorophyll.

(i) Radiationless de-excitation. Energy may be lost as heat and the molecule returns to the ground state. The energy is lost in a series of small de-excitation steps, each one emitting a quantum of low energy which may be envisaged as heat.

(ii) Fluorescence. Energy may be re-emitted as a quantum of light. Usually this is at a slightly longer wavelength and thus represents a quantum of slightly lower energy. Following absorption of the exciting quantum, some rearrangements within the molecule associated with small radiationless transitions to slightly lower energy states precede emission of the quantum as fluorescence. Only a small percentage of the excited states decay by fluorescence.

(iii) Photochemistry. The molecule may return to the ground state through involvement in a photochemical reaction.

These three processes are in competition for the excited state of the molecule. Thus in a population of excited molecules in which all three processes are occurring, inhibition of the photochemical reaction increases the fluorescence. Thus fluorescence measurements can be a useful guide to the photochemical activity of a population of pigment molecules.

(b) Higher excited states

So far we have considered only the consequences of absorbing a quantum of red light. However, chlorophyll also has a conspicuous blue absorption peak and the absorption of a quantum of blue light will raise the molecule to a higher excited state. The only fate of this higher excited singlet state is transition to the first excited singlet state by a number of small steps releasing small amounts of energy, i.e. through a series of sub-states (see Fig. 13.2). Because this pathway is an easy one, the higher excited states have very short lifetimes (of the order of 10^{-12} s) and conversion to the lower state occurs before fluorescence or a photochemical reaction can occur. Thus blue quanta, although possessing more energy, do not contribute more energy for photochemical purposes than red quanta.

(c) Energy transfer

A further possible fate of the excited state is transfer of the energy to an adjacent pigment molecule. The molecule to which energy is transferred may be one of the same species or of a different species. For example, irradiation of a concentrated solution of chlorophylls *a* and *b* at a wavelength absorbed primarily by chlorophyll *b* will give chlorophyll *a* fluorescence.

For efficient transfer between molecules of different types, the absorption band of the acceptor should overlap the fluorescence band of the donor and the molecules should be close together. Efficiency of transfer varies inversely with the sixth power of the distance between donor and acceptor. Energy will transfer from a donor pigment to a recipient with an absorption maximum at longer wavelength, requiring quanta of slightly lower energy value for excitation. 'Uphill' transfer can occur when the difference in absorption maxima is not great and can be met by thermal energy, but this process is less likely. It should be noted that transfer is far too efficient to be explained in terms of fluorescence of the donor and absorption of the fluorescent quantum by the acceptor.

(d) Triplet states

We have so far examined excited singlet states where all the electrons are in pairs of opposite spin quantum number (antiparallel spin). De-excitation of excited singlet chlorophyll can occur through a metastable triplet state with a long life (of the order of 10^{-5} s) where one pair of electrons have more or less parallel spins. Such a long-lived state may be able to undergo other reactions such as the transfer of energy to oxygen. Oxygen can be said to quench the excited triplet state of chlorophyll and singlet oxygen, a very reactive species of oxygen, is formed. Oxygen is somewhat unusual in that its ground state is a triplet; reaction with triplet chlorophyll gives singlet oxygen. This type of reaction is highly undesirable in a chloroplast as it leads to a destructive oxidation of membrane lipids and other chloroplast components including chlorophyll itself. A major function of carotenoids in chloroplasts is to quench the triplet state of chlorophyll, and also singlet oxygen, the energy being lost as heat.

13.3 Pigments

(a) Pigment distribution

As shown in Table 13.1, the light-absorbing pigments of higher plants are chlorophylls *a* and *b* (Fig. 12.9; see Gloe *et al.*, 1975) and the carotenoids. The main components of the latter group are β-carotene, lutein, neoxanthin and violaxanthin. A similar pigment composition is found in green algae (Chlorophyta) and in bryophytes and ferns. The other algae, however, show a variety of pigment compositions.

Reference to Table 13.1 shows that all groups of plants capable of oxygen-evolving photosynthesis possess chlorophyll *a* together with other pigments. The loss of chlorophyll *a* (by mutation) is associated with a loss of photosynthetic ability and this substance therefore has a primary function in photosynthesis. This is not true of the other pigments and they have been described as accessory pigments whose role is to absorb light mainly in regions of the spectrum where chlorophyll *a* absorption is poor and to transfer the energy to chlorophyll *a*. Thus in green algae (Chlorophyta) and higher plants, chlorophyll *b* and carotenoids fulfil this function, while in some other groups (e.g. brown algae, Phaeophyta) the carotenoid fucoxanthin transfers its energy with high efficiency to chlorophyll *a*. The biliprotein pigments, phycocyanin, phycoerythrin and allophycocyanin, act as accessory pigments in red (Rhodophyta) and blue-green algae (Cyanophyta or Cyanobacteria). Red algae possess a predominance of phycoerythrin, while most blue-greens possess phycocyanin, but this is not an invariable rule. In red algae such as *Porphyridium*, phycoerythrin is the major light-absorbing pigment and contributes energy to both pigment systems. The biliproteins are globular proteins which have tetrapyrrole prosthetic groups (see Fig. 13.3). These pigments, unlike the chlorophyll–protein complexes discussed below, are readily extracted by water or dilute salt solutions. The bilin pigments are located in large pigment–protein complexes or particles attached to the surface of the thylakoid membranes. These particles are termed phycobilisomes and are found in red and blue-green algae. The question of how energy is passed

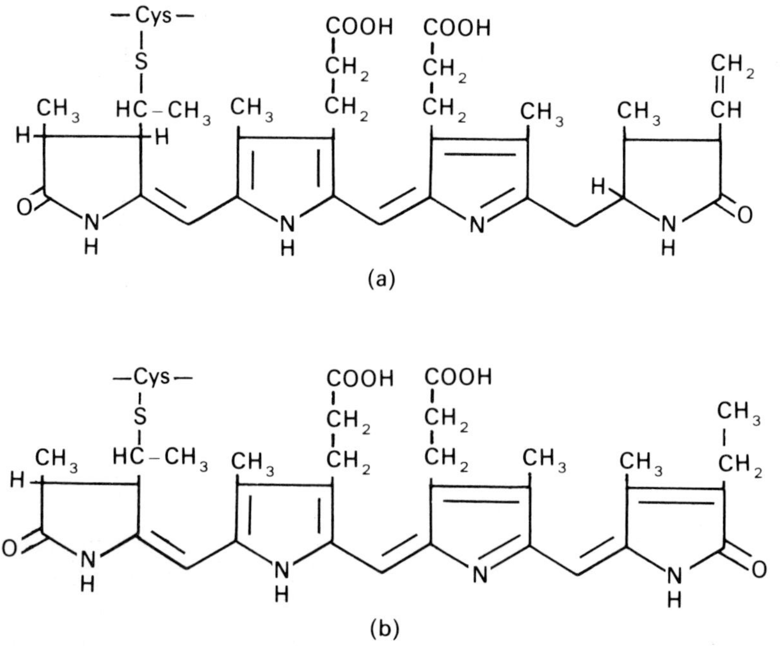

Fig. 13.3 Prosthetic groups of biliproteins: (a) phycoerythrobilin; (b) phycocyanobilin. The prosthetic groups are probably attached to a cysteine residue of the protein.

Table 13.1 Pigment distribution in photosynthetic eukaryotes

	Higher plants Bryophytes Pteridophytes	Algae					
		Chlorophyta Euglenophyta	*Phaeophyta*	*Rhodophyta*	*Cyanophyta*	*Pyrrophyta*	*Chrysophyta*
Chlorophylls *a*	+	+	+	+	+	+	+
b	+	+	−	−	−	−	−
c	−	−	+	−	−	+	+
d	−	−	−	+	−	−	−
Phycobilins							
Phycoerythrin	−	−	−	+	(+)	+	−
Phycocyanin	−	−	−	(+)	+	+	−
*Carotenoids							
Fucoxanthin	−	−	+	−	−	−	+
Peridinin	−	−	−	−	−	+	−

* The carotenoids are universally present in naturally occurring photosynthetic cells. Those found in higher plants and green algae are β-carotene, lutein, neoxanthin and violaxanthin. Other groups contain β-carotene (or occasionally α-carotene) and xanthophylls of varying structure (see Goodwin, 1976).

from discrete particles to chlorophylls embedded in the membrane has been puzzling, but a possible close relationship between phycobilisomes and the membrane-bound pigments is suggested in Fig. 13.4. The transfer of energy to chlorophyll *a* in *Porphyridium* has been shown to be highly efficient, the sequence being: phycoerythrin → phycocyanin → allophycocyanin → chlorophyll *a*.

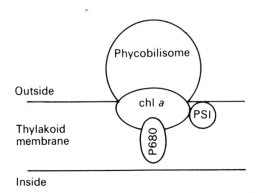

Fig. 13.4 Relation of phycobilisomes to thylakoid membranes (modified from Mimuro and Fujita, 1978).

(b) Forms of chlorophyll

Higher plants and green algae are characterised by two chlorophylls of very similar structure – *a* and *b*. In solution in organic solvents such as ethanol, chlorophyll *a* has a red absorption maximum at 662 nm and chlorophyll *b* at 644 nm. However, spectroscopy of chloroplast membranes or membrane particles gives evidence of several peaks of chlorophyll *a* absorbance at low temperature. Each of these peaks is thought to represent a species of the pigment in a particular environment or with a specific binding to protein. The latter appears unlikely in view of the nature of the chlorophyll proteins discussed below. Thus there is evidence for Chl a_{660}, Chl a_{670}, Chl a_{680}, Chl a_{685}, Chl a_{690} and Chl $a_{700-720}$. Chlorophyll *b* absorbs at about 650 nm although there has also been a suggestion for the existence of Chl b_{640}.

The main pigments in PS II are Chl *b*, Chl a_{660}, Chl a_{670}, Chl a_{680} and Chl a_{685}. The same forms of chlorophyll are found in PS I, but this system also includes Chl a_{690} and Chl $a_{700-720}$. Distribution between the pigment system is such that PS I absorbs at longer wavelengths than PS II. The action spectrum for PS II activity shows a maximum at 677 nm with a shoulder at 650 nm (due to substantial amounts of Chl *b*), while PS I shows a maximum at 681 nm and extends into the infra-red (above 700 nm), unlike PS II.

During the greening process, the two main forms of chlorophyll *a* are synthesised successively. Chl a_{680} is formed first and Chl a_{670} is formed second, possibly from the former. Longer wavelength forms are found later with the onset of photosynthetic activity.

The fluorescence of the chloroplast pigments of higher plants and green algae has a main band at 685 nm and is attributed to Chl a_{680}. Most of the emission originates from PS II but weak fluorescence in the 710–715 region is attributed to PS I (Fig. 13.5). Carotenoid pigments do not fluoresce.

(c) Chlorophyll proteins

Extraction of thylakoid membranes and fractionation has resulted in the isolation of two spatially defined chlorophyll–protein complexes. CP I (chlorophyll–protein I), which is found in a wide variety of oxygen-evolving photosynthetic organisms including red and blue-green algae, has a molecular weight of the order of 100 000 to 150 000, possesses the reaction centre for PS I (P700), about 30 molecules of chlorophyll, one molecule of β-carotene and one or two polypeptides of molecular weight about 64 000. The Chl a_{660}, Chl a_{670}, Chl a_{680} and Chl a_{685} spectral forms of chlorophyll can be detected. If the complex is so prepared that it retains a non-haem iron centre, the photo-oxidation of the reaction centre, P700, may be observed. The polypeptides are synthesised on thylakoid ribosomes.

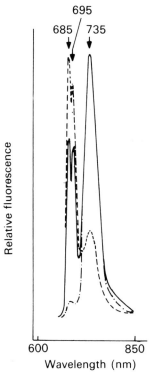

Fig. 13.5 Chloroplast fluorescence at 77 K. (Chloroplast fragments ——, PS II particles −−−, PS I particles −·−·−·) (after Bril *et al.*, 1969).

CP II (chlorophyll *a/b* complex) has a minimum molecular weight of 27 000–32 000 in higher plants and may be present as a dimer. The complex represents up to about 60 per cent of the total chlorophyll, chlorophylls *a* and *b* being present in approximately equal amounts. Spectrally Chl b_{650}, Chl a_{660}, Chl

a_{670}, Chl a_{680} and Chl a_{685} appear to be involved. Unlike CP I, the polypeptides (MW 20000–25000) are synthesised in the cytoplasm. Mutants lacking CP II possess photosynthetic activity and it is therefore concluded that CP II represents an antenna pigment complex (light harvesting) in contrast to CP I which includes the essential reaction centre. In the alga *Acetabularia*, a variant chlorophyll a/b protein complex is found of molecular weight 67000, containing similar but not identical polypeptides of 21500 and 23000 molecular weight.

More recently a third pigment–protein complex, containing only chlorophyll a and associated with a polypeptide of molecular weight 42000, has been described by Henriques and Park (1978).

In etioplasts (plastids of plants grown in the dark), a precursor of chlorophyll, protochlorophyllide a (see Fig. 12.9), is found associated with a protein. Illumination of plants grown in the dark results in chlorophyll synthesis and the conversion of protochlorophyllide a to chlorophyllide a. It has been suggested that the protochlorophyllide-a–protein complex is composed of 10 protein subunits (MW = 60000) and 10 pigment molecules.

13.4   Development of concepts of pigment function

Early work by Emerson and Arnold (1932b) using algae (*Chlorella*) showed that only a very small proportion of the chlorophyll molecules were involved in photochemical reactions. These workers used short flashes of light (10^{-5} s) with adequate dark periods between the flashes to ensure that the dark reactions were not rate-limiting, and measured the oxygen evolved for each flash. They found that the maximum oxygen yield per flash was 1 molecule for every 2480 chlorophyll molecules. Although not properly understood at the time, these experiments, confirmed later by de Kouchkovsky and Joliot (1967) for maize chloroplasts, laid the basis for the concept of the pigment system. A large number of pigment molecules act as a means for harvesting light energy for transfer to one chlorophyll molecule so located that it can carry out a photochemical oxidation–reduction reaction.

In the 1940s, Emerson commenced an investigation of the efficiency with which light of various wavelengths is used by photosynthesising systems. Emerson found that far-red light on its own was very inefficiently used for photosynthetic oxygen evolution, although at these wavelengths chlorophyll a was the only light absorbing pigment (see Fig. 13.6a). In the late 1950s, Emerson and coworkers showed that although far-red light was relatively inefficient when used on its own, efficient photosynthesis could be obtained if far-red light were supplemented with weak white light (the enhancement effect). The white light could be replaced by monochromatic light at the blue end of the spectrum or even red light at 644 nm (Emerson and Rabinowitch, 1960). These experiments were interpreted as showing the necessity of activating two pigment systems for normal photosynthesis, one being activated by far-red light and the other by shorter wavelengths. Oxygen evolution was found to be more closely associated with the shorter wavelength system, PS II.

Later the use of dyes such as dichlorophenol indophenol (DCPIP) made possible a clearer demonstration of two separate pigment systems. Red light (rather than far-red light) could be shown to activate a reduction of the dye coupled with oxygen evolution in chloroplast preparations.

$$H_2O \xrightarrow{\quad e \quad} PS\ II \xrightarrow{\quad e \quad} DCPIP$$
$$\searrow \qquad \begin{bmatrix} red \\ light \end{bmatrix}$$
$$O_2$$

If DCMU, 3-(3,4-dichlorophenyl)1,1-dimethyl urea, which inhibits PS II, is added to chloroplasts, oxidation of dichlorophenol indophenol coupled to NADP reduction can be demonstrated. Such a system can be activated effectively by far-red light (Losada *et al.*, 1961)

$$ascorbate \xrightarrow{\quad e \quad} DCPIP \xrightarrow{\quad e \quad} PS\ I \xrightarrow{\quad e \quad} NADP$$
$$\begin{bmatrix} far\text{-}red \\ light \end{bmatrix}$$

Thus the thylakoid pigments are seen as being distributed in two pigment systems with slightly different absorption properties. In each system, most of the pigment molecules will be involved not in catalysing a photochemical reaction, but in gathering light energy for a reactive pigment, the reaction centre. Since the evolution of a molecule of oxygen from water (requiring 2480 chlorophyll molecules, according to Emerson and Arnold, 1932b) will involve four electron transfers, each electron being transferred through two light reactions, a pigment system would contain on average $2480/(4 \times 2) = 310$ chlorophyll molecules. That is to say about 300 chlorophyll molecules compose a pigment system capable of catalysing an electron transfer if activated with a quantum of light of suitable wavelength. Such a figure may be a little high, but it indicates the approximate size of the chlorophyll pool in a photosystem.

13.5 Thylakoid photochemistry

(a) Antenna molecules and reaction centres

Emerson and Arnold's early experiments led to the conclusion that large numbers of chlorophyll molecules were grouped together for the photochemical reactions of photosynthesis which were measured as oxygen evolution. Numerous other experimental approaches confirm this view. In fact almost all the chlorophyll molecules in complexes of 200–300 molecules serve as a means of absorbing light energy and then transferring the energy to specific molecules which can undergo a photochemical reaction. The former molecules, which absorb light energy from the environment, are referred to as antenna chlorophylls and the latter as the reaction centre. Collectively, the chlorophylls with their ancillary molecules make up the pigment system. The antenna chlorophylls are the main source of fluorescence. Most of the observed fluorescence comes from PS II at normal temperatures and has a maximum at 685 nm, although a small amount at about 740 nm is attributed to PS I. At low temperatures (77 K) fluorescence comes from both PS II and PS I (Fig. 13.5). The first steps in the photosynthetic process are:

(a) the absorption of a quantum of light by a pigment molecule;
(b) the transfer of the absorbed energy to other pigment molecules;
(c) the receipt of the energy by a reaction centre molecule capable of undergoing de-excitation through a photochemical reaction.

We have already considered absorption of the quantum and we turn now to the question of energy transfer within the pigment system.

(b) Energy transfer

De-excitation of one chlorophyll can be achieved by excitation of another. Thus a quantum of light energy can wander through the mass of similar, spatially related, pigment molecules. We noted earlier that transfer from one pigment to another of a different type is more efficient when the absorption maximum of the recipient is at a longer wavelength. Thus transfer from Chl a_{670} to Chl a_{680} is more likely than the reverse, although 'uphill' transfer has been demonstrated in photosynthetic systems.

Thus a quantum being transferred among the pigments of the thylakoid will tend to move to those of longer wavelength absorption. If the molecule capable of carrying out a photochemical reaction has a long wavelength absorption maximum, energy absorbed by other pigments will migrate eventually to such a reaction centre molecule. Thus on theoretical grounds we would expect to find reaction centre molecules among those of long wavelength absorption; as we will see, this is in general true. Failure to transfer energy to the reaction centre will result in increased fluorescence of the antenna molecules.

(c) Action spectra

An action spectrum, a measure of photochemical activity against wavelength, can be defined as the reciprocal of the relative number of light quanta needed to bring about an effect of a certain size and is plotted against wavelength. If whole systems are used, photosynthesis can be measured as oxygen evolution. An example of an action spectrum for photosynthesis is shown in Fig. 13.6 where a unicellular green alga with a pigment composition similar to that of the higher plant chloroplast, has been used. An absorption spectrum is also included and the two curves are normalised at 675 nm. It can be seen that the absorption curve and action spectrum match between 580 and 680 nm. In the carotenoid region (blue end of the spectrum) light is not as efficient in promoting photosynthesis showing that carotenoids are not efficient in the transfer of their energy to chlorophyll. Above 680 nm, the efficiency is also low because far-red light activates PS I and not PS II, which lacks long wavelength chlorophylls. This 'red drop' was the basis for the Emerson enhancement effect discussed earlier.

An action spectrum for a red alga, *Porphyra*, which possesses phycoerythrin as its main accessory pigment, is shown in Fig. 13.6. The absorption spectrum of the thallus (the whole multicellular alga) is given, together with the absorption spectrum of phycoerythrin. It is clear that the most effective wavelengths are those activating phycoerythrin, while chlorophyll activation seems relatively ineffective. Here it is necessary to assume that light absorbed by phycoerythrin can activate both pigment systems.

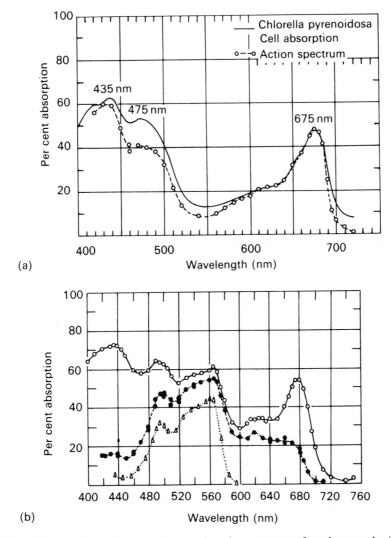

Fig. 13.6 (a) Absorption spectrum and action spectrum for photosynthesis of the unicellular green alga *Chlorella* (after Haxo, 1960). (b) The absorption spectrum of the thallus (——), the action spectrum for photosynthesis (– – –) and the absorption of an aqueous extract of the phycoerythrin (. . . .) in a red alga (*Porphyra*) (after Haxo and Blinks, 1950).

(d) Evidence for energy transfer

The transfer of energy from accessory pigments was first demonstrated in a diatom by Dutton *et al.* (1943). In this organism fucoxanthin is the main accessory pigment. Activation of fucoxanthin at 546 nm gives chlorophyll fluorescence which is as intense as that obtained by activating chlorophyll itself. Studies such as these, relying on chlorophyll fluorescence have been pursued with a variety of systems. It has been concluded that chlorophyll *b* transfers energy to chlorophyll *a* with almost 100 per cent efficiency while the phycobilins are less efficient but may still transfer up to 90 per cent. The carotenoid fucoxanthin found in some groups of algae also has a high efficiency of transfer. In higher plants other oxygen-containing carotenoids (xanthophylls) transfer energy weakly. The hydrocarbon carotenoid, β-carotene, is relatively efficient in this respect. However, it should be noted that a proportion of the xanthophylls are located in the envelope of the chloroplast where there is no chlorophyll.

(e) Transfer of energy in pigment systems

As noted earlier, there are two pigment systems, PS I and PS II. Their composition is shown diagrammatically in Fig. 13.7. Energy taken up by pigments absorbing shorter wavelengths will be distributed among those pigments or be transferred to those absorbing longer wavelengths, progressively moving to the reaction centres P700 or P680. A computer calculation of the rate of energy transfer to the reaction centre has shown that the existence of multiple spectral forms of chlorophyll substantially increases the rate of energy transfer to the reaction centre (Seeley, 1973). There are two main ways in which models have been developed for the process of energy migration through a pigment system. Firstly, we will consider a simple resonance energy transfer process similar to that which can occur in solutions; this will be efficient in the thylakoid membranes where the pigment molecules are closely packed together, since the efficiency of transfer is inversely proportional to the sixth power of the distance between the molecules. Excitation energy can randomly wander through the pigment system until it reaches a reaction centre. Here the coupling between molecules can be described as weak.

An alternative model is based on the assumption that the pigment molecules are so closely packed that there is strong intermolecular coupling and the excitation energy cannot be regarded as localised in a specific molecule, but delocalised over the aggregate. The absorption of a quantum of light by such an aggregate of organic molecules results in a distribution of the excitation energy in much the same way as the energy absorbed by a single isolated molecule is delocalised over that molecule. The model considers energy transfer in terms of 'excitons', delocalised excitations. However, the aggregate may be quenched by a reaction centre, the energy being now localised at the reaction centre. The second model gives a faster rate of energy transfer, the rate depending on the intermolecular coupling strength.

A variety of other models have been explored including transfer via triplet states and by semiconductor mechanisms (see Borisov and Godik, 1973). However, those dependent on singlet excitation and resonance energy transfer are accepted most widely.

The above considerations have assumed a more or less homogeneous distribution of chlorophylls and related pigments through which energy is transferred (Fig. 13.7). However, the picture is clearly more complex since the pigments are aggregated in complexes such as CP II, each complex possessing chlorophylls of various absorption maxima. In this case energy transferred between the pigments would reach the forms absorbing longer wavelengths by a

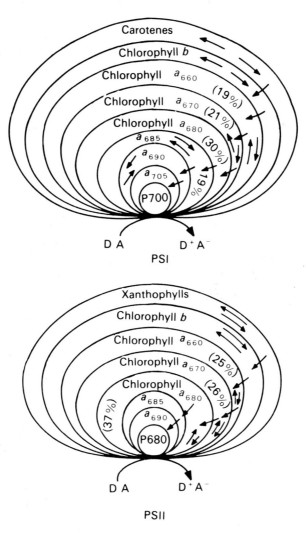

Fig. 13.7 Diagrammatic representation of chloroplast pigment systems. Absorption of a quantum of light by any pigment in the complex will be followed by transfer of the energy to adjacent pigment molecules. When transfer to a pigment absorbing at longer wavelengths occurs, there is a lower probability of transfer back to the pigment absorbing at shorter wavelengths. This process will result in energy transfer to pigments of long wavelength absorption and finally to the reaction centre (P700 or P680) where an oxidation–reduction will take place. Note that the overall absorption spectrum of PS I includes longer wavelengths than does that of PS II. The approximate percentage composition of the spectral forms of chlorophyll *a* is shown in brackets.

much quicker route. Energy transfer has been demonstrated with pigment–protein complexes. In algae containing the carotenoid peridinin as a prosthetic group, a complex containing four peridinin molecules, one chlorophyll and one protein molecule has been isolated. Here it has been demonstrated that excitation energy is transferred efficiently from peridinin to chlorophyll. A detailed knowledge of energy transfer will require, among other things, a much better understanding of the detailed structure of the thylakoid membrane.

(f) 'Puddles' and 'lakes'

The photosystems of the thylakoid can be related in two opposing ways. Firstly the energy absorbed by antenna molecules in a complex may be available to that pigment system and to no other. Alternatively, the energy trapped in the antenna molecules of a pigment system may be available to a number of reaction centres, that is a reaction centre may be thought of as being set in a 'lake' of antenna molecules. The first alternative envisages a number of discrete 'puddles'.

It is of course possible to combine these theories in intermediate models in which, for example, there could be discrete units each containing several reaction centres. Although at present there is uncertainty as to which models reflect the *in vivo* situation, there is evidence to support the notion of a small number of reaction centres sharing the excitation energy from antenna complexes.

(g) The 'spillover' or 'separate package' hypothesis

As noted earlier, the absorption maximum of pigment system II is at a shorter wavelength than pigment system I. Can excitation energy be transferred from PS II to PS I? Since selective activation of PS II by red light and PS I by far-red light has been widely and successfully used in photosynthetic studies, random transfer of energy from one pigment system to another seems unlikely. Thus the separate package hypothesis (no energy transfer between pigment systems) was proposed. However, it now seems probable that when the PS II system is strongly excited and the PS I trap (reaction centre) is open, excitation energy can reach the PS I centre from PS II. Hence there is an apparent spillover of excitation energy from PS II units to PS I.

We may look at this problem in another way by asking about the distribution of energy between the two pigment systems. Experiments with chloroplast preparations over the last decade show that the distribution is not constant, but influenced by conditions, particularly the cation concentration of the medium. Addition of Mg^{2+} decreases the transfer of energy to the PS I reaction centre, whereas low concentrations of monovalent cations have been reported to have the reverse effect. Mg^{2+} is known to have several effects on chloroplast membranes (e.g. promotion of the stacking of membranes in grana) and the mechanism whereby energy distribution between reaction centres is regulated, is not understood. However, regulation of energy transfer between photosystems does not occur in the absence of the chlorophyll a/b complex, CP II (Lieberman *et al.*, 1978).

13.6 Photochemical reactions

The photochemical process in the chloroplast is initiated by the excitation of a pigment molecule to a first excited singlet state. The energy is then transferred to a specific chlorophyll molecule in an environment where the photochemical reaction can take place. Study of chlorophyll chemistry has shown that a number of reactions can occur when the pigment is excited under various conditions. In the chloroplast, the reaction may be represented as:

$$Chl \xrightarrow{hv} Chl^* \qquad \text{excitation}$$

$$Chl^* \longrightarrow Chl^+ + e \qquad \text{oxidation of centre}$$

$$Chl^+ + e \longrightarrow Chl \qquad \text{reduction of centre}$$

where Chl^* represents the excited molecule.

This sequence of events can more realistically be expressed in terms of a donor (D) and an acceptor (A), the excited chlorophyll catalysing the oxidation of the donor and the reduction of the acceptor.

$$D \ Chl \ A \longrightarrow D \ Chl^* \ A \longrightarrow D \ Chl^+ A^- \longrightarrow D^+ \ Chl \ A^-$$

In effect this involves a charge separation and the process will not be efficient if the back reaction occurs to a significant extent, i.e.

$$D^+ \ Chl \ A^- \longrightarrow D \ Chl^* \ A$$

Clearly the structure of the reaction centre and the surrounding region must be such that the photochemical 'dipole' is stabilised. However, when the back-reaction does occur, there will be a possibility of fluorescence. PS II accounts for most of the fluorescence observed normally; fluorescence due to back reactions may be distinguished by delay in its emission which may be observed up to several minutes after illumination ceases. The process is described as delayed fluorescence, luminescence, afterglow or delayed light.

The first reaction centre to be discovered was in PS I and was named P700. Kok in 1956 observed an absorbance change at 700 nm after activation by a flash of light. Later a corresponding change observed at 433 nm led to the conclusion that P700 was probably a chlorophyll *a*. Various studies of the reaction centre pigment suggest that it is probably a dimer, although as yet it has not been completely isolated.

It proved more difficult to detect a reaction centre for PS II. In the period 1967–72, however, work in Witt's laboratory showed the existence of absorbance changes at 435 and 682–690 nm after a flash of light and these were attributed to the PS II reaction centre. This second centre is also oxidised on illumination and is referred to as P680.

13.7 Summary

The consequence of absorbance of a single quantum of light can be seen as follows. The absorbing pigment (or pigment–protein complex) which is located in a relatively large pigment complex is raised to an excited state. The excitation energy migrates through the pigment complex eventually arriving at the reaction centre, where a specific excited molecule undergoes an oxidation–reduction reaction. The radiant energy is now converted to chemical energy. An electron carrier of relatively low (oxidation–reduction) potential is reduced by one of relatively high (more positive) potential. A quantum of light of wavelength 700 nm is equivalent to 169 kJ/mol and this energy is in turn equivalent to a potential of about 1.8 volts per electron.

Further reading

Boardman, N. K., Anderson, J. M. and Goodchild, D. J. (1978) Chlorophyll–protein complexes and structure of mature and developing chloroplasts. *Current Topics in Bioenergetics*, **8**, 35–109.

Clayton, R. (1970) *Light and Living Matter*, Vol. I. McGraw-Hill, New York.

Chapter 14

The chloroplast electron transport chain

14.1 Development of the idea of a chloroplast electron transport chain

(a) Oxygen evolution and the Hill reaction

The study of the oxidation–reduction systems of the chloroplast can be traced back to the end of the nineteenth century when Ewart observed that dead or broken cells retained a limited capacity to evolve oxygen in the light. Molisch (1925) used dried leaf powders which, if ground in water, yielded a preparation capable of oxygen evolution in the light. These results later encouraged Hill at Cambridge to study oxygen evolution of chloroplast preparations (Hill, 1937). He found that oxygen evolution could be obtained from chloroplasts, particularly if a suitable electron acceptor was added to the preparation (Hill, 1939). Ferric oxalate was used as the oxidant, but in later experiments ferricyanide and a wide range of electron acceptors were used. It was concluded that the reduction of the oxidant and the evolution of oxygen represented a photolysis of water:

$$2H_2O \xrightarrow{\text{light}} 4H^+ + 4e + O_2$$

Initially doubt was expressed as to whether this reaction was related to photosynthetic oxygen evolution of the chloroplast. No obvious requirement for CO_2 could be demonstrated although Warburg subsequently showed a requirement for very low levels of CO_2 (see sect. 14.3(c)). There was, furthermore, no specificity of electron acceptor.

Support for the idea of water as the source of oxygen evolved in photosynthesis came from two sources. Earlier studies of comparative microbial biochemistry suggested that oxygen might well be derived from water. Van Niel from a study of photosynthetic bacteria in the 1930s (see van Niel, 1941) had drawn attention to the similarities between the photosynthesis of sulphur bacteria (see sect. 15.1(b)) and that of higher plants. The latter evolve oxygen:

$$CO_2 + 2H_2O \longrightarrow (CHOH) + H_2O + O_2$$

whereas the former do not:

$$CO_2 + 2H_2S \longrightarrow (CHOH) + H_2O + 2S$$

or

$$4CO_2 + H_2S \longrightarrow (CHOH)_4 + SO_4^{2-} + 2H^+$$

This centred attention on the requirement for reducing equivalents (from H_2O or H_2S) in the conversion of CO_2 to carbohydrate (CHOH).

Using the heavy isotope of oxygen, ^{18}O, Ruben *et al.* (1941) showed that the oxygen came primarily from water. These workers prepared $H_2^{18}O$ and $C^{18}O_2$ and found that in photosynthesis with whole algal cells $^{18}O_2$ was evolved when $H_2^{18}O$ was the substrate but not when $C^{18}O_2$ was used.

(b) NADP reduction

The problem of a physiological electron acceptor for the Hill reaction was solved when Vishniac and Ochoa (1951) showed that the pyridine nucleotides could be reduced by broken chloroplasts in the light when low oxygen tensions were maintained

$$2H_2O + 2NAD(P) \xrightarrow{\text{light}} 2NAD(P)H_2 + O_2$$

In their early experiments, Hill and Scarisbrick (1940) had found that the Hill reaction could readily be assayed with a system in which ferric oxalate was reduced by chloroplast preparations but reoxidised by methaemoglobin (haemoglobin in which the iron is oxidised to the ferric state). They noted that their chloroplasts could not reduce methaemoglobin directly but only in the presence of ferric oxalate. In 1952 Davenport *et al.* showed the existence of a protein (normally washed out of their chloroplasts) which was capable of reducing methaemoglobin in the presence of illuminated chloroplasts and in the absence of ferric oxalate. This was termed the methaemoglobin reducing factor (MRF). A few years later, San Pietro and Lang (1958) isolated a protein from spinach which catalysed NADP reduction by illuminated chloroplasts. This protein, the photosynthetic pyridine nucleotide reductase (PPNR), was later shown to be the same as the MRF and was capable of undergoing oxidation and reduction which could be demonstrated spectrophotometrically. In 1962, Tagawa and Arnon isolated a protein which they called ferredoxin. This protein was identical to MRF and PPNR and also very similar to a protein previously described as ferredoxin and shown to be involved in redox reactions in the anaerobic bacterium *Clostridium pasteurianum* from which it had earlier been isolated. Ferredoxin is involved in pyruvate metabolism (pyruvate synthase) in this organism:

$$\text{pyruvate} + \text{coenzyme A} + \text{ferredoxin}_{ox} \rightleftharpoons \text{acetyl CoA} \\ + \text{ferredoxin}_{red} + CO_2$$

The new photosynthetic ferredoxin could be obtained from chloroplasts and photosynthetic bacteria and had an $E_{m\,7.0}$ of -420 mV. Among the properties of chloroplast ferredoxin demonstrated by Tagawa and Arnon, was its ability to couple with a hydrogenase (obtained from bacteria) to evolve hydrogen in the light:

It was possible to reduce ferredoxin with molecular hydrogen and hydrogenase, but NADP reduction also required the presence of broken chloroplasts suggesting the existence of a chloroplast membrane-bound enzyme. Subsequently, such a flavoprotein enzyme, ferredoxin–NADP reductase (EC 1.6.7.1) was isolated from chloroplast membranes. Thus the reduction of NADP can be represented:

$$H_2O \xrightarrow{\quad e \quad} chloroplast \xrightarrow{\quad e \quad} ferredoxin \xrightarrow{\quad flavoprotein \quad} NADP$$
$$\searrow$$
$$\tfrac{1}{2}O_2 + 2H^+ \quad \boxed{light}$$

(c) Cytochromes

The oxido–reduction system studied by Hill suggested the possible existence of other components in the chloroplast, particularly cytochromes. Cytochrome c (of mitochondrial origin) had been shown to act as an electron acceptor for chloroplast oxygen evolution. This led Hill and Scarisbrick (1951) to search for cytochromes in the chloroplast. Cytochrome f, a non-autoxidisable pigment eventually shown to be a c-type cytochrome with an α-band at 555 nm, was found associated with chloroplasts. A second cytochrome, b_6, with an α-band at 563 nm, was also identified by Hill (1954) in etiolated barley leaves. Both cytochromes were found to be widely distributed in photosynthetic systems, being present at a concentration of about 1 cytochrome per 400 chlorophyll molecules. The mid-point potentials at pH 7.0 were 365 mV (f) and -40 mV (b_6).

An additional b-cytochrome was also found in chloroplasts, initially designated b_3[1] but later b_{559}. Subsequently, b_{559} was found to exist in low and high potential forms (rather like b_T of complex III in the mitochondrion). The function of the b-cytochromes in the chloroplast has been problematic and will be discussed later.

(d) Electron transport between the photochemical systems

The studies of action spectra by Emerson and by Blinks outlined in the last chapter led to the concept of two pigment systems. Studies of the oxidation and reduction of cytochrome f in the red alga *Porphyridium* by Duysens et al. (1961) provided evidence of two related photochemical reactions. These workers found that whereas 680 nm light was most effective in cytochrome f oxidation, 560 nm light brought about a reduction (Fig. 14.1a). 560 nm light, which activated the accessory pigment phycoerythrin, was also most effective in promoting chlorophyll fluorescence. It was concluded that 560 nm activated a pigment system reducing cytochrome f while 680 nm light activated a system oxidising the cytochrome. Similar conclusions were reached with higher plant chloroplasts. At the same time Losada et al. (1961) showed that the dye dichlorophenol indophenol could be reduced by red light (664 nm) with oxygen evolution in a Hill-type reaction which was sensitive to 3 (3,4-dichlorophenyl)-1,1-dimethylurea (DCMU) (see sect. 13.4). Photo-oxidation of the dye could be coupled to NADP reduction, but here longer wavelengths were effective and the system was insensitive to DCMU (Fig. 14.1b).

---

[1] b_3 was first described by Hill and Scarisbrick (1951) as a soluble cytochrome. The name b_3 was later used by Lundegårdh (1962) to describe the tightly membrane-bound pigment.

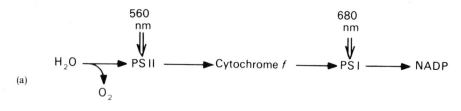

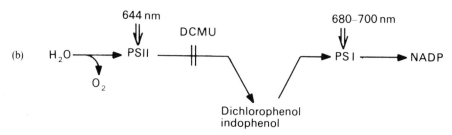

Fig. 14.1 Evidence for two light reactions. (a) Cytochrome *f* reduced by light absorbed by phycoerythrin (560 nm) and oxidised by far-red light (680 nm) in the red alga *Porphyridium*. (b) Reduction in red light and oxidation in far-red light of dichlorophenol indophenol. The oxidation occurs in the presence of DCMU.

These experiments laid the basis for wide-ranging studies of oxidation–reduction in the chloroplast. Those electron carriers which, like cytochrome *f*, can be reduced by activating PS II with red light and oxidised by PS I activated with far-red light, are regarded as mediating the transfer of electrons from PS II to PS I.

In addition to the components already discussed, a blue copper protein, plastocyanin, was discovered in 1960 by Katoh (see Katoh *et al.*, 1961). Plastocyanin is closely associated with PS I, has a MW about 10 000 and contains one atom of copper. Its oxidation–reduction is inhibited by cyanide, $HgCl_2$ and polycations such as polylysine. The $E_{m\ 7.0} = +370$ mV approximately.

A second type of carrier mediating electron transport between the pigment systems is plastoquinone. Chloroplasts could be shown to lose their ability to carry out a Hill-type reaction after extraction with light petroleum; the activity could be restored by addition of the extract. Bishop (1961) identified the active component of the extract as plastoquinone. Plastoquinones (Fig. 14.6) have an $E_{m\ 7.0}$ close to zero and are reduced readily by activation of PS II.

An outline of the chloroplast electron transport chain is shown in Fig. 14.2, where it will be noted that the carriers appear in order of standard red-ox potential. The experimental techniques used to analyse the electron transport chain of the chloroplast have included difference spectra (light-treated minus dark-treated chloroplasts) and the analysis of spectral changes consequent on a brief flash of light sufficient to bring about single electron transfers at most reaction centres in the preparation. This latter approach is capable of a number of refinements. Preferential activation of either PS I or PS II reaction centres is achieved with far-red and red light respectively. The spectral changes can be subjected to detailed kinetic analysis. The use of low temperatures restricts the

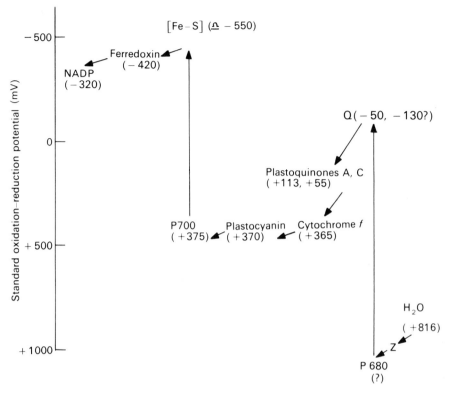

Fig. 14.2 Electron transport chain of the chloroplast (standard oxidation–reduction potentials in mV).

number of individual oxidation–reduction reactions that occur after photosystem activation. It should be added, however, that difficulties have been experienced in assigning specific spectral changes to the appropriate chloroplast components.

(e) Photophosphorylation

Early studies in this field not only developed an understanding of ATP synthesis in the chloroplast but also introduced the idea of cyclic electron flow. The first demonstration of photophosphorylation in the chloroplast was made by Arnon and coworkers (1954). They showed that illuminated chloroplasts would incorporate ^{32}Pi into ATP. The initial rates of phosphorylation were very low, but a number of catalysts were found which substantially increased activity, including vitamin K (menadione, etc.), riboflavin, riboflavin phosphate and NADP. Catalysts such as vitamin K and riboflavin promoted phosphorylation not associated with oxygen evolution or reduction of an electron acceptor. Arnon explained these observations as being due to a cyclic electron flow. He named the process 'cyclic photophosphorylation' (see Fig. 14.3).

$$ADP + Pi \xrightarrow[\text{chloroplast}]{\text{light}} ATP + H_2O$$

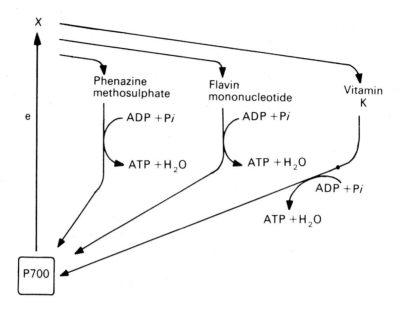

Fig. 14.3 Cyclic phosphorylation in the presence of non-physiological electron carriers.

In contrast, phosphorylation promoted by NADP was associated with oxygen evolution, was sensitive to the inhibitor DCMU which inhibits electron flow close to PS II, and was apparently associated with the flow of electrons from water to NADP. This 'non-cyclic photophosphorylation' (Arnon, 1959) may be represented as:

$$H_2O + NADP \longrightarrow \tfrac{1}{2}O_2 + NADPH_2$$
$$ADP + Pi \longrightarrow ATP + H_2O$$

$$NADP + ADP + Pi \xrightarrow[\text{chloroplast}]{\text{light}} \tfrac{1}{2}O_2 + NADPH_2 + ATP$$

The equation assumes a P/2e (or P/O) ratio of unity. A third type of photophosphorylation was also described by Arnon, pseudocyclic photophosphorylation. This may be differentiated from cyclic by sensitivity to DCMU. It is in fact a form of non-cyclic phosphorylation in which one electron carrier, ferredoxin, for example, is oxidised by oxygen after its reduction; hydrogen peroxide is formed but this is converted to water and oxygen by catalase. It may be represented as:

$$4 \text{ ferredoxin}_{ox} + 2ADP + 2Pi \xrightarrow[\text{chloroplasts}]{\text{light}} 4 \text{ ferredoxin}_{red}$$
$$+ 4H^+ + O_2 + 2ATP$$
$$4 \text{ ferredoxin}_{red} + 4H^+ + 2O_2 \longrightarrow 2H_2O_2 + 4 \text{ ferredoxin}_{ox}$$
$$2H_2O_2 \xrightarrow{\text{catalase}} 2H_2O + O_2$$

$$2ADP + 2Pi \longrightarrow 2ATP + 2H_2O$$

Overall there is no net oxidation or reduction and no net oxygen evolution, although there is electron transport in which water is the electron donor and the non-cyclic pathway is coupled to ATP synthesis. However, unlike the cyclic pathway, PS II activity is involved. It is not clear whether pseudocyclic phosphorylation has any significance *in vivo*.

In the early studies, photophosphorylation systems were explained in terms of a single photochemical reaction. Later it was shown that while cyclic phosphorylation could be activated by far-red light (PS I only), the non-cyclic process (and also the pseudocyclic) required both pigment systems to be activated.

These early studies imply that ATP synthesis is coupled to electron transport and consequently that the mechanism for photophosphorylation will be similar to that for oxidative phosphorylation. However, they also imply that in addition to the linear flow of electrons through electron carriers from oxygen to NADP, there is also a cyclic pathway of electron transport in which there is no net oxidation–reduction. Direct evidence for such a pathway was gained in the 1960s when cytochrome b_6 was found to be both reduced and oxidised by far-red light activating PS I.

14.2   Oxidation–reduction reactions associated with PS I

(a) P700

In the last chapter the arguments for the mechanisms associated with the reaction centre were outlined. In 1956, Kok reported changes in the absorption spectrum of photosynthetic systems at 700–705 nm which could be observed in a variety of organisms when illuminated (Fig. 14.4). Subsequently bleaching at 700 nm was found to correlate with a similar loss of absorption at 433 nm. The spectral changes and disappearance of the spectral properties during graded chlorophyll extraction suggested that they represented a chlorophyll pigment which was termed P700 (Kok, 1961). (Note that this is also referred to as Chl a_I.) P700 may be oxidised by ferricyanide and has $E_{m\,7.0} = +430$ mV. Assuming the pigment to be a chlorophyll, about 1 P700 is present for every 400 chlorophyll molecules. It therefore became the obvious candidate for the reaction centre for PS I, the long wavelength absorbing system. P700 has not so far been isolated, although considerable progress has been made in obtaining small complexes containing the compound. It seems likely from physical evidence that it is a dimer. The basic reaction may be represented:

$$P700 \xrightarrow{\text{light}} P700^*$$
$$P700^* \longrightarrow P700^+ + e$$
$$P700^+ + e \longrightarrow P700$$

The dye methyl viologen has been widely used as an electron acceptor. The $P700^+$ (bleached) state can be generated in PS I digitonin particles (lacking an electron donor) by illumination in the presence of methyl viologen. P700 is restored by addition of ascorbate-TMPD (tetramethylphenylenediamine) as reducing agent.

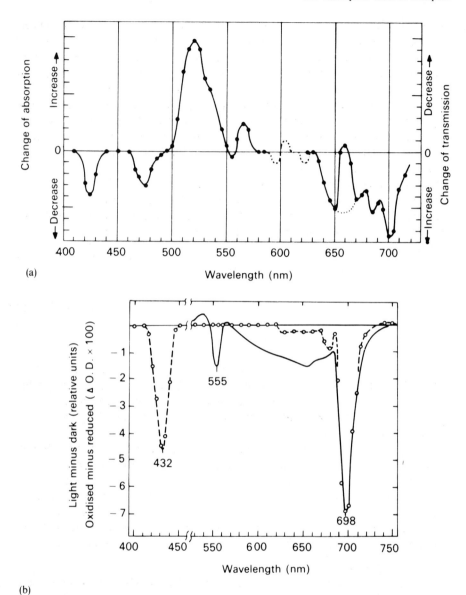

Fig. 14.4 (a) Absorption changes of a suspension of whole algal cells (*Scenedesmus*) consequent on 500 μs flashes of white light (after Kok, 1957). (b) Oxidised minus reduced spectrum of acetone-extracted chloroplasts which have a low chlorophyll content (solid line) and light minus dark spectrum (dotted line). (after Kok, 1961).

(b) Primary acceptor for P700

The nature of the primary acceptor for P700 is poorly understood. Arnon's group demonstrated that ferredoxin could be reduced by PS I and ferredoxin was proposed as the primary acceptor in the early 1960s (Whatley *et al.*, 1963). However, PS I was shown to be capable of reducing dyes considerably more electronegative than ferredoxin ($E_{m\,7.0} = -420\,\text{mV}$) down to a potential of about $-550\,\text{mV}$ or lower. In 1971, Malkin and Bearden obtained evidence for the reduction of an iron–sulphur centre by PS I in EPR studies at low temperatures with preparations from which soluble ferredoxin had been removed. It was shown that this centre (initially referred to as bound ferredoxin and later as centre A) was irreversibly reduced by P700. If centre A was reduced before illumination, a photo-oxidation of P700 was observed due to reduction of a second Fe–S centre (centre B) (Evans *et al.*, 1974). If both centres A and B were reduced, a reversible oxidation of P700 could still be observed, associated with reduction of an intermediate X, an electron carrier between P700 and the iron–sulphur protein complex. X may also be detected by EPR spectroscopy. The EPR data may suggest the sequence of redox reactions shown below (see Williams-Smith *et al.*, 1978). The centres A and B have E_m values of -550 and $-585\,\text{mV}$ respectively.

Iron–sulphur protein complex

Donor — P 700 / P 700⁺ —e→ X → [Fe–S]ₐ / [Fe–S]ᵦ —e→ Ferredoxin → Ferredoxin–NADP Reductase (bound flavoprotein) → NADP

An absorbance decrease at 430 nm detectable at room temperature has been attributed to a pigment P430 with an apparent $E_{m\,7.0} = -530\,\text{mV}$ and spectral characteristics of an iron–sulphur protein. P430 appears to correlate with centres A and B. The reduction of P430 involves centres A_1 and A_2 presumably equivalent to X (Sauer *et al.*, 1978).

Ferredoxin itself is the soluble iron–sulphur protein referred to earlier (sect. 9.4 (c)). It has a molecular weight of 12 000 and has two iron atoms and two equivalents of labile sulphide. It acts as a one electron carrier. As previously indicated it is easily lost from chloroplast preparations when the envelope is ruptured. It is probably loosely bound to the outside of the thylakoid membrane. Ferredoxin is the substrate for the membrane-bound ferredoxin–NADP reductase and also for cytochrome b_6 reduction (see sect. 14.4 (b)).

(c) Electron donor to P700

Experiments in the early 1960s suggested that cytochrome *f* was the donor to P700. Oxidation of the cytochrome appeared to be linked to the reduction of P700⁺. However, the copper protein plastocyanin was also known to act, like cytochrome *f*, between PS II and PS I, being reduced by the former and oxidised by the latter. The $E_{m\,7.0}$ of plastocyanin is $+370\,\text{mV}$ and that of cytochrome $f + 365\,\text{mV}$, suggesting a preferred sequence:

cytochrome $f \longrightarrow$ plastocyanin $\longrightarrow$ P 700

Plastocyanin and cytochrome *f* may nevertheless act as alternative donors to PS I. Under some special circumstances, *f* is oxidised directly by P700. However, detergent extraction of chloroplasts (or sonication) can give plastocyanin-depleted particles which lack the ability to photoreduce NADP. Activity can be restored by addition of plastocyanin. It can also be shown that plastocyanin stimulates oxidation of cytochrome f by P700. Cyanide and polylysine, both of which inhibit plastocyanin redox reactions, also inhibit oxidation of cyto-chrome f.

(d) Algal c-type cytochromes

There is a major difference between the c-type cytochromes of algae and those of higher plants. Algae appear to have a cytochrome comparable with the cytochrome f(λ_{max} = 555 nm) of higher plants and this is membrane-bound. In addition, algae also possess a soluble acidic cytochrome c with a maximum at shorter wavelengths (552 – 553 nm) and a molecular weight (10 000) well below that of cytochrome f. However, although most green algae and the more primitive blue-green algae (cyanobacteria) contain plastocyanin, other algal chloroplasts lack this electron donor to PS I. It has therefore been suggested that c_{552} can function in place of plastocyanin and evidence in support of this view has been obtained. For many algae the electron donor system to PS I may thus be:

$$\text{cyt } f \longrightarrow \text{cyt } c_{552} \longrightarrow \text{P700}$$

14.3 Electron transport associated with photosystem II

(a) P680

The study of photosystem II has been hampered by the difficulty in identifying with certainty any electron donor other than water and any electron acceptor other than the plastoquinone pool. More recently, detailed kinetic studies have made possible an analysis of the entities involved, although their chemical nature remains uncertain.

Following wide acceptance of P700 as the reaction centre of PS I it was expected, by analogy, that PS II would function in a similar way with a reaction centre to the pigment complex but here the absorption maximum might be expected at a wavelength 10 to 15 nm shorter than P700. Unfortunately, this region is difficult to analyse spectroscopically after a flash of light, owing to high chlorophyll absorbance and fluorescence, the latter not being constant (see below). Nevertheless Witt's group (Doering *et al.*, 1967) succeeded in observing a flash-induced absorption change at about 680–690 nm which decayed with a half-life of 200 μs. This was distinguishable from the P700 change which showed a 20 ms decay time. The 680 nm change was sensitive to the PS II inhibitor DCMU but not to far-red light; it was attributed to a pigment P680. Although considerable scepticism about the identity of P680[2] with the PS II reaction centre existed for many years, this pigment has now been widely accepted as the PS II equivalent of P700. The 680 nm change may be correlated with one at about 435 nm and the pigment is assumed to be a form of chlorophyll a. The $E_{m\,7.0}$ of P680 has so far proved too high to measure but must be greater than + 820 mV. Energy absorbed by the antenna pigments is transferred to the reaction centre where the excited P680 undergoes an oxidation–reduction reaction.

[2] P680 is also referred to as P690 owing to variable estimates of its absorption maximum, and also as Chl a_{II}.

$$P680 \xrightarrow{\ h\nu\ } P680^* \longrightarrow P680^+ + e$$

$$P680^+ + e \longrightarrow P680$$

$P680^+$ would be expected to give an EPR signal and such a signal has been detected. An absorption band at 835 nm has also been attributed to this radical.

Neither the immediate electron donor nor the acceptor to the reaction centre is known with certainty and they have therefore been designated Z and Q respectively. Thus the overall reaction, which can occur even at 77 K, may be represented:

$$Z.P680.Q \xrightarrow{\ \text{light}\ } Z.P680^*.Q \longrightarrow Z.P680^+.Q^- \longrightarrow Z^+.P680.Q^-$$

Z^+ is then reduced by electrons derived from water. An alternative suggestion is that P680 does not participate directly in the redox reaction but acts as a photosensitiser. This may be represented:

$$ZQ \xrightarrow{\ P680^*\ } Z^+Q^-$$

(b) Fluorescence studies; Q the primary electron acceptor for PS II

Most of the chlorophyll fluorescence observed when chloroplasts are illuminated arises from PS II antenna chlorophyll *a* and has a maximum around 693–698 nm with a second band around 740 nm. PS I fluorescence, which is observed at around 710–715 nm at room temperature, is weak and shows none of the induction effects observed in PS II fluorescence.

About 1930 Kautsky noted an induction of fluorescence lasting only a few seconds when photosynthesis of a green leaf was initiated with illumination; there was a rapid rise to a maximum followed by a slow decline. The 'Kautsky' effect has been investigated by several workers. In broken chloroplasts fluorescence is low when illumination begins but rises to a peak, declines slightly before rising substantially to a plateau (Fig. 14.5). In intact chloroplasts and whole cells a rather similar rise in fluorescence is seen followed by a slow decline. The shape of the curves obtained experimentally is dependent on conditions.

The rapid rise in fluorescence is interpreted along the lines proposed by Duysens and Sweers (1963). They postulated an electron carrier Q as the primary electron acceptor for PS II (P680). In the oxidised state but not the reduced state, Q is an effective quencher of fluorescence. Thus when illumination is initiated, Q is mainly oxidised and able to quench fluorescence. As Q becomes reduced by photochemical activity, quenching is less and fluorescence rises to a steady state. As electron transfer through PS II declines, so does oxygen evolution. The slow rise in fluorescence from O to P (Fig. 14.5) implies the reduction of a substantial pool of electron acceptors beyond Q:

$$P680 \xrightarrow{\ e\ } Q \xrightarrow[\text{DCMU}]{\ e\ \ \ } A$$

Since fluorescence reaches the plateau P slowly, there appears to be an equilibrium between Q and A. Full reduction of Q requires substantial reduction of the secondary pool of electron carriers A. Addition of DCMU prevents oxidation of Q^- (reduced Q) and consequently Q^- accumulates much more

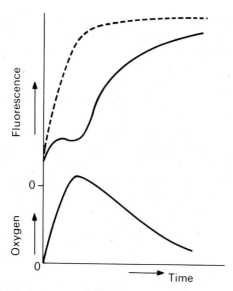

Fig. 14.5 Diagram of fluorescence and oxygen evolution in broken chloroplasts. Illumination commences at zero time. The inclusion of an electron transport inhibitor (DCMU) results in a rapid rise in fluorescence (dotted line).

rapidly together with a much faster rise in fluorescence. Thus DCMU inhibits electron transport beyond Q. Maximal fluorescence can also be obtained by reducing Q with dithionite. The small initial plateau in fluorescence seen under the same conditions is attributed to quenching effects of the electron donor to P680, Z.

Why do the antenna chlorophylls fluoresce when Q becomes reduced? Normally excitation energy absorbed by antenna pigments is transferred to the reaction centre which may result in a photochemical reaction, radiationless de-excitation, fluorescence or return to the antenna pigments. If Q is in the oxidised state, fluorescence will be quenched and a photochemical oxidation–reduction will be a very probable event. As Q becomes reduced, a photochemical reaction will cease to be a possibility. With reduced photochemical activity, the energy will be transferred back to the antenna molecules where fluorescence may occur. Thus the increase in fluorescence will come from P680, but particularly from the antenna pigments.

The fall in fluorescence seen in intact systems is in reality a complex process but reflects in part the oxidation of Q by the intermediate electron transport chain and PS I.

Assuming the properties described above, Q is a one-electron carrier with an $E_{m\ 7.0}$ about -50 mV. However, the $E_{m7.0}$ is pH sensitive and more recent studies have led to the conclusion that the effective value may be as low as -130 mV. There is one Q per PS II reaction centre.

(c) Plastoquinones, secondary electron acceptors for PS II

The pool of electron acceptors, A, discussed above, has been equated with the plastoquinone pool. The quinone content of the chloroplast is shown in

Table 14.1. The function of phylloquinone (vitamin K_1) and the tocopherol quinones is uncertain; only the plastoquinones are known to function directly in the photosynthetic process. Of the three types, A (Fig. 14.6) is much the most significant, being present at up to 20 to 40 molecules per cytochrome f. However, the functional plastoquinone pool has been estimated to be of a much lower magnitude, about 7 quinones per chain. The $E_{m\,7.0}$ values of the major quinones A and C have been estimated at $+113$ and $+55$ mV respectively and they are therefore substantially more positive than the primary acceptor Q. As noted earlier, the quinone is necessary for the PS-II-dependent Hill reaction; it is reduced by activation of PS II and oxidised by activation of PS I, the reduction but not the oxidation, being DCMU-sensitive. The oxidation and reduction may be observed optically at 265 nm. Small concentrations of HCO_3^- are necessary for plastoquinone reduction by PS II; this explains the observation of Warburg that low levels of CO_2 are necessary for the Hill reaction.

Table 14.1 Quinone content of spinach chloroplasts

	µmol/mmol chlorophyll
Plastoquinone A	108
Plastoquinone B	5
Plastoquinone C	18
α-Tocopherol quinone	8
β-Tocopherol quinone	0.5
Phylloquinone, vitamin K_1	14

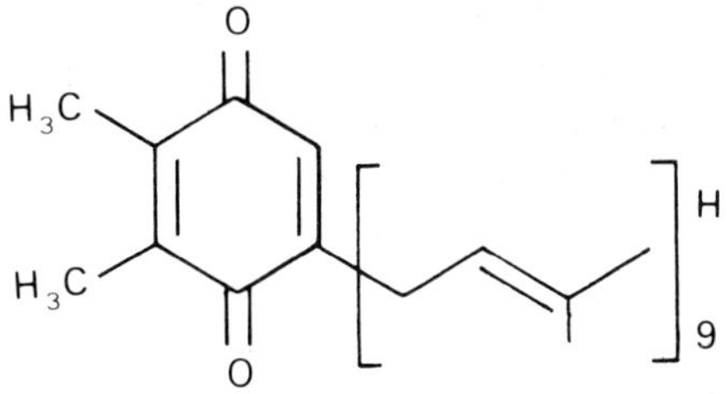

Fig. 14.6 Plastoquinone A.

Dibromothymoquinone (2,5-dibromo-3-methyl-6-isopropyl-*p*-benzo-quinone) has proved a valuable inhibitor in studies of electron transport. It acts as an antagonist of plastoquinone, inhibiting oxidation of the reduced form. Thus cytochrome f reduction by PS II may be inhibited by dibromothymo-quinone but the inhibition may be overcome by addition of plastoquinone.

The quinone pool appears to interact with several electron transport chains. Addition of an inhibitor, DCMU, at a concentration sufficient to inactivate 90 per cent of the PS II activity, leaves almost all the PS I systems active if there is an adequate supply of reducing equivalents under otherwise suitable conditions. The plastoquinones may thus be envisaged as mobile lipid-soluble carriers located in the membrane lipid phase, capable of being reduced by several PS II systems and oxidised by several PS I systems (see Fig. 14.7). Such a situation resembles the model suggested earlier for ubiquinone in the mitochondrial membrane, where quinone is capable of oxidising a variety of flavoprotein complexes and is also capable of reducing one of several cytochrome chains.

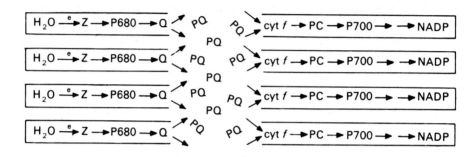

Fig. 14.7 The plastoquinone pool in electron transport from PS II to PS I. The pool of quinones oxidises any one of a number of PS II electron transport chains and reduces any one of a number of PS I chains.

The oxidation of plastoquinones by cytochrome f appears to be a rate-limiting step. In, for example, uncoupled chloroplasts in strong white light, plastoquinone accumulates in the reduced and cytochrome f in the oxidised state. The association of a high-potential iron–sulphur protein ($E_{m\ 7.0} = +290$ mV) with this step has been proposed but not substantiated.

(d) The identities of Q, X320 and C550

The search for the identity of Q has resulted in the discovery of closely associated absorption changes at 550 nm and 320 nm.

Knaff and Arnon (1969) described a decrease in absorption at 550 nm associated with PS II activation and distinct from the absorption changes due to known cytochromes. The optical effect could be observed at 77 K and was therefore closely associated with the primary photochemical reaction of PS II. It could also be induced without illumination by reduction with dithionite. The absorbance change has been attributed to a substance C550, which became a candidate for the role of primary electron acceptor in the PS II photoreaction. Later work has suggested that the absorption change is due to a carotenoid which does not itself undergo oxidation–reduction. However, the optical properties of C550 do appear to reflect the state of Q under most conditions and it has been used experimentally as an indicator of Q reactions.

Also in 1969, Stiehl and Witt described an increase in absorbance at 320 nm closely associated with PS II activity. The compound responsible for this, X320 (X335), is formed very rapidly (less than 1 μs) at room temperature and under most conditions in the presence of DCMU. It decays with a half-life of 600 μs. X320 may also be formed at 77 K. The spectrum corresponds to that of plastosemiquinone (PQ^-) and the ratio of PQ^- to P680 is estimated at 1 : 1.

Extraction of plastoquinone to low endogenous levels results in loss of P680 oxidation, but a substantial proportion of the original activity can be recovered if plastoquinone is returned to the preparation, which again suggests that the primary acceptor of PS II is a specialised plastoquinone molecule.

Many workers are now disposed to identify Q with X320 and with a special plastoquinone which acts as a one-electron acceptor of the PS II reaction, forming the semiquinone. It is then oxidised by the plastoquinone pool possibly via another specialised plastoquinone which also acts as a one-electron carrier. Kinetic evidence for such an intermediate has been obtained both from fluorescence and from spectral studies. The reducing side of PS II is summarised in Fig. 14.10.

(e) Oxygen evolution and the electron donor to PS II

The ultimate electron donor to PS II is water. Thus the sequence of carriers is:

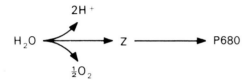

The system is relatively unstable, requires chloride ions and is inhibited by high concentrations of Tris (Tris(hydroxymethyl)aminomethane) at pH 8.0. Diphenylcarbazide and hydroxylamine will act as electron donors to P680 in systems which have lost the ability to evolve oxygen.

A role of manganese in photosynthesis has been known since the 1930s. From studies with algae grown in manganese-deficient media and from preparations from which manganese was removed by temperature shock or washing with Tris, it is clear that the primary role is in relation to the oxygen-evolving side of PS II. Manganese-deficient preparations can be shown to use diphenylcarbazide or hydroxylamine but not water as electron donors to PS II. There are probably about 6 Mn^{2+} per PS II reaction centre. A manganese protein has been postulated as an essential part of the water-splitting system.

Oxygen evolution following flashes of light shows an induction phase. Allen and Franck (1955) observed that a single flash of light did not give oxygen evolution in algae which were dark-adapted (kept in darkness for the preceding period; 40 minutes will serve as an adequate period for dark adaptation) although oxygen evolution would occur after pre-illumination or if there were weak background light. Joliot and Joliot (1968) used a sensitive oxygen electrode system to demonstrate that with dark-adapted chloroplasts, oxygen evolution per flash showed a periodicity of four (see Fig. 14.8). A first flash gave no oxygen evolution while the yield from the second was low; the third flash, however, gave a substantial evolution but the fourth was again low. The flashes were brief (10 μs) and were adequate to excite a single electron transfer in most active centres; electron transfer could be demonstrated in terms of variable fluorescence.

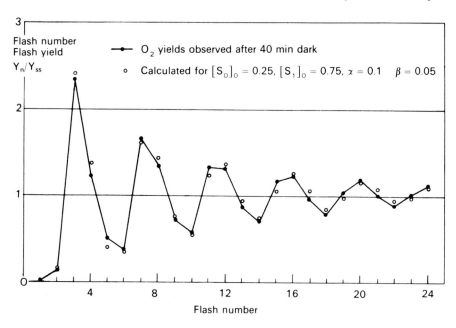

Fig. 14.8 Oxygen evolved per flash by spinach chloroplasts which have been previously kept in the dark (·——·). The open circles represent a calculated yield assuming the model shown in Fig. 14.9 with 25 per cent of the oxygen-evolving centres in state S_0 and 75 per cent in state S_1. It is also assumed that a single flash activates 5 per cent of the centres twice and fails to activate 10 per cent of the centres. $Y_n = O_2$ yield per flash, Y_{ss} = the steady-state yield per flash (after Radmer and Kok, 1975). Reproduced from the *Annual Review of Biochemistry* 44 © 1975 by Annual Reviews Inc.

The explanation for this system proposed by Kok *et al.* (1970) assumes for the Z.P680.Q system, four photoactive states and an additional state which is non-photoactive, S_0, S_1, S_2, S_3 and S_4 (Fig. 14.9). Each flash results in a single electron transfer creating an increasing number of positive charges in Z. The accumulation of four positive charges is necessary for oxygen evolution. To accumulate four charges would require the reoxidation of Q which carries only one electron. If this oxidation is partially inhibited by DCMU, the periodicity of 4 remains but the oxygen yields are much reduced. Since it is the third flash that results in major oxygen evolution, it is necessary to assume that S_1 is the stable state and that this must be the major form in dark-adapted chloroplasts. If it is also assumed that while most systems will undertake a single electron transfer for each flash, a few will undertake two transfers and a few none, the data are consistent with about 75 per cent of the systems being in the S_1 state and 25 per cent in S_0 in dark-adapted chloroplasts. Thus in the first flash, S_4, necessary for oxygen evolution, cannot be reached, although in two flashes, one of which involved two electron transfers, S_4 could be attained by a small number of systems. The model implies proton release in association with oxygen evolution and this is in fact the case.

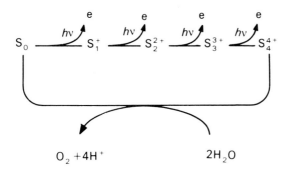

Fig. 14.9 Scheme for oxygen evolution, showing the states of Z.

It was noted earlier that the state of the electron donor Z influenced the curves obtained for the increase in fluorescence. Each flash would be expected to bring about a reduction of Q and hence an increase in fluorescence. It has been found that while flashes received by S_0 and S_1 states do have this effect, those absorbed by S_2 and S_3 result in reduced fluorescence, implying that these states have quenching properties. Electron flow associated with PS II is summarised in Fig. 14.10.

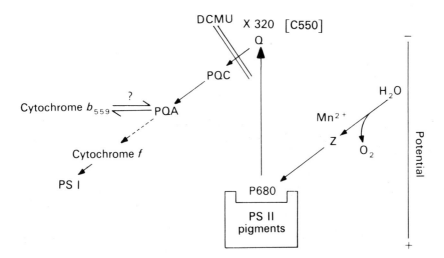

Fig. 14.10 Electron transport chain associated with PS II.

(f) Delayed fluorescence

This has been observed in leaves, isolated chloroplasts and also photosynthetic bacteria. Light emission occurs up to several minutes after illumination and therefore cannot be due to direct excitation.

Delayed fluorescence is attributable to the lowest excited singlet state of chlorophyll *a*, the antenna pigments of PS II being primarily responsible. However, both PS II and PS I will show delayed fluorescence. Although several explanations have been put forward, recombination of the primary products of the photoact is now widely accepted as the mechanism. Thus with $P680^+ . Q^-$, the primary charge pair gives rise to excited P680.

$$P680^+ . Q^- \longrightarrow P680^* . Q$$

The energy may then migrate to antenna molecules giving rise to fluorescence. In support of this explanation, it has been shown that hydroxylamine, an electron donor to P680, suppresses delayed fluorescence, but reduction of Q by dithionite enhances it.

14.4   The *b*-cytochromes

(a) Cytochrome b_{559}

The account of PS II above omits a major unsolved problem, the function of the *b*-cytochrome associated with this system, b_{559}. This cytochrome, like cytochrome b_T of the mitochondrion, has been found to exist in at least two potential forms. Cytochrome b_{559} HP (high potential) has an $E_{m\ 7.0}$ at about $+70$ mV. There are at least two molecules of cytochrome b_{559} per PS II reaction centre, of which one and probably both are normally in the HP form. Several treatments are known to convert HP to LP cytochrome including extraction of plastoquinone; adding back the quinone restores the HP form. Detergent treatment, heating, ageing and addition of antimycin or hydroxylamine all promote the HP to LP conversion. The existence of the cytochrome b_{559} LP in intact unaltered chloroplasts is doubted by several workers, who regard the LP form as a product of manipulation.

There have been three major proposals for the function of cytochrome b_{559}. These are: (*a*) as an electron acceptor for at least one of the four electrons associated with the water-splitting reaction, although it is doubtful whether the potential of b_{559} can be high enough for this role. The cytochrome can, however, be oxidised by PS II, but only at low temperature (77 K). This also implies a close structural relationship between PS II and the cytochrome; (*b*) as a component for a cyclic reaction round PS II. Since the cytochrome can be reduced by PS II at room temperatures and oxidised at low temperature, it has been proposed that it mediates a cyclic electron flow. Evidence for such a pathway is lacking; (*c*) as a secondary electron acceptor for PS II. In the dark the cytochrome is in the reduced state. If it is oxidised, the reduction by PS II, which is DCMU-sensitive, can be demonstrated at room temperature, although the rate is too low for the cytochrome to participate in linear electron flow from PS II to PS I as originally proposed. However, oxidation by PS I of cytochrome b_{559} can also be demonstrated. Electron flow from plastoquinone to cytochrome *f* is inhibited by dibromothymoquinone (DBTQ) and this quinone also inhibits the oxidation but not the reduction of cytochrome b_{559}. Since there is some evidence to suggest competition between cytochrome b_{559} and cytochrome *f* for electrons, it has been suggested that the reduction of the cytochrome and its oxidation might be represented:

$$P680 \longrightarrow Q \xrightarrow{\text{DCMU}} PQ \xrightarrow{\text{DBTQ}} \underset{f}{\text{cyt}} \longrightarrow \text{plastocyanin} \longrightarrow P700$$
$$\updownarrow$$
$$\text{cyt } b_{559}$$

The function of cytochrome b_{559} in the chloroplast is not clear and remains to be elucidated.

(b) Cytochrome b_6

This b-cytochrome, also known as cytochrome b_{563}, is associated with PS I, catalyses cyclic electron flow round PS I and has an $E_{m\,7.0}$ at about $+5$ mV. Both reduction and oxidation by far-red light can be demonstrated. A rapid transient reduction is observed on illumination of PS I followed by oxidation. Oxidation is sensitive to dibromothymoquinone and cyanide, implying the participation of plastoquinones and plastocyanin in this process (Fig. 14.11). The cytochrome accompanies PS I in the fractionation of treated chloroplasts and a cytochrome f–cytochrome b_6 complex has also been obtained, in accordance with the view that this pigment is primarily associated with PS I photoreactions. It is unlikely that cytochrome b_6 reacts directly with P700 without the intervention of the primary electron acceptor for PS I. Antibodies to ferredoxin inhibit reduction of cytochrome b_6 by PS I, which is consistent with the belief that the cyclic pathway involves ferredoxin (Bohme, 1977). There is also some evidence in the literature for a reduction of cytochrome b_6 by PS II, but this is not regarded as a major pathway and could reflect a reverse electron flow from plastoquinone.

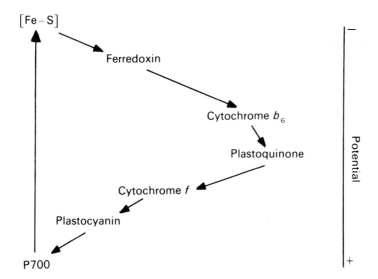

Fig. 14.11 The role of cytochrome b_6 in cyclic electron flow.

14.5 Orientation of photosystems in the thylakoid membrane

(a) Transmembrane orientation of the electron transport chain

The balance of the evidence strongly favours the view that the water-splitting reaction, the oxidising part of PS II, is located on the inside of the thylakoid membrane. Electron donors to P680 (such as diphenylcarbazide) which replaces the water-splitting reaction itself as an electron donor, react with PS II in fragmented systems but only poorly in intact thylakoids. The proton release shown in Fig. 14.12 can only be measured in the presence of an uncoupler, such as methylamine, which allows release of protons from the thylakoid interior. Furthermore, the membrane potential (see sect. 14.7) created by activation of PS II is positive inside and negative outside. It should nevertheless be noted that several experiments have shown inhibition of the PS II donor system by non-permeating agents in intact thylakoids, particularly when PS II is illuminated. For example, *p*-(diazonium)-benzenesulphonate (DABS), a water-soluble compound unable to penetrate the lipid region of the membrane, is highly reactive towards lipid and protein functional groups (α-amino, ε-amino, sulphydryl, imidazole and guanidino groups). [^{35}S]DABS will hence label chloroplast membranes rapidly. DABS inhibits reduction of PS II by water, but only if added in the light, suggesting that illumination of PS II causes a conformational change which exposes interior parts of the membrane. The enzyme lactoperoxidase in the presence of KI and H_2O_2 will iodinate exposed groups of membrane proteins and such treatment has been shown to inhibit close to the reaction centre of PS II in intact thylakoids.

These results taken together argue for a transmembrane orientation of the PS II reaction centre with the cleavage of water,

$$2H_2O \longrightarrow 4H^+ + 4e + O_2$$

occurring inside.

(b) Location of PS II electron acceptors

A transmembrane orientation of the PS II oxidation–reduction system would, by the above argument, require that the reducing side of PS II was on the outside of the membrane. Cytochrome b_{559} in intact thylakoids is readily oxidised in the dark by ferricyanide which does not permeate membranes. If the earlier argument for the function of this cytochrome is accepted, it will imply that Q must also be located close to the outside of the membrane. Electron acceptors that react in a Hill-type reaction involving only PS II are lipophilic substances, such as oxidised phenylenediamines and benzoquinones, suggesting that the reducing side of PS II is buried in the membrane. Further, protons are taken up from outside the thylakoid during plastoquinone reduction (see later). Hence, although it is buried in the membrane, the reducing side of PS II is close to the outside surface. The PS II system is summarised in Fig. 14.12.

(c) Electron donors to PS I: plastocyanin and cytochrome *f*

Antibodies against plastocyanin only inhibit PS I activity after sonication of chloroplast thylakoids, so that they must probably enter the loculus (internal space) of the thylakoid to be effective. Similar results are obtained with antibodies to cytochrome *f*. These results support the location of the plasto-cyanin and cytochrome *f* on the inside of the membrane. However, contrary results for plastocyanin have also been obtained more recently and it has been suggested that some part of the molecule is accessible from the external surface of the thylakoid (see Smith *et al.*, 1977).

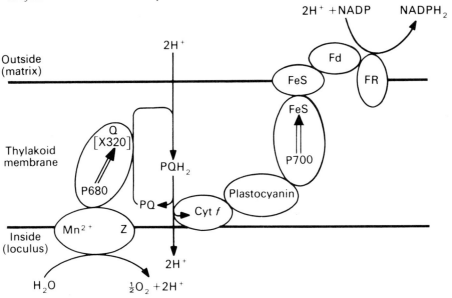

Fig. 14.12 Distribution of electron carriers across the thylakoid membrane (Fd = ferredoxin, FR = ferredoxin–NADP reductase).

(d) Electron acceptor for PS I

NADP, the ultimate electron acceptor of the electron transport chain, does not permeate the membrane and hence ferredoxin–NADP oxidoreductase must be located on or near the outside of the thylakoid membrane. This view is supported by experiments with antibodies to the oxidoreductase and to ferredoxin. Many artificial electron acceptors unable to penetrate the membrane, such as ferricyanide, are also readily reduced by PS I. Such evidence suggests that the electron acceptor system for PS I is located near the outside of the thylakoid membrane. A summary of the distribution of electron carriers in the membrane is shown in Fig. 14.12.

14.6 Artificial electron donors and acceptors for electron transport

The study of electron transport and its associated reactions has been greatly facilitated by the use of artificial electron acceptors and donors. However, a major problem in this field arises from the possibility that a given donor or acceptor may interact with the chain at more than one site. There is also a possibility of side effects.

A series of donors have been used to donate electrons to PS II in place of the water-splitting reaction. These do not involve the charge-accumulating component Z. They include hydroquinones, phenylenediamines benzidine semicarbazide, diphenylcarbazide, hydrazobenzol, hydrazine, hydroxylamine and ascorbate, etc. The hydroquinones, phenylenediamines and ascorbate will also act as electron donors to PS I. The preferred electron acceptors for PS II are the lipophilic group (Class III acceptors of Saha *et al.*, 1971) such as phenylenediamines and benzoquinones. The reduction of these substances is sensitive to DCMU but not to the inhibitor of plastoquinone oxidation, dibromothymoquinone, nor to inhibitors of plastocyanin such as KCN and $HgCl_2$. They therefore accept electrons close to the site of plastoquinone reduction.

The best known donor system to PS I is ascorbate–dichlorophenol indophenol, the latter being the immediate donor while the ascorbate maintains the dye in the reduced state. Phenylenediamines such as diaminodurene and TMPD (2,3,5,6-tetramethyl-p-phenylenediamine) have also proved to be a useful group of electron donors used in conjunction with ascorbate.

The artificial electron acceptors for PS I (Class I acceptors) include methyl viologen, flavin mononucleotide, ferricyanide, oxidised dichlorophenol indophenol and diaminodurene. Reduction of these electron acceptors is normally sensitive to plastocyanin inhibitors (KCN, $HgCl_2$, etc.) and in many cases will show enhancement effects when H_2O is the electron donor. In fragmented chloroplasts, however, ferricyanide and DCPIP act as PS II acceptors. Consequently methyl viologen ($E_{m\,7.0} = -440\,mV$) and related dyes (benzyl viologen $E_{m\,7.0} = -379\,mV$), which have low mid-point potentials, are more specific for PS I systems. Thus PS I may be readily assayed with ascorbate–DCPIP as an electron donor and methyl viologen as an electron acceptor.

Work with artificial carriers is complicated by the fact that some of them have unexpected side effects in relation to protons. For example, TMPD and diaminodurene are both lipid-soluble, have similar mid-point potentials and will both donate electrons to PS I. However, only the diaminodurene crosses the membrane in a protonated reduced form. Its oxidation by PS I on the inside of the thylakoid releases the protons. TMPD also crosses the membrane to reach the oxidising side of PS I, but is only an electron carrier. Thus a proton gradient is created by using diaminodurene as a donor to PS I; the gradient is not formed with TMPD. Phenazine methosulphate acts in a similar way to diaminodurene (see Fig. 14.15).

14.7   Proton translocation and membrane potentials

The widely accepted explanation for proton translocation in the chloroplast, as with the mitochondrion outlined earlier, is that it results from an electron transport chain arranged in loops across the membrane. The distribution of components outlined in Fig. 14.12 supports this conclusion. As noted above, the primary reactions of photosynthesis involve a charge separation. Both reactions appear to be arranged so as to give a transmembrane potential, positive inside and negative outside. Experimental evidence for this has been obtained by Witt's group (see Schliephake *et al.*, 1968 and also Junge, 1977). They have used the 515 nm optical absorbance change as an indicator of membrane potential. This change, attributed primarily to carotenoids but also to other pigments, is a broad band in the region 515 to 530 nm and first observed by Duysens. The absorbance increase has been shown to reflect an increase in the membrane potential across the membrane[3]. A flash of light gives rise to a rapid increase in

[3] The argument used here applies to higher plant chloroplasts rather than to algae, where the situation is more complex.

the 515 nm absorbance which then decays. Increasing the permeability of the membrane to ions by addition of ionophores or osmotic shock increases the decay rate as expected. Using this approach, Witt's group have observed the formation of a membrane potential. This is attributed to the initial photochemical reaction because it is established in less than 20 ns and is therefore much faster than secondary reactions and because it is also set up at low temperatures ($-50°C$). Artificially induced membrane potentials influence the delayed fluorescence described above. The polarity of the potential has been shown to be positive inside and negative outside. If chloroplast preparations capable of either PS II or PS I reactions only are examined, about half the membrane potential is formed as compared with preparations in which both photochemical reactions can occur. These experiments support the transmembrane orientation of the reaction centres as shown in Fig. 14.12.

Unlike the mitochondrion, the chloroplast on activation by light, takes up protons which are released in the interior (loculus) of the thylakoid. Estimates of the protons translocated per electron transferred from water to NADP are variable, but a ratio $H^+/e = 2$ is probable. Since the cleavage of water releases protons internally and NADP reduction requires protons externally in the matrix of the chloroplast, these terminal reactions effectively transfer one proton per electron. Schliephake *et al.* (1968) observed the uptake of one proton per electron for each photosystem. For PS I this can be attributed to the reduction of the terminal electron acceptor. The most probable proton carrier across the membrane is the plastoquinone which would require protons for its full reduction near the outer surface of the thylakoid membrane and release them if oxidised by cytochrome *f* at the inner surface of the thylakoid membrane. The kinetics of the release of protons into the thylakoid loculus (indicated by the dye neutral red) after a flash of light matched the kinetics of water oxidation (with a half-life of 300 μs) and plastoquinone oxidation ($t_{\frac{1}{2}} = 20$ ms). The latter can be shown to be sensitive to dibromothymoquinone. The uptake of protons can be made to match plastoquinone reduction by PS II kinetically if a permeability barrier on the outside of the thylakoid membrane is disrupted. The translocation of protons is summarised in Fig. 14.12.

14.8 The chloroplast ATPase

(a) Discovery

The chloroplast electron transport chain is coupled to a phosphorylating enzyme, the ATPase. Although in principle this enzyme is very similar to the mitochondrial proton-translocating ATPase, there are important differences. In an early study, it was concluded that unlike the mitochondrial system, the chloroplast did not possess an ATPase. However, Petrack and Lipmann (1961), using a cyanobacterium, noticed that ATP hydrolysis in fragments capable of photophosphorylation, could be observed in the presence of high concentrations of a thiol reagent (glutathione or cysteine) and light. Later, it was found that the ATPase activity could be removed from EDTA-treated chloroplasts and restored by addition of the washings. Vambutas and Racker (1965) isolated the

EDTA-removable factor, CF_1 (chloroplast factor 1) and showed that it possessed ATPase activity after trypsin treatment. An antibody to the purified CF_1 inhibits inducible ATPase activity and photophosphorylation, showing that the CF_1 functions in both processes. CF_1 can be shown to be attached to the outside of the thylakoid membrane (see sect. 12.3) and is therefore accessible to solutes in the stroma. This location is comparable with that in the mitochondrion (where F_1 is on the inside), if it is realised that the polarity of the thylakoid membrane is the reverse of the inner mitochondrial membrane with respect to proton translocation and membrane potential.

(b) Catalytic activity

The latent membrane-bound ATPase may be activated by thiol compounds and light and is Mg^{2+}-dependent. The light and thiol treatment can be given before assay, but the activation effect decays slowly in the dark. Light can be replaced by an acid–base transition (see sect. 14.9(a)), suggesting that the major role of light in activation is to drive proton translocation. Since ATP hydrolysis also drives proton translocation inwards, this will maintain the activated state. Trypsin treatment induces a Mg^{2+}-dependent ATPase activity which is not light-dependent.

The ATPase activities of the thylakoids and of isolated CF_1, and of photophosphorylation itself, are inhibited by the antibiotic Dio-9 and the glucoside, phloridzin. Oligomycin and aurovertin do not inhibit the chloroplast ATPase. In contrast, DCCD will inhibit photophosphorylation and membrane-bound ATPase, but not CF_1 activity. Since DCCD also stimulates the formation of a light-driven proton gradient in CF_1-depleted thylakoids (but not in coupled systems), it appears to act in the membrane part of the ATPase complex rather than on CF_1 and may block a proton-translocating pore uncovered by CF_1 removal. A DCCD-binding protein (MW = 8000) has been isolated. Removal of CF_1 increases the permeability of the thylakoid membrane to protons. Uncouplers such as CCCP stimulate the ATPase at low concentrations, but inhibit activity at high concentrations.

Early attempts to demonstrate an ATP/Pi exchange similar to that seen in mitochondria failed, until it was realised that activation of the preparation by light and a thiol reagent was necessary. Other exchanges similar to the mitochondrial ones have also been demonstrated (see sect. 9.1(a)).

(c) CF_1

CF_1 is a cold-labile ATPase with a molecular weight of 325 000. It is a spherical protein (9 nm diameter) which can be removed from thylakoid membranes by EDTA or silicotungstate treatment. ATPase activity is very low until it is activated by trypsin, heat or dithiothreitol. The activated enzyme in contrast to the membrane-bound form is Ca^{2+}-dependent and inhibited by Mg^{2+}. Five subunits have been detected, designated α, β, γ, δ and ε with molecular weight 59 000, 56 000, 37 000, 17 500 and 13 000 respectively but the number of subunits of each type in the complex is uncertain (possibly $\alpha_2\beta_2\gamma\delta\varepsilon_2$). The ε-subunit can be shown to inhibit active CF_1 and to be sensitive to trypsin. It is the subunit responsible for inhibition of the bound ATPase and is released from its inhibitory site by the membrane potential established during activation. Reconstitution of coupling activity can be achieved by adding CF_1 to partially depleted membranes in the presence of Mg^{2+}.

The membrane-bound ATPase has been shown to possess bound adenine nucleotides like mitochondrial and bacterial ATPases. Removal of CF_1 from the membrane releases the nucleotides. It is estimated that approximately 2 ATP and 1 ADP per ATPase are bound (Harris and Slater, 1975). In dark non-energised membranes these nucleotides are firmly bound and do not readily exchange with added nucleotides. However, in an energised system (as is obtained by illumination in the presence of the phosphorylation catalyst pyocyanine – see sect. 14.10(d)), exchange occurs under conditions (e.g. when Pi is omitted) where net ATP synthesis does not occur. ATP will then exchange with bound ADP or bound ATP. The level of ATP on the ATPase remains constant under full activation but the ADP level falls to near zero. These exchange reactions are sensitive to uncouplers and can also be induced by acid–base transitions. The fact that free adenine nucleotides can exchange with those bound to the ATPase only in the light, implies a conformational change in the enzyme when it is energised (see below).

14.9 The ATPase and theories of photophosphorylation

The two major current theories concerned with the mechanisms of oxidative phosphorylation have both been applied to photophosphorylation. In principle, the problem is the same, since ATP synthesis is in both cases driven via an unknown intermediate state by oxidation–reduction reactions in a membrane-bound system. The theories themselves have been discussed earlier and here some pertinent experimental observations will be discussed.

(a) Chemiosmotic evidence
As briefly noted above, activation of the ATPase results in the translocation of protons inwards across the thylakoid membrane. The ATPase (CF_1) appears to be attached to the membrane at a point where there is a 'pore', enabling the externally located ATPase to communicate with the protons of the interior of the thylakoid. DCCD, which inhibits photophosphorylation, reduces the proton permeability of the membrane when the CF_1 particle is removed. ATP synthesis can be shown to correlate in a predictable manner with the pH gradient and membrane potential, when a series of light flashes are used to drive proton translocation (see Graber and Witt, 1976).

A useful experimental approach has been the acid–base wash. Here an artificially induced pH gradient across the thylakoid membrane can be shown to drive ATP synthesis. Broken chloroplasts are first exposed to a medium of pH 4 containing an organic acid (e.g. succinate, which permeates the thylakoid) and then transferred to pH 8. This results in a low internal pH (due to the organic acid) and a pH gradient.

In broken chloroplasts the kinetics of ATP synthesis and proton efflux can be followed using an appropriate pH indicator. A correlation between the pH gradient across the membrane and ATP synthesis has been observed. The

H^+/ATP ratio has been estimated by many groups and a figure between 2 and 3 is obtained (see Reeves and Hall, 1978, for a summary).

Racker's group have carried out an ingenious reconstitution to demonstrate ATPase activity. Vesicles were formed in which hydrophobic thylakoid proteins and CF_1 were incorporated. Such vesicles would carry out an uncoupler-sensitive ATP/Pi exchange. If bacteriorhodopsin was also added, the light-dependent ATP synthesis could be demonstrated (Winget *et al.*, 1977). Bacteriorhodopsin (see sect. 15.9) is a readily purified protein which translocates protons across membranes in the light. Thus the ATP synthesis is shown to be a response to light.

These experiments show that the chloroplast ATPase is a proton-translocating enzyme which *can* synthesise ATP in response to a pH gradient.

(b) Conformational evidence

Some evidence that conformational changes are associated with activation of the ATPase in the light has already been given. For example, adenine nucleotide exchange in the light but not the dark is consistent with Boyer's conformational theory of energy coupling. Interesting evidence for conformational changes was obtained by Ryrie and Jagendorf (1971). Chloroplasts were illuminated in a medium containing tritiated water, ^3H_2O, allowing some hydrogen exchange with groups on the protein. The chloroplasts were then washed in ordinary water in the dark, thus removing the exchangeable tritium from groups still exposed in the dark. Isolation, purification and assay of CF_1 from this preparation showed the presence of 50–100 atoms of tritium per CF_1. Clearly these atoms are exchangeable only in the light and therefore remain labelled when the preparation is washed in the dark. This indicates a substantial conformational change in CF_1. The tritium incorporation could be prevented by uncouplers. Labelling, indicating the formation of the 'energised' conformational state, could be obtained with light, with ATP hydrolysis in the dark or with an acid–base wash.

Such results as these lead to the conclusion that the proton motive force, phosphorylation by the ATPase and conformational changes in the ATPase are all related, although the full nature of the relationship is a matter of debate. Some conformational changes associated with the photosystems will be outlined below (sect. 14.10(c)).

14.10 Photophosphorylation

(a) Estimates of the P/2e ratio

Earlier it was seen that measurements of the P/O ratio in oxidative phosphorylation assisted understanding of the process. Attempts to obtain comparable precise data for non-cyclic photophosphorylation have been less successful. The literature contains a wide range of values, but it now seems that for the transfer of two electrons from water to NADP the probable value is about 1.5 to 2.0 (see Reeves and Hall, 1978). If the electron acceptors ferricyanide and methyl viologen are used in place of NADP, slightly lower values are obtained. The two photosystems can be assayed separately (see Fig. 14.13). Thus PS II

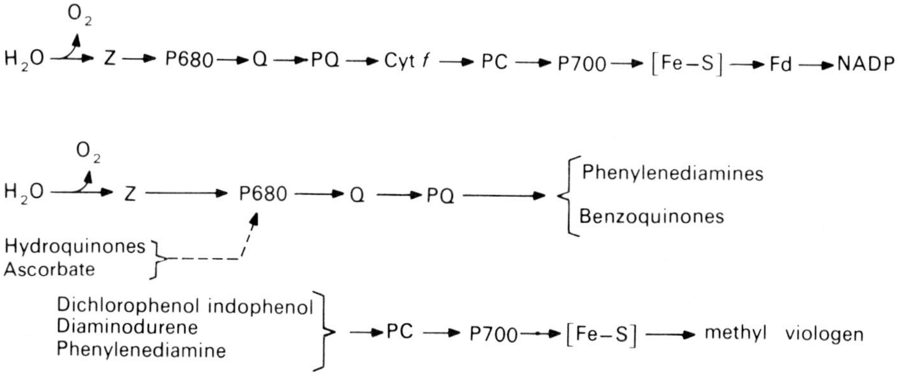

$$O_2$$
$$H_2O \xrightarrow{\quad} Z \rightarrow P680 \rightarrow Q \rightarrow PQ \rightarrow Cyt\ f \rightarrow PC \rightarrow P700 \rightarrow [Fe-S] \rightarrow Fd \rightarrow NADP$$

$$O_2$$
$$H_2O \xrightarrow{\quad} Z \xrightarrow{\qquad} P680 \xrightarrow{\quad} Q \xrightarrow{\quad} PQ \xrightarrow{\quad} \begin{cases} \text{Phenylenediamines} \\ \\ \text{Benzoquinones} \end{cases}$$

Hydroquinones ⎱
Ascorbate ⎰

$$\left.\begin{array}{l} \text{Dichlorophenol indophenol} \\ \text{Diaminodurene} \\ \text{Phenylenediamine} \end{array}\right\} \rightarrow PC \rightarrow P700 \rightarrow [Fe-S] \rightarrow methyl\ viologen$$

Fig. 14.13 Assay of photophosphorylation using artificial electron donors and acceptors.

activity using water as electron donor and phenylenediamine, diaminodurene or dimethyl-*p*-benzoquinone as electron acceptor, together with dibromothymoquinone to prevent reduction of PS I, gives a P/2e ratio of about 0.6. PS I activity with methyl viologen as electron acceptor and DCPIP, diaminodurene or phenylendiamine as electron donor gives a P/2e ratio between 0.5 and 1.0. While the values obtained are very variable, they are consistent with the proposition that phosphorylation is associated with both PS I and PS II, and that the sum of phosphorylations associated with the two photosystems is close to that obtained when full non-cyclic photophosphorylation is assayed with water as donor and methyl viologen as acceptor.

There has been a desire to determine the 'true' P/2e ratio (comparable to the 3.0 obtained from mitochondria with NADH as substrate and oxygen as electron acceptor). The figure 2.0 has been widely considered as the most likely value. This can be interpreted to mean that one phosphorylation is associated with each pigment system, the transfer of two electrons through each photosystem being coupled to the synthesis of one molecule of ATP. There are, however, two reasons why this theoretical figure might not be obtained experimentally.

Firstly, when NADP is used as the electron acceptor, it is possible that 'pseudocyclic phosphorylation' occurs in addition to non-cyclic, that is ATP synthesis not associated with either NADP reduction or *net* oxygen evolution (see Fig. 14.14). This will tend to give a high P/2e ratio, since the estimate of electron transport through the photosystems will be too low. Evidence in support of such pseudocyclic phosphorylation has been obtained (see Allen, 1975).

Secondly, even carefully prepared chloroplasts show a low rate of uncoupled electron transport. This low rate may be subtracted from the electron transport obtained under phosphorylation conditions in order to obtain the true P/2e ratio. If this procedure is justified, higher P/2e ratios are obtained approaching a value of 2 for the oxidation of water and reduction of NADP.

The prime function of the light reaction in photosynthesis is to provide cofactors for the operation of the Calvin cycle. This requires three molecules of

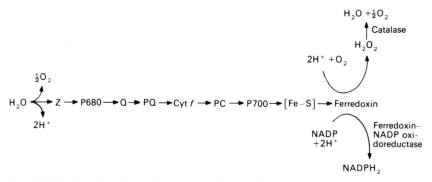

Fig. 14.14 Pseudocyclic and non-cyclic electron flow.

ATP and two molecules of $NADPH_2$ for each molecule of CO_2 fixed and is equivalent to a P/2e ratio of 1.5. Although the Calvin cycle is not the only system in the chloroplast requiring ATP and reduced NADP, it is quantitatively the most important. Thus the metabolic system of the particle has a cofactor requirement equivalent to a P/2e ratio of 1.5 which broadly corresponds to that obtained experimentally. While this correspondence between metabolic requirement and the photophosphorylating electron transport chain is satisfying, it takes no account of the cyclic processes (see below). However, if an ATP-exporting shuttle were operative (sect. 12.5), a higher ratio might be expected.

(b) Sites of phosphorylation

The implication of the foregoing argument is that two sites of phosphorylation might be expected in the non-cyclic pathway, one associated with each pigment system and with the synthesis of one molecule of ATP for two electrons transferred through the site. The P/2e ratio deviates from this ideal figure of 2.0 for reasons not fully understood but in part related to uncoupled electron transport and pseudocyclic processes.

The location of these sites has been problematic. Phosphorylation has not been detected in association with the reducing side of PS I and NADPH reduction. Evidence for phosphorylation coupled to the oxidation of water by PS II has been published, but its significance has been questioned (see, however, (c) below). There has been almost universal agreement that ATP synthesis is associated with electron transport between PS II and PS I. The location of sites in mitochondrial phosphorylation was elucidated in part by the study of cross-over points in relation to the availability of ADP (see sect. 2.3(g)). Comparable studies have been made with chloroplasts. Thus Bohme and Cramer (1972) showed that ADP was necessary for electron transport from the plastoquinone pool to cytochrome f. In the absence of an electron acceptor, cytochrome f can be oxidised by far-red light and reduced by red light. However, with an electron acceptor (methyl viologen) and red light (643 nm), cytochrome f is oxidised (presumably by partial activation of PS I), but now addition of ADP is necessary for reduction by red light. The ATPase inhibitor Dio-9, will inhibit the ADP effect, indicating that ADP is acting through the ATPase. Addition of uncoupler, NH_4Cl, results in strong reduction of cytochrome f, since electron transport from PS II to the cytochrome now no longer requires ADP.

Thus the coupling of ATP synthesis to the oxidation of plastoquinone and reduction of cytochrome f can be demonstrated, although it is much more difficult to demonstrate a second coupling site. Indeed, if the chemiosmotic theory were correct, only one clear coupling site could be demonstrated, since the second ATP would be synthesised as a consequence of release of protons internally during water splitting and uptake externally during NADP reduction. It should be noted that numerous other claims for phosphorylation sites have been made but not adequately substantiated.

(c) Mechanism of phosphorylation

The theories of the mechanism of phosphorylation were outlined earlier. Most workers in the field of photosynthesis have shown a strong preference for the chemiosmotic mechanism of Mitchell, presumably because of the demonstration of the necessary proton gradients and their effects on the ATPase. The chloroplast electron transport chain was shown to translocate four protons for two electrons transferred from water to NADP. The proton translocating loops are: (a) the oxidation of water and reduction of plastoquinone with an $H^+/e = 1$; and (b) the oxidation of plastoquinone and reduction of NADP (see Fig. 14.12) also with an $H^+/e = 1$. If it is assumed that the ATPase translocates two protons per phosphorylation ($P/H^+ = 0.5$), then the P/2e ratio would be 2. Such a figure is consistent with experimental work.

Evidence in support of a conformational theory has also been obtained, particularly in the case of the ATPase outlined earlier. Conformational changes in membrane structure associated with the light reaction have been demonstrated. The labelling of membrane proteins by diazoniumbenzene sulphonate shows that illumination exposes parts of PS II not accessible in the dark (see sect. 14.5(a)). This change correlates with proton release (Giaquinta *et al.*, 1975). Detailed studies of changes in membrane conformation suggest that they may be too slow to be directly involved in the process of photophosphorylation. Cytochrome b_{559} is known to be capable of existing in two different redox forms, but most of the evidence suggests that only the high-potential form is found in unaltered chloroplasts.

(d) Cyclic photophosphorylation

The study of the cyclic process has been hampered by the difficulty of measuring accurately the electron flow; most investigators have relied primarily on measurements of phosphorylation under conditions where there is no net oxidation–reduction. Initial work stemmed from the discovery that the addition of appropriate catalysts to chloroplasts greatly stimulated phosphorylation which was independent of PS II. It was noted that optimal conditions required the catalyst to be partly but not fully reduced. The catalysts included phenazine methosulphate and pyocyanine, which were capable of inducing very high rates of ATP synthesis, while DCPIP, menadione, TMPD, diaminodurene, ferredoxin and FMN gave a lower rate. In each case examined, it is clear that the cyclic process is dependent on far-red light, is insensitive to DCMU[4] and is driven by

[4] DCMU can influence the functioning of cyclic photophosphorylation by affecting the redox state of the system, which can be critical for the cyclic process.

326

the PS I photoact. Analysis of the reactions involved shows that they do not all function in the same way. Thus Hauska *et al.* (1974) noted that the plastoquinone antagonist, dibromothymoquinone, inhibited photophosphorylation associated with the lipid-insoluble catalysts such as ferredoxin and lipophilic quinones with low redox potentials such as menadione. It is therefore believed that these compounds catalyse a cyclic electron flow involving the plastoquinones (see Fig. 14.15). In contrast, phenazine methosulphate and pyocyanine are insensitive to dibromothymoquinone and to the plastocyanin inhibitor KCN. They catalyse a high rate of phosphorylation which only shows light saturation at high intensities, suggesting a very short cycle around PS I without the rate-limiting steps associated with the other catalysts (Fig. 14.15a). A third type of catalyst is represented by DCPIP and the phenylenediamines. These promote a cyclic photophosphorylation which is inhibited by KCN (and therefore involves plastocyanin) but is only partly inhibited by dibromothymoquinone. All these catalysts will not only promote a cyclic flow of electrons coupled to phosphorylation but also participate in the transfer of protons across the membrane. Their function is therefore readily interpreted in terms of the chemiosmotic theory.

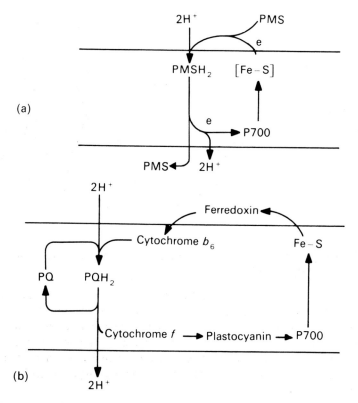

Fig. 14.15 Cyclic electron flow and proton transport. (a) Cyclic flow round P700 catalysed by phenazine methosulphate (PMS). (b) Ferredoxin-catalysed cyclic flow through cytochrome b_6.

It has frequently been questioned whether the cyclic process has a role in photosynthesis, especially since it serves only to synthesise ATP and not to reduce NADP, whereas the non-cyclic process appears to synthesise both cofactors at about the required relative proportions. Arguments in favour of cyclic flow *in vivo* are as follows. Firstly, a number of PS I systems in the chloroplasts of higher plants appear to be unlinked to PS II systems and therefore to have no function other than cyclic phosphorylation. These systems occur outside the grana. Secondly, the addition of dihydroxyacetone phosphate to chloroplasts blocks CO_2-dependent oxygen evolution (non-cyclic electron flow), but does not block CO_2 fixation which, under suitable conditions, can occur anaerobically. This CO_2 fixation must therefore be dependent on cyclic photophosphorylation to provide the necessary ATP. The use of dihydroxy-acetone phosphate avoids a requirement for $NADPH_2$ (Fig. 14.16). Thirdly, full fixation of CO_2 in intact chloroplasts has been found to require the cyclic process. Study of other energy-dependent systems (i.e. anaerobic assimilation of glucose in cells requiring ATP which can be provided only by the cyclic process) has shown that they are light-driven under conditions where non-cyclic electron flow is blocked (by DCMU, for example).

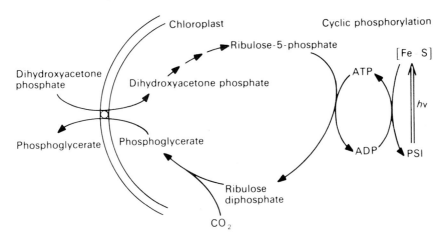

Fig. 14.16 Carbon dioxide fixation in the presence of dihydroxyacetone phosphate driven by cyclic photophosphorylation (see text).

Thus in the intact systems (intact chloroplasts, cell suspensions of algae) cyclic photophosphorylation occurs in the absence of added cofactors. What is the endogenous cofactor? The most likely candidate is ferredoxin, since this substance catalyses phosphorylation at reasonable rates and the process is sensitive to antimycin A and dibromothymoquinone, which are known to inhibit phosphorylation. Further, ferredoxin appears to be necessary for cytochrome b_6 reduction by PS I.

It has been proposed, though not fully demonstrated, that pseudocyclic phosphorylation may also have a physiological role in ATP synthesis under conditions where NADP becomes reduced.

14.11 Energetics

(a) Quantum yield

The quantum yield (or quantum efficiency) for photosynthesis may be expressed as the number of oxygen molecules evolved per quantum of light absorbed. However, it has frequently proved more convenient to use the reciprocal of the yield, the quantum requirement, i.e. the number of quanta required for the evolution of one molecule of O_2. Measurements of the quantum yield or requirement have been made by many workers with variable results. Such measurements are attractive in that they allow an assessment of the overall photosynthetic process in whole algal cells such as *Chlorella*. Early studies produced a controversy between Emerson who obtained values of 8 to 10 quanta per molecule of O_2 evolved at maximum efficiency and Warburg who obtained values as low as 2.8 to 4. More recent evidence has supported Emerson; for example Govindjee *et al.* (1968) obtained a value of 8.3 quanta at maximum efficiency. It is now accepted that values somewhat higher than 8 represent maximum efficiency.

Since the evolution of one molecule of O_2 is normally coupled to the fixation of one molecule of CO_2, these figures also represent the quantum requirement for CO_2 fixation. The operation of the non-cyclic pathway will evolve one molecule of O_2 and reduce the two molecules of NADP required for the fixation of one molecule of CO_2. This will require 8 quanta to transfer the four electrons through each of the two photosystems. Theoretically, if the P/2e ratio is 1.5 or more for non-cyclic photophosphorylation, adequate ATP for CO_2 fixation will be formed. If, however, there is some ATP wastage, or if ATP is also required simultaneously for other processes, ATP synthesis by cyclic photophosphorylation will occur, resulting in more than 8 quanta being required for fixation of one molecule of CO_2. These data support the schemes outlined earlier.

(b) Thermodynamic considerations

The fixation of CO_2 in the chloroplast can be summarised by:

$$CO_2 + H_2O \longrightarrow O_2 + (CH_2O) + 477 \text{ kJ}$$

The total energy, provided as light energy, required for fixation of CO_2 is 477 kJ/mol CO_2. Each quantum of light at 700 nm is equivalent to 171 kJ/mol and therefore the theoretical quantum requirement is $471/171 = 2.8$. If we accept an approximate experimental figure of 8.5 quanta per C atom fixed, the *maximum* efficiency of the energy conversion process $= 2.8/8.5 = 33$ per cent. In other words, about two-thirds of the radiation energy absorbed by chlorophyll is lost during the photosynthetic process. A substantial part of this will be lost in the metabolic processes. Can we expect losses in the conversion of light energy into chemical energy in the photosystem reaction centres? To put the question in another way, can we expect the full 1.8 eV (equivalent to 171 kJ/mol and to 1 quantum of far-red light at 700 nm) to be available? The answer is clearly no. Firstly, there must be an immediate energy loss in the process of absorption since this is not fully reversible; the absorber is not in equilibrium with the radiation falling on it and consequently it does not radiate at the same rate as it absorbs (if it did, the quantum yield would be zero!). Thus the entropy of the absorber must be greater than that in the radiant field. Secondly, the energy conversion reaction in which radiant energy is converted to chemical energy also involves an increase in entropy.

Several investigators have attempted to estimate the free energy (ΔG) available from a quantum of red light absorbed by a photosystem and hence the proportion of the 1.8 eV available to do chemical work in the oxidation–reduction reaction. The simplest approach taken by Duysens (see Duysens and Amesz, 1959) and by Spanner (1964; see also Knox, 1969) is to consider the photosynthetic system as a heat engine of classical thermodynamics. The maximum efficiency of such an engine, as derived from the second law of thermodynamics, is given by

$$\eta_{max} = \left(1 - \frac{T_2}{T_1}\right)$$

where T_2 is the absolute temperature of the photosynthetic system (say 300 K) and T_1 is the temperature attributable to the monochromatic radiation (in practice a finite band between λ and $\lambda + \delta\lambda$) of known intensity.

The significance of T_1 has to be understood in terms of the temperature of a black body radiating energy according to the relationship shown in Fig. 14.17 which shows the intensity of radiation against wavelength and also how this varies with change in temperature. It is assumed that with monochromatic light, the black body radiator is surrounded by a filter transmitting fully the monochromatic radiation of interest and totally absorbing all other wavelengths. The temperature T_1 is that temperature of the black body which

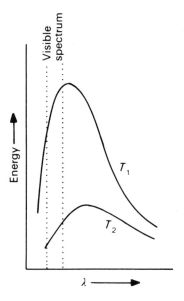

Fig. 14.17 Energy emitted from a black body radiator at two temperatures, T_1 and T_2.

radiates the appropriate intensity of the monochromatic light under consideration. In Duysens's study, T_1 was estimated at 1100 K.

$$\text{Hence} \quad \eta_{max} = 1 - \frac{300}{1100} = 0.73$$

Maximum efficiency is thus 73 per cent, giving rise to free energy equivalent to 1.31 eV. It should be noted that this will vary with intensity.

A more thorough treatment by Ross and Calvin (1967) produced a maximum figure of 1.19 eV for PS I and 1.23 eV for PS II assuming an effective incident intensity of white light of 1000 lux, giving an efficiency of about 67 per cent. Thus it is clear that only about two-thirds of the energy absorbed by the photosystems themselves is available as free energy. Such a potential is adequate but not excessive for the reactions outlined in Fig. 14.2 where, for example, the standard mid-point potential span is probably about 0.93 eV between PS I and its primary acceptor.

(c) Proton gradients and membrane potentials

A quantitative approach to the bioenergetics of photophosphorylation is possible if the chemiosmotic theory is assumed to be valid. This measures the energy conserved by the proton pump as the proton motive force (see sect. 8.3 and sect. 9.11).

$PMF = \Delta\psi + Z\Delta pH$ where $Z = 59$ at $25\,^\circ C$ when potentials are expressed in millivolts. Numerous estimates of $\Delta\psi$ have been made, frequently based on shifts in carotenoid absorption (see sect. 14.7), but very variable values have been obtained. However, using flashes of light, Graber and Witt (1976) have observed a potential in excess of 100 mV as a consequence of illumination. Their results are in agreement with those obtained by methods based on luminescence measurements. With a pH difference of 2.0, Witt's group found that ATP synthesis in flashing light was proportional to the membrane potential, being zero at zero $\Delta\psi$. Under continuous illumination with the external phase maintained at pH 8, the internal phase falls to pH 5. At steady state, the membrane potential is probably about 100 mV, although there are reports of lower values. Assuming these conditions,

$$PMF = 100 + 59 \times 3$$
$$= 277 \text{ mV}$$

since $\Delta G = \Delta E \cdot n \cdot F$, this is equivalent to 53.5 kJ.

The energy requirement for ATP synthesis has been determined by Kraayenhof (1969) who allowed intact (class I) chloroplasts to carry out photophosphorylation. These chloroplasts showed a state 3 to state 4 transition as in mitochondria. In state 4, the phosphate potential at pH 7.6 gave an approximate value of $\dfrac{(ATP)}{(ADP)(Pi)} = 40\,000 \text{ M}^{-1}$. Using a $\Delta G^* = 30 \text{ kJ/mol}$ for the synthesis of ATP (Rosing and Slater, 1972), the energy required for ATP synthesis at $25\,^\circ C$ is given by

$$\Delta G = \Delta G^{\ominus} + RT \ln \frac{(\text{ATP})}{(\text{ADP})(\text{P}i)}$$

$$= 30\,000 + 8.314.298 \ln 40\,000$$

$$= 56.3 \text{ kJ/mol}$$

Taking into account the difficulties of measurement, such a value, within about 6 per cent of the value obtained from estimations of the proton motive force, suggests that in state 4, the concentrations of adenine nucleotides correlate with the proton motive force.

Further reading

Baird, B. A. and Hammes, G. G. (1979) Structure of oxidative- and photophosphorylation coupling factor complexes. *Biochim. Biophys. Acta*, **549,** 31–53.

Blankenship, R. E. and Parson, W. W. (1978) The photochemical electron transfer reactions of photosynthetic bacteria and plants. *Annu. Rev. Biochem.*, **47,** 635–53.

Crofts, A. R. and Wood, P. M. (1978) Photosynthetic electron transport chains of plants and bacteria and their role as proton pumps. *Current Topics in Bioenergetics*, **7,** 175–244.

Harris, D. A. (1978) The interaction of coupling ATPases with nucleotides. *Biochim. Biophys. Acta*, **463,** 245–73.

Junge, W. M. (1977) Membrane potentials in photosynthesis. *Annu. Rev. Plant Physiol.*, **28,** 503–36.

Malkin, R. and Bearden, A. J. (1978) Membrane-bound iron–sulphur centres in photosynthetic systems. *Biochim. Biophys. Acta*, **505,** 147–81.

McCarty, R. E. (1979) Roles for a coupling factor for photophosphorylation in chloroplasts. *Annu. Rev. Plant Physiol.*, **30,** 79–104. (see also Ch. 15, Further reading).

Reeves, S. G. and Hall, D. O. (1978) Photophosphorylation in chloroplasts. *Biochim. Biophys. Acta*, **463,** 275–98.

Trebst, A. (1974) Energy conversion in photosynthetic electron transport of chloroplasts. *Annu. Rev. Plant Physiol.*, **25,** 423–58.

Witt, H. T. (1979) Energy conversion in the functional membrane of photosynthesis. Analysis of light pulse and electric pulse methods. The central role of the electric field. *Biochim. Biophys. Acta*, **505,** 355–427.

Also:

Shavit, N. (1980) Energy transduction in chloroplasts: Structure and function of the ATPase complex *Annu. Rev. Biochem.*, **49,** 111–38.

Velthuys, B. R. (1980) Mechanisms of electron flow in photosystem II and toward photosystem I. *Annu. Rev. Plant Physiol.*, **31,** 545–68.

Chapter 15

Bacterial photosynthesis

15.1 Groups of photosynthetic prokaryotes

There are three fundamentally different types of photosynthesis[1] exhibited by three separate groups of bacteria. The study of each group has made a very significant contribution to our understanding of photosynthesis and hence of chloroplast biochemistry and cell energetics.

(a) Cyanobacteria

These are filamentous or unicellular organisms exhibiting oxygen-evolving photosynthesis of a type discussed in previous chapters. They have also been described as blue-green algae (Cyanophyta) because of their general and biochemical similarity to other algae. Like them they possess chlorophyll *a* as the main photosynthetic pigment, together with phycobilin and carotenoid pigments. They are classified as prokaryotes primarily because of their cell structure (see Fig. 15.1d). They lack a well-defined nucleus with a nuclear membrane, mitochondria, clearly defined chloroplasts and other eukaryotic membranous structures. However, these bacteria possess oxygen-evolving photosynthesis (with two pigment systems) and may therefore be considered as the possible origin of the chloroplast in the endosymbiotic theory. The primitive characteristics, together with the ability to evolve oxygen photosynthetically and the capacity of many strains to fix atmospheric nitrogen, are among the most interesting features of the group. As oxygen-evolving photosynthesis has been considered earlier, the cyanobacteria will not be discussed further.

(b) Rhodospirillales

The classical photosynthetic bacteria are now all included in this group of unicellular organisms capable of anaerobic photosynthesis in which oxygen is not evolved. They possess only one pigment system and cannot use water as an electron donor. H_2S and reduced inorganic sulphur compounds, reduced organic compounds or even hydrogen may be used as reductants. The photosynthetic pigments are the bacteriochlorophylls together with carotenoids. There are two suborders, the Rhodospirillineae or purple bacteria and the Chlorobiineae, the green (and brown) sulphur bacteria. The main organisms in the second suborder

---

[1] The word photosynthesis is used here to denote the conversion of light energy into chemical energy and is consistent with the definition given in the *Shorter Oxford English Dictionary* 'Chemical combination caused by the action of light'. A more restrictive definition would exclude the third group discussed, the Halobacteria.

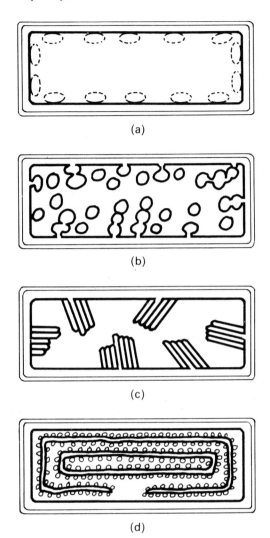

(a)

(b)

(c)

(d)

Fig. 15.1 Diagrammatic representation of photosynthetic prokaryotic cell structure. (a) *Chlorobium* cells with simple plasma membrane and attached non-membranous pigment-containing vesicles. (b) Purple bacterial cells (*Rhodospirillum rubrum, Rhodopseudomonas sphaeroides, Chromatium vinosum*) showing invagination of cell membranes to form vesicular membrane structures. (c) *Rhodospirillum molischianum* showing thylakoid-like invagination of the plasma membrane. (d) Simple cyanobacterial type with intracellular thylakoids with attached phycobilisomes (see sect. 13.3(a) and Fig. 13.4).

are *Chlorobium limicola* (a straight rod-shaped organism) and *C. vibrioforme* (a curved rod). These bacteria are strictly anaerobic and their metabolism is CO_2-, light- and H_2S-dependent. Elementary sulphur is formed as an intermediate in the oxidation of H_2S and is excreted into the medium where sulphur granules are formed. Under suitable conditions the sulphur may be oxidised to sulphate

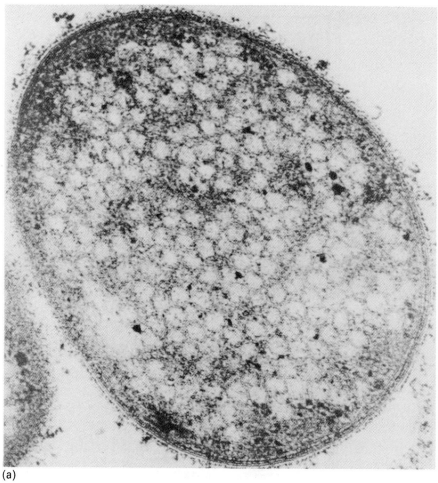

(a)

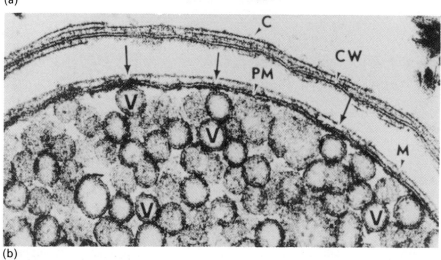

(b)

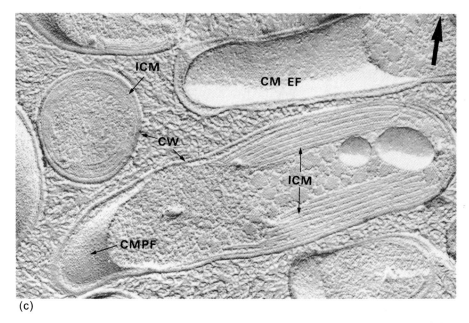

(c)

Fig. 15.2 Electron micrographs of photosynthetic bacteria, (a), (b) Thin sections of the purple sulphur bacterium *Thiocapsa floridana*, (a) showing whole cells, × 108 000, and (b) detail of the membranes in a preparation from which the ribosomes have been released. The site of association of the intracellular vesicles, V, (chromatophores) with the plasma membrane (PM) is arrowed. The outer layer represents the cell wall (CW) with some capsular material (C) attached, × 167 000. (Courtesy of Dr. B. J. Takacs, from Takacs, B. J. & Holt, S. C. (1971) *Biochim. Biophys. Acta* **233**, 258–77. (c) Freeze-etched electron micrograph of the purple non-sulphur bacterium *Rhodopseudomonas palustris*. To the left a fracture face across the long axis of a rod may be seen showing the cell wall (CW) and the intracytoplasmic membranes (ICM). A fracture face along a rod may be seen in the centre. The EF and PF of the plasma membrane (cytoplasmic membrane, CM) are also seen, × 51 500. (Courtesy of Dr. J. R. Golecki, Albert-Ludwigs Universität, Freiberg, W. Germany).

almost completely; CO_2 is fixed, but probably not by the pentose phosphate pathway. The genus *Chloroflexus*, a green filamentous gliding organism, is also included in this group, although it can grow aerobically on organic media (like purple bacteria).

The Rhodospirillineae (purple bacteria) may be divided into the purple sulphur bacteria (Chromatiaceae), of which *Chromatium* and *Thiocapsa* are characteristic organisms, and the purple non-sulphur bacteria (Rhodospirillaceae) characterised by *Rhodospirillum*, *Rhodopseudomonas* and *Rhodomicrobium*, etc. The purple sulphur bacteria will metabolise H_2S, but the sulphur is deposited intracellularly and can be oxidised further to sulphate. The purple non-sulphur bacteria grow best under anaerobic conditions on organic media in the light. Most species can also grow aerobically on organic media, but under these conditions the photosynthetic pigments are repressed.

In general photosynthetic bacteria have offered rather simpler systems for

study and have contributed very considerably to our understanding of photosynthetic reaction centres.

(c) Halobacterium

In recent years considerable interest has been shown in this salt-loving
organism which possesses a system for the conversion of light energy into
chemical energy in which chlorophylls are not involved. The organism possesses
a light-absorbing protein, bacteriorhodopsin, similar to the mammalian eye
pigment, rhodopsin. Illumination of the pigment results in ATP synthesis and the
mechanism is of particular interest in relation to our understanding of oxidative
phosphorylation and photophosphorylation.

15.2   Organelles of photosynthetic bacteria (Rhodospirillales)

(a) Purple bacterial chromatophores

In 1952, Schachman *et al.* isolated pigmented particles from the purple
non-sulphur bacterium, *Rhodospirillum rubrum*. These particles had a diameter
of about 60 nm and were subsequently shown to be vesicles bounded by a unit
membrane. Photophosphorylation by such particles (chromatophores) was first
reported by Frenkel (1954) who thus demonstrated that they possessed the
photochemical apparatus. Similar structures are found in other purple bacteria.

Chromatophores can be prepared from broken cells by differential
centrifugation and purified by density gradient centrifugation. It is also possible
to separate chromatophore membranes from cytoplasmic or plasma membranes
by the same means.

Electron micrographs of sections of purple bacterial cells show that
chromatophores are not discrete membranous vesicles but interconnected
structures. They are also connected with the plasma membrane (Figs. 15.1 and
15.2).

Although this is not universally accepted, it seems that the intracytoplasmic membranes are probably formed by invagination of the cytoplasmic
membrane (see Oelze, 1976). Support for this view comes from studies of purple
non-sulphur bacteria. As noted above, these organisms will grow photosynthetically in the light under anaerobic conditions but chemo-organotrophically in the
presence of oxygen which, above a critical partial pressure, inhibits bacteriochlorophyll synthesis and cytoplasmic membrane formation. The transition
from the chemo-organotrophic state with only the plasma membrane, to the
photosynthetic state with intracytoplasmic organelles, proceeds by the progressive invagination of the plasma membrane. Transition back to aerobic
metabolism involves the reverse process. Grown under aerobic conditions, the
plasma membrane of *Rhodospirillum rubrum* possesses an NADH and succinate
oxidase system with flavoproteins, ubiquinone, a *b*-type cytochrome and
cytochrome *o*. *Rhodopseudomonas sphaeroides* possesses a *c*-type cytochrome
and cytochrome oxidase; a transhydrogenase is also found. When intracytoplasmic membranes are present, the NADH oxidase system is found in both
cytoplasmic and plasma membranes. In organisms capable of photosynthesis,
light inhibits respiration.

Table 15.1 Lipid composition of light-grown cells of purple bacteria

	Percentage composition	
Lipid	*Rhodospirillum rubrum*	*Chromatium*
Phosphatidylcholine	6	
Phosphatidylethanolamine	57	47
Phosphatidylglycerol	29	53
Diphosphatidylglycerol (cardiolipin)	8	

The lipid composition of the bacterial membrane is shown in Table 15.1. The major phospholipids are phosphatidylethanolamine and phosphatidylglycerol. The major fatty acids have chain lengths of C_{16} and C_{18} with only one double bond.

(b) Organelles of green sulphur bacteria
The major difference between these organisms and the purple bacteria is that the photosynthetic apparatus is exclusively located in the cytoplasmic membrane, while the chlorobium vesicles contain only the light-harvesting bacteriochlorophylls and not the reaction centres (Fig. 15.1a). Unlike the chromatophores of purple bacteria, the vesicles do not have an organised membrane.

(c) Photosynthetic pigments
Several types of bacteriochlorophyll have been identified (Fig. 15.3). Bacteriochlorophylls *a*, *c*, *d* and *e* are found in the green sulphur bacteria. By contrast bacteriochlorophyll *a* is characteristic of purple bacteria although a few species have bacteriochlorophyll *b* instead. A wide variety of carotenoids are associated with the bacteriochlorophylls as protective agents and as light-harvesting pigments. The pigments associated with the photosynthetic apparatus vary greatly. The carotenoids spirilloxanthin and lycopene are found in many purple non-sulphur bacteria, okenone in purple sulphur bacteria and other aromatic carotenoids such as chlorobactene and isorenieratene in green sulphur bacteria.

15.3 Photochemistry

The overall arrangement of pigments in photosynthetic bacteria is comparable to that found in chloroplasts. Photosystems are composed of about 35 bacteriochlorophyll molecules in *Rhodospirillum*, rather more in *Rhodopseudomonas*, with a reaction centre capable of carrying out a light-driven oxidation–reduction reaction. In contrast with oxygen-evolving organisms, there is only one photosystem. The carotenoids can transfer their excitation

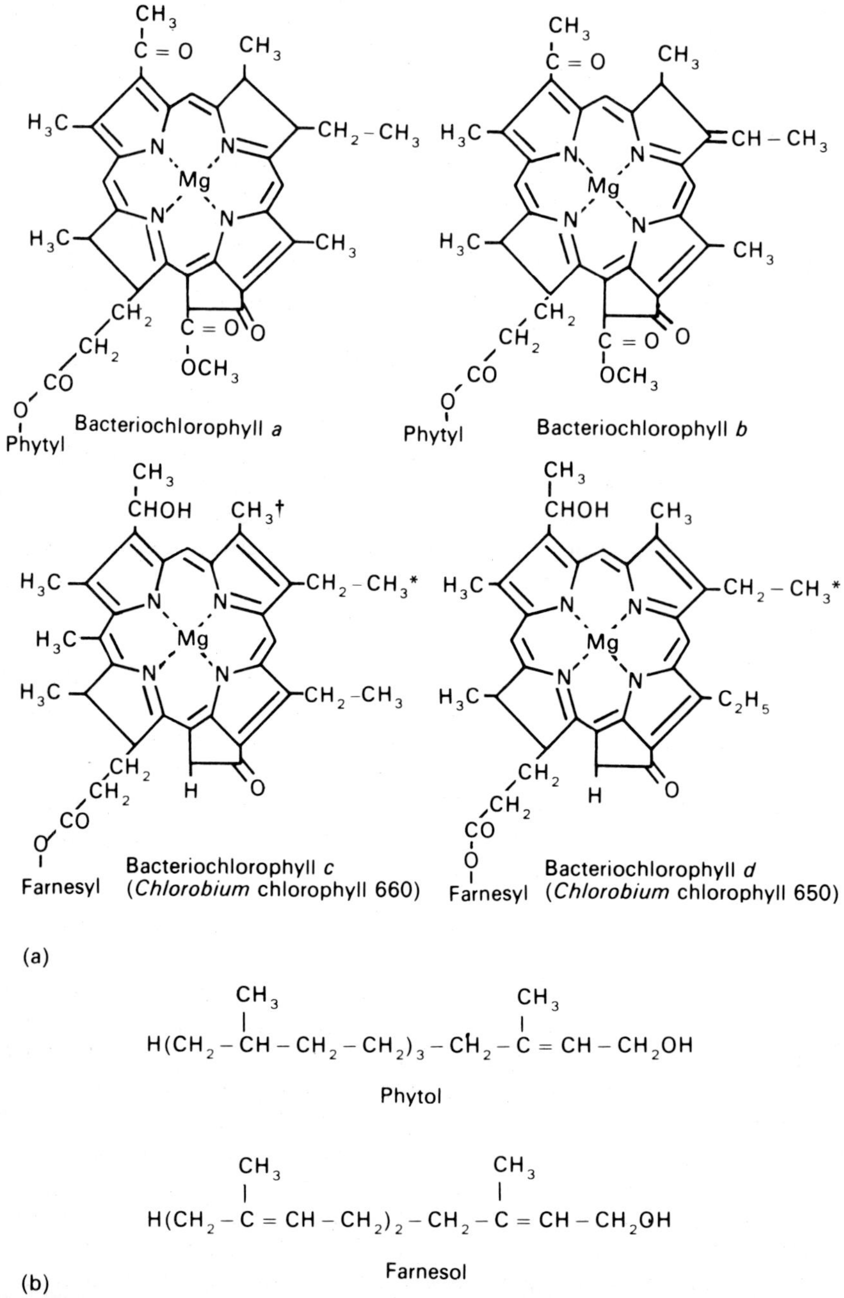

Fig. 15.3 (a) Bacteriochlorophylls (see Caple *et al.*, 1978, Gloe, *et al.*, 1975). (b) Phytol and farnesol. Note that (i) bacteriochlorophyll *a* may possess either phytyl or geranylgeranyl side chains; (ii) bacteriochlorophylls *c* and *d* may possess in place of $-CH_2-CH_3$*, either $-C_3H_7$ or $-C_4H_9$ (iii) bacteriochlorophyll *e* (not shown) is similar to bacteriochlorophyll *c* except that the methyl group ($-CH_3$†) is replaced by $-CHO$ and may possess either farnesyl or phytyl side chains.

energy to bacteriochlorophyll, but estimates of the efficiency of transfer have varied. Antenna pigment–protein complexes have been isolated from several organisms. From green sulphur bacteria, a water-soluble complex (MW = 150 000 approx.) has been isolated. This can be dissociated into three identical subunits each containing seven strongly interacting molecules of bacteriochlorophyll *a*. This complex functions in the transfer of energy from the antenna pigments to the reaction centres (Fenna and Matthews, 1975). From purple bacteria several pigment–protein complexes have been isolated. A polypeptide (MW about 10 000) is associated with a complex I which has an absorption maximum at about 875 nm. Complex II with absorption maxima at about 800 and 850 nm has three polypeptides with a small number of bacteriochlorophyll molecules. The molecular structure of these complexes is, however, not yet clear (see Feick and Drews, 1978).

In the early 1960s, Clayton developed the idea of a reaction centre in studies with *Rhodopseudomonas* where bleaching in the infra-red (even at very low temperature) was attributable to such a centre (see Clayton, 1963). The reaction centre has an absorption maximum at 870 nm and is therefore referred to as P870. The position of the absorption band varies between species (especially those with bacteriochlorophyll *b* such as *Rhodopseudomonas viridis* with a centre P960). The existence of two different reaction centres in some organisms has been proposed, but this is now largely discounted.

In 1968, Reed and Clayton succeeded in isolating a reaction centre complex free of antenna molecules from *Rhodopseudomonas sphaeroides* by extracting chromatophores with detergents. The purest preparations have a particle weight of 70 000, with one or two molecules of ubiquinone and one non-haem iron atom per molecule of P870. The complex is free of cytochromes and copper but possesses one molecule of carotenoid. The carotenoid requirement is not specific and one of several carotenoids is able to function at the reaction centre. The photo-oxidation of P870 in the reaction centre complex can be demonstrated by exposure to a single short flash of light; photo-oxidation is followed by relatively slow reduction which can be accelerated by addition of a reduced soluble *c*-type cytochrome, the cytochrome becoming oxidised as the P870 is reduced. Analysis of the reaction centre shows that it is composed of four molecules of bacteriochlorophyll *a* and two of bacteriophaeophytin. Three very hydrophobic polypeptides are associated with the complex (MW approx. 21 000, 23 000 and 27 000) in a 1 : 1 : 1 stoicheiometry. The heaviest of the three can be removed together with iron but the residual complex of the two smaller polypeptides retains the pigments and photochemical activity.

A study of absorption spectra of reaction centres has suggested three pigment groups. P870 is associated with two bacteriochlorophyll molecules which are bleached on illumination. P800 is also associated with bacteriochlorophylls but undergoes only small shifts in absorbance when P870 is bleached while P760 also shows small changes and is associated with bacteriophaeophytin (Prince *et al.* 1977). P870 apparently exists in comparable complexes in purple and green sulphur bacteria, although the degree of purification described for *Rhodopseudomonas sphaeroides* cannot be so readily attained.

15.4   Reaction centres

The oxidation of the P870 complex is associated with reduction of a component X which in turn reduces the secondary electron acceptor, ubiquinone (Fig. 15.4). Chromatophore membranes contain 5–10 molecules of ubiquinone per reaction centre and the quinone is reduced in the light. When the bulk of the quinone has been extracted, the P870 can still be photo-oxidised with a first flash of light (reducing X), but a second flash is without effect. Addition of ubiquinone to the preparation restores reactivity with the second flash. Thus ubiquinone is the secondary electron acceptor, although in some organisms (*Chromatium vinosum*) it is replaced by menaquinone (Romijn and Amesz, 1977).

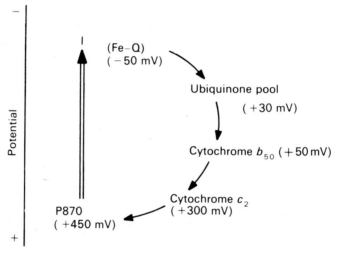

Fig. 15.4 Bacterial cyclic electron flow

The identity of the primary electron acceptor in the chain, X, has been a matter for considerable debate. There have been two rival candidates for this role, a specialised ubiquinone molecule or a non-haem iron protein. Optical measurements support the formation of a semiquinone in the primary photo-chemical reaction. Furthermore, *complete* removal of quinone causes a loss of photochemical activity which can be be restored by titrating back one quinone per reaction centre. Evidence for non-haem iron as an electron acceptor comes from the demonstration of an EPR signal which can be detected after a single flash of light. Various studies suggest that this signal is derived from a non-haem iron of the same mid-point potential as the primary acceptor. Since either a non-haem iron or an ubiquinone molecule appears to act as primary acceptor, it has been suggested more recently that some form of iron–ubiquinone complex could be the elusive electron carrier acting between the P870 reaction centre and the quinone pool.

The nature of the reaction at the centre itself has also been problematic. However, several approaches (see Parson and Cogdell, 1975) suggest that the reaction P870 $\longrightarrow$ P870$^+$ results in an unpaired electron being shared by only two of the four bacteriochlorophylls at the reaction centre. Prior to the reaction, the four bacteriochlorophylls interact strongly and the two bacterio-phaeophytins much less strongly. The formation of P870$^+$ breaks the interaction with two of the bacteriochlorophylls. A study of the kinetics of bacteriochlorophyll *a* oxidation show that BChl$^+$ is formed regardless of the oxidation–reduction state of the proposed Fe–Q complex. This has been interpreted as requiring an intermediate carrier to accept the electron from BChl. Such a component would function between the reaction centre bacteriochlorophyll and the primary acceptor. Oxidation of the bacteriochlorophyll dimer, moreover, takes about 10 ps whereas reduction of X requires 100–200 ps. Bacterio-phaeophytin has been identified as the intermediate.

The formation of P870$^+$ is followed by the oxidation of a *c*-type cytochrome as first observed by Duysens (1954). This oxidation can occur at low temperatures, suggesting an intimate relationship between the cytochrome and the reaction centre.

The sequence of events occurring at the reaction centre can be sum-marised as follows where *c* represents the electron donor (cytochrome c_2), BChl–BChl the bacteriochlorophyll dimer, I the intermediate carrier (bacterio-phaeophytin) and X the primary acceptor, probably an iron–ubiquinone complex.

$$c[\text{BChl-BChl I}]X \xrightarrow{h\nu} c[\text{BChl-BChl* I}]X \longrightarrow c[\text{BChl-BChl}^+ \text{I}^-]X$$
$$\longrightarrow c[\text{BChl-BChl}^+ \text{I}]X^- \longrightarrow c^+[\text{BChl-BChl I}]X^-$$

The quantum efficiency of the reaction centre appears to be close to 1.

15.5   Electron transport chain

(a) Electron donor to P870

In *Rhodopseudomonas sphaeroides*, two *c*-type cytochromes (c_2) act as electron donors to P870. Excitation of the reaction centre results in the photo-oxidation of first one and then the other cytochrome. These are loosely bound to the chromatophore membrane. Cytochrome c_2 has an E_m of about $+300$ mV and can be solubilised and purified. The structure is similar to that of the mitochondrial cytochrome *c*. In *Rhodospirillum rubrum*, two distinguishable *c*-type cytochromes are found, the high-potential pigment being of the c_2 type and photo-oxidised by P870, while the low-potential one, *c'* ($E_{m7.0}$ about 0 V), may also be photo-oxidised by P870. In *Chromatium*, cytochrome c_{555} is firmly bound to the membrane, has an $E_{m7.0}$ of about 300 mV and is photo-oxidised by P870 directly. A second *c*-type cytochrome, c_{553} (with two haems per molecule and an $E_{m7.0}$ probably around 165 mV) can also be photo-oxidised under conditions where c_{555} is already oxidised. The green bacterium *Chlorobium vinicola* has two identical *c*-type cytochromes ($E_{m7.0}$ about $+250$ mV) in reaction centre preparations. In the purple bacteria, soluble *c*-type cytochromes also appear to act as electron donors to the reaction centre complex.

(b) Cyclical electron transport: cytochrome b_{50}

Nishimura (1963) showed that antimycin A added to illumi-
nated chromatophores increased the oxidation of cytochrome c_2 and reduced a
b-type cytochrome (b_{50}) in *Rhodopseudomonas sphaeroides*. Cytochrome b_{50}
($E_{m\,7.0} = +50\,mV$) can be rapidly reduced by a flash of light in the presence of
antimycin. The rate of oxidation of this cytochrome can be shown to match the
rate of cytochrome c_2 reduction under suitable conditions. Thus this
cytochrome, found in green and purple sulphur bacteria as well as in non-sulphur
bacteria, acts in a cyclical pathway between the electron-donating (reducing) and
electron-accepting (oxidising) side of P870 (Fig. 15.4). More recently, a further
component of the electron transport chain (Z) has been proposed. Z, the major
donor to cytochrome c_2, is a hydrogen carrier and is reduced by ubiquinol (see
Prince and Dutton, 1977 and Fig. 15.7).

(c) Proton translocation

The existence of a proton gradient and membrane potential were
ingeniously demonstrated by Jackson and Crofts (1968). Chromatophores, on
illumination, showed an uncoupler-sensitive uptake of protons. In the presence
of valinomycin (which is a specific K^+ ionophore, see sect. 7.4) a light-dependent
efflux of K^+ was observed together with proton uptake. The addition of nigericin
(which promotes K^+/H^+ exchange, see sect. 7.4) in the presence of K^+
apparently inhibited light-dependent proton uptake. These results can be
interpreted as showing both a proton gradient and a membrane potential formed
in the light as a consequence of a proton pump (Fig. 15.5).

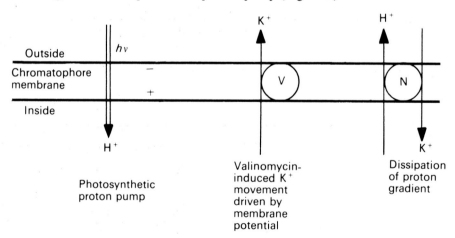

Fig. 15.5 Use of valinomycin (V) and nigericin (N) to demonstrate proton gradient and
membrane potential (see text).

Evidence has been obtained for an asymmetric distribution of the electron
transport system across the membrane. Antibodies to cytochrome c_2 only inhibit
photo-oxidation of c_2 after chromatophores have been disrupted suggesting that
this cytochrome is on the inside face of the membrane. Several experiments have

led to the conclusion that the reducing side of P870 is on the outside of the chromatophore; for example, the secondary electron acceptor takes up protons from the outside on photoreduction (Fig. 15.6).

More recent experiments have shown that there are two phases of proton uptake (see Petty *et al.*, 1977), a rapid phase associated with the initial photochemical reaction and a second phase which may be associated with the reduction of the ubiquinone pool. The second phase of uptake, but not the first, is sensitive to antimycin A and this has led to the formulation of the model of electron transport shown in Fig. 15.7. It is not certain whether the proton release on the inside of the chromatophore occurs when the quinone is oxidised or when cytochrome b is oxidised. It is suggested that proton release may also occur in two phases.

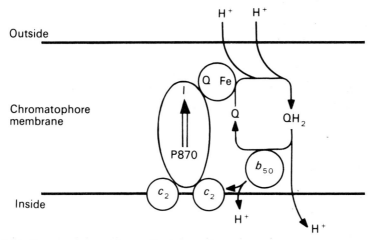

Fig. 15.6 Proton transport by cyclic electron flow

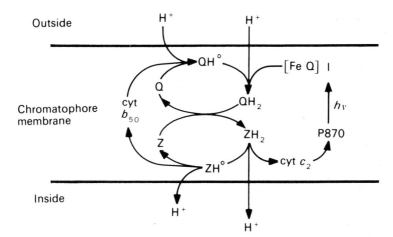

Fig. 15.7 Cyclic electron transport and the proton pump. A modification of Fig. 15.5 to accommodate an additional component, Z, and transport of two protons per electron (modified from Petty *et al.* 1977).

15.6   Photophosphorylation

Following the early studies in chloroplasts which differentiated cyclic and non-cyclic phosphorylation, the main pathway in chromatophores was seen as cyclic phosphorylation, ATP synthesis driven by the operation of the pathway shown in Fig. 15.4. Although non-cyclic phosphorylation has been proposed on many occasions, sometimes involving a second pigment system, the validity of the proposals is difficult to assess.

The proton translocating ATPase is located on the outside (c-side) of the chromatophore membrane and is similar to other ATPases discussed earlier. The soluble ATPase consists of five types of subunit and has a molecular weight of about 300 000. In organisms which can grow photosynthetically or hetero-trophically, the same ATPase will function in both oxidative phosphorylation and photophosphorylation.

Measurement of the phosphate potential in *Rhodospirillum rubrum* together with the proton motive force suggests that the latter is of sufficient magnitude to drive ATP synthesis by a chemiosmotic mechanism as long as the H^+/ATP ratio is 2 (Leiser and Gromet-Elhanan, 1977).

An unusual coupling of the energy of electron transport is found in *Rhodospirillum rubrum*. This organism can couple the synthesis or hydrolysis of pyrophosphate, $PP + H_2O \rightleftharpoons 2Pi$, through a membrane-bound pyrophos-phatase, to electron transport, ATP hydrolysis or synthesis and transhydrogen-ation; the enzyme also translocates protons (Rao and Keister, 1978).

15.7   Carbon dioxide fixation

The Calvin cycle is operative in purple sulphur bacteria and purple non-sulphur bacteria. In the green bacteria, however, the pathway of carbon fixation is not clear and may involve a separate metabolic system. Incubation of these organisms with $^{14}CO_2$ results in rapid labelling of malate, oxoglutarate, pyruvate, oxaloacetate, glutamate and aspartate (Sirevag, 1974). These results are consistant with the functioning of a reductive tricarboxylic acid pathway (Fig. 15.8).

The $NADH_2$ needed for carbon fixation in purple photosynthetic bacteria can be formed photosynthetically using appropriate electron donors such as succinate or thiosulphate (or non-photosynthetically in the case of those reductants of sufficiently low mid-point potential, e.g. molecular hydrogen). However, a direct reduction by the primary electron acceptor of P870 (X or Fe–Q) seems improbable as, unlike the state of affairs in the chloroplast PS I system, the mid-point potential of the acceptor is much more positive than that of the NAD/NADH couple (−320 mV). Reduction of NAD can, however, be demonstrated in the light or in the dark if ATP is added. In the former case, it is sensitive to oligomycin and, in both cases, to uncouplers. It therefore seems probable that NAD is being reduced through reverse electron flow driven by an energised state set up by ATP hydrolysis or light-driven cyclic electron flow.

A possible exception to the above is found in *Chlorobium* where CO_2 is fixed by a reductive tricarboxylic acid cycle (Fig. 15.8) and where the potential of

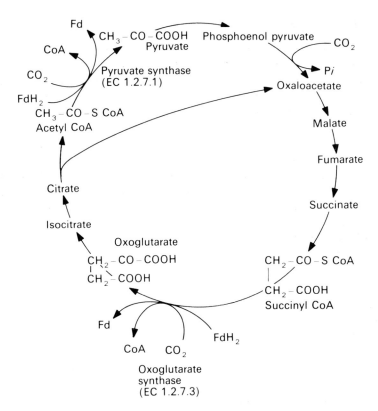

Fig. 15.8 Proposed pathway for CO_2 fixation in *Chlorobium* (reductive tricarboxylic acid cycle). Fd and FdH_2 are oxidised and reduced ferredoxin respectively.

the primary acceptor is sufficiently low (possibly -450 mV) for reduction of NAD and ferredoxin to occur. Here a light-driven reduction of NAD, mediated by ferredoxin, has been demonstrated.

15.8   Respiratory metabolism

The purple non-sulphur bacteria can be grown either photosynthetically (anaerobically) or aerobically when the photosynthetic pigments are repressed and metabolism is supported by respiratory energy. Respiratory metabolism has been poorly characterised. In photosynthetically grown cells of several species, two terminal oxidases are found, cytochrome *o* and a second oxidase not fully described. The function of these oxidases under anaerobic conditions is not clear. In *Rhodopseudomonas sphaeroides*, aerobic growth leads to the formation of an *a*, a_3 type oxidase with properties very similar to those of the mitochondrial enzyme.

Have the pathways of photosynthetic and respiratory electron transport common components? In one species, *Rhodopseudomonas capsulata*, an attempt has been made to answer this question. Study of mutants suggests that a *b*-type cytochrome ($E_{m\ 7.0} = 47$ mV) and a *c*-type cytochrome ($E_{m\ 7.0} = 342$ mV) are involved in both pathways. A high potential b-type cytochrome ($E_{m\ 7.0} =$ 413 mV) is involved only in respiratory metabolism and could be a terminal oxidase (see Fig. 15.9).

It should be noted that the photosynthetic bacteria have a number of potential electron or hydrogen carriers, the functions of which are at present unknown. These include iron–sulphur proteins and cytochromes.

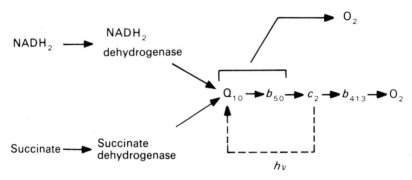

Fig. 15.9 Proposed pathway for respiration and photosynthetic cyclic electron flow in *Rhodopseudomonas capsulata* (modified from Zannoni *et al.*, 1976).

15.9   Photosynthesis without chlorophyll

(a) The purple membrane

Organisms of the genus *Halobacterium* show a unique form of photophosphorylation which requires neither an electron transport chain nor any form of chlorophyll but involves a purple pigment related to rhodopsin, the visual pigment of the mammalian retina. The pigment has been called bacteriorhodopsin.

The organisms live in solutions of salt above about 12 per cent (2 M), such as pools where sea water evaporates. They have atypical walls composed of glycoproteins. The cells lyse in water and the cell membrane is disrupted. As shown originally by Stoeckenius (see Stoeckenius and Kunau, 1968) the resultant membrane fragments can be separated into red and purple fractions. The purple fraction is only a minor component of lysates from well-oxygenated cells, but in cells grown at low oxygen tensions, it can account for up to 50 per cent of the membrane. Oesterhelt and Stoeckenius (1971) showed that the purple membrane contained only one polypeptide (MW = 26 000; Reynolds and Stoeckenius, 1977) and that the prosthetic group of the protein was retinal (vitamin A aldehyde) which is also the prosthetic group of the eye pigment, rhodopsin. The retinal was bound to a lysine residue by a Schiff base linkage (see Fig. 15.10). Later studies have established a structure of the purple membrane and particularly its role in photophosphorylation as first proposed by Oesterhelt and Stoeckenius (1973). Purple membrane has also been shown to transport protons outwards in the light.

Purple membrane isolated as a result of cell lysis is derived from specialised areas or patches of the plasma membrane. These have a structure

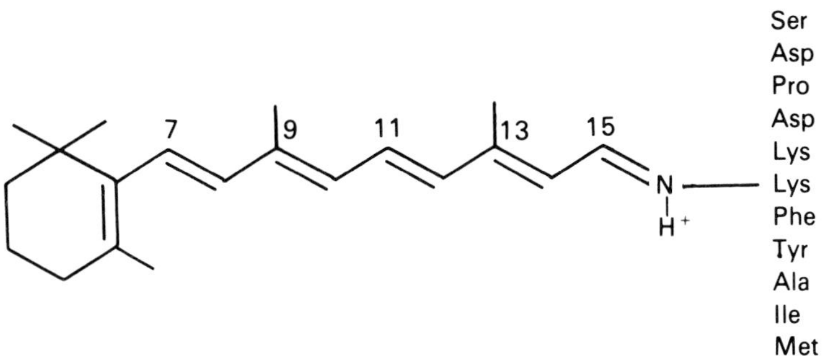

Ser
Asp
Pro
Asp
Lys
Lys
Phe
Tyr
Ala
Ile
Met

Fig. 15.10 Retinal linked by a Schiff's base to a lysine residue

quite different from the red membrane which is of a conventional bacterial type and contains carotenoids (bacteriorubrin, etc.) giving the fraction its colour. The purple membrane patches are composed of 25 per cent lipid and 75 per cent protein. The protein fraction has only one component (in contrast to the red membrane which contains a variety of different proteins). The major lipids are a diether of phosphatidylglycerophosphate (about 50 %, Fig. 15.11), triglycosyl diether (about 20 %), sulpholipids (about 15 %), neutral lipids including squalene and phosphatidylglycerol. Dihydrophytol ether-linked to glycerol is a major component of these lipids (Fig.15. 11). The proteins are arranged in a hexagonal lattice which spans the membrane as shown in Fig. 15.12, three proteins being incorporated into each cell of the lattice.

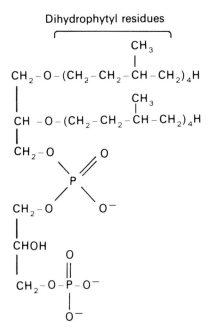

Dihydrophytyl residues

Fig. 15.11 Diether phosphatidylglycerophosphate

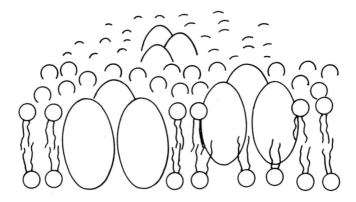

Fig. 15.12 Purple membrane. Three molecules of protein are associated with about 12–14 molecules of lipid.

(b) Photochemistry

Bacteriorhodopsin in the 'dark-adapted' form (bR_{560}) has an absorption maximum at 560 nm, but in the light this is converted to the 'light-adapted' form with a maximum at 570 nm (bR_{570}). If left in the dark, bR_{570} slowly returns to bR_{560} (dark-adapted). Upon further illumination bleaching occurs and a pigment with a maximum at 412 nm is formed (M_{412}). This second light reaction is independent of the first ($bR_{560} \rightarrow bR_{570}$). M_{412} has a short life and is rapidly converted back to the 570 nm form. The formation of M_{412} is associated with the release of protons, while the conversion to bR_{570} is accompanied by proton uptake. The bleached M_{412} form can be trapped by cooling illuminated membranes. Other intermediate forms have also been detected in the cycle (Fig. 15.13). Extraction of M_{412} gives exclusively 13-*cis* retinal whereas the dark-adapted pigment (bR_{560}) gives a mixture of 13-*cis* and all-*trans* isomers. The light-adapted purple pigment (bR_{570}) possesses all-*trans* retinal as its prosthetic group. The apoprotein, bacterio-opsin (protein from which the retinal has been removed) will combine with either the all-*trans* or the 13-*cis* but not the 9-*cis* or 11-*cis* isomers to give the purple pigment.

(c) Function of the purple membrane

As with other bacteria, respiration of *Halobacterium* results in an electrochemical proton gradient across the cell membrane. Under anaerobic conditions, however, cells possessing the purple membrane will also form an electrochemical proton gradient in the light without involving electron transport. Uncouplers abolish the gradient. Under aerobic conditions, light inhibits respiration of such cells. About 24 quanta of light energy inhibit the uptake of one molecule of oxygen. Since the quantum efficiency of proton transport is measured at between 0.5 and 0.8 protons per quantum and one bacteriorhodopsin undergoing one cycle translocates one proton, 24 quanta may be expected to bring about translocation of 12–19 protons. This would be equivalent to about one molecule of oxygen, assuming a H^+/O_2 ratio of 12 or possibly 16 (cf. sect. 10.4).

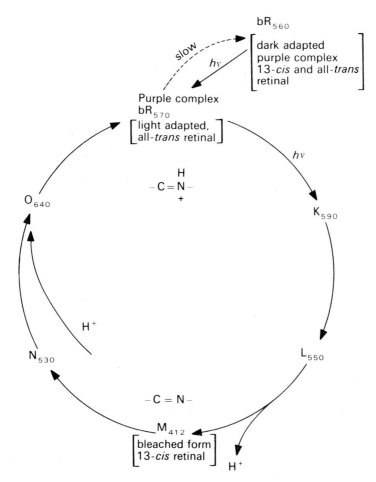

Fig. 15.13 Bacteriorhodopsin cycle. There are uncertainties about the sequence of intermediates between M_{412} and bR_{570} and also about the point at which a proton is taken up.

Measurements with intact cells have shown the establishment of a proton motive force of about 150 mV ($\Delta pH = 1.0$, $\Delta \psi = 90$ mV) under aerobic conditions in the dark. In the light the value can be raised to about 190 mV ($\Delta pH = 1.2$, $\Delta \psi = 110$ mV) and to 280 mV ($\Delta pH = 2.2$, $\Delta \psi = 160$ mV) in the presence of the ATPase inhibitor DCCD.

Illumination also leads to phosphorylation. Measurement of the intracellular ATP level shows that high ATP (and very low ADP and AMP) levels are found under aerobic conditions (in light or dark) and under anaerobic conditions in light. Thus either light or respiratory activity can lead to the formation of an electrochemical proton gradient and to ATP synthesis. The action spectrum for ATP synthesis in the absence of oxygen shows that light absorbed by bacteriorhodopsin drives ATP synthesis. This photophosphorylation is sensitive to DCCD (an ATPase inhibitor) and to uncouplers, but not to

(a)

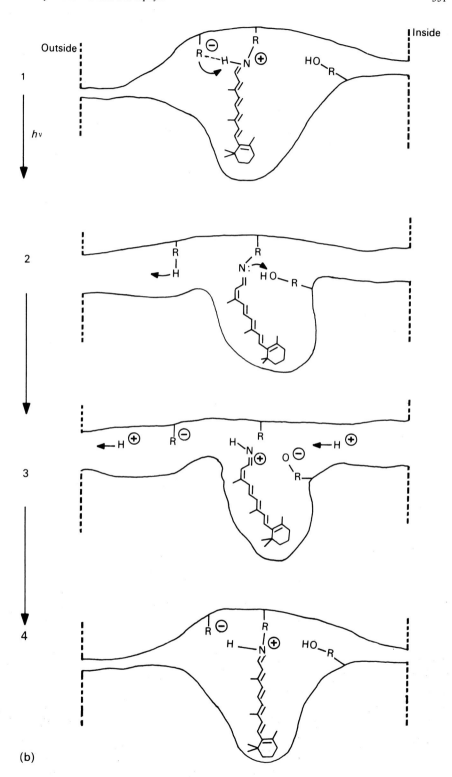

(b)

inhibitors of respiration. Illumination has also been shown to drive the active uptake of cations such as potassium. The light-driven protonation and deprotonation of bacteriorhodopsin must occur vectorially, taking up protons on one side of the membrane and releasing them on the other. Two proposals on how this might occur are shown in Fig. 15.14. Protons could be transferred through the membrane via either a grated pore (Packer and Konishi, 1978) or a chain of hydrogen exchange reactions in the protein (Stoeckenius *et al.*, 1979, ref. cit. below.)

It may be concluded that in *Halobacterium* illuminated in the absence of oxygen, the purple membrane uses light energy to pump protons across the membrane. Apparently only the protein bacteriorhodopsin, and no electron transport system, is involved in the proton pump. The electrochemical proton gradient thus set up is able to drive phosphorylation by a proton-translocating ATPase. Further evidence for this view has been obtained by using vesicles containing purple membrane. These can be shown to translocate protons in the light and to synthesise ATP if an ATPase is incorporated into the vesicle.

(d) Ecological and theoretical significance of bacteriorhodopsin-catalysed photophosphorylation

The solubility of oxygen in strong solutions of sodium chloride is substantially less than that in the media where bacteria normally grow (the solubility is five times greater in water than in 4 M NaCl). Halobacteria are adapted to grow in solutions of high salt concentration and hence low oxygen. The bacteriorhodopsin system enables photophosphorylation to occur at low oxygen levels where oxidative phosphorylation would be relatively weak and may therefore be regarded as an adaptation to a high salt environment.

Cations, particularly Na^+, have been shown to play an important role in membrane energetics. Thus a Na^+/H^+ exchanging translocator has been described and the formation of Na^+ gradients in the light have been demonstrated (Wagner *et al.*, 1978). The Na^+ gradient appears to be the immediate energy source for the active uptake of amino acids, especially glutamate. Potassium uptake has also been shown, suggesting the presence of a K^+ translocator responding to the membrane potential (Eisenbach *et al.*, 1977).

The energetics of *Halobacterium* are summarised in Fig. 15.15. The proton gradient and its accompanying potential are seen as the driving force of phosphorylation. These studies have substantially supported the chemiosmotic hypothesis of Peter Mitchell.

Fig. 15.14 Possible mechanisms for proton transport by purple membranes. (a) Transfer of protons through membrane protein (bacteriorhodopsin). A series of groups in the protein are able to transfer protons from one to another. Such a chain would accept a proton at one end of the chain and release one at the other as a consequence of the photoact. An increase in the pK of the Schiff base as a consequence of excitation would be sufficient to initiate the chain of proton transfers. The arrows represent shifts of electrons. (b) Transfer of protons through a gated pore. The photoactivation of the chromophore leads to a conformational change accompanying the release of protons towards the outside. During return to the initial conformation, a proton is taken up from the outside.

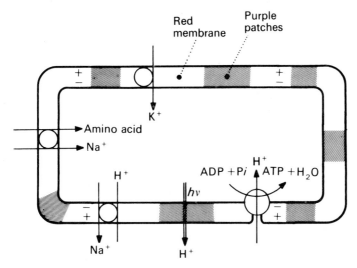

Fig. 15.15 The energetics of the *Halobacterium* membrane

Further reading

Blankenship, R. E. and Parson, W. W. (1978) The photochemical electron transfer reactions of photosynthetic bacteria and plants. *Annu. Rev. Biochem.,* **47,** 635–53.

Clayton, R. K. and Sistrom, W. R. (1977) *The Photosynthetic Bacteria.* Plenum Press, New York.

Drew, G. (1979) Structure and development of the membrane system of photosynthetic bacteria. *Current Topics in Bioenergetics,* **8,** 161–207.

Jones, O. T. G. (1977) Electron transport and ATP synthesis in photosynthetic bacteria. In *Microbial Energetics* (Haddock, B. A. and Hamilton, W. A., eds.) Soc. General Microbiology Symposium 27, Cambridge University Press.

Lanyi, J. K. (1978) Light energy conversion in *Halobacterium halobium. Microbiol. Rev.,* **42,** 682–706.

McCarty, R. E. (1978) The ATPase complex of chloroplasts and chromatophores. *Current Topics in Bioenergetics,* **7,** 245–78.

Parson, W. W. and Cogdell, R. J. (1975) The primary photochemical reaction of bacterial photosynthesis. *Biochim. Biophys. Acta,* **416,** 105–49.

Pfenning, N. (1977) Phototrophic green and purple bacteria. A comparative systematic survey. *Annu. Rev. Microbiol.,* **31,** 225–74.

Stoeckenius, W., Lozier, R. H. and Bogomolni, R. A. (1979) Bacteriorhodopsin and the purple membrane of halobacteria. *Biochim. Biophys. Acta,* **505,** 215–278.

Stanier, R. Y. and Cohen-Bazire, G. (1977) The cyanobacteria. *Annu. Rev. Microbiol.,* **31,** 225–74.

References

Albracht, S. P. J., Dooijewaard, G., Leeuwerik, F. J. & van Swol, B. (1977) *Biochim. Biophys. Acta*, **459**, 300–17.

Alexandre, A., Reynafarje, B. & Lehninger, A. L. (1978) *Proc. Natl. Acad. Sci. USA*, **75**, 5296–300.

Allen, F. L. & Franck, J. (1955) *Arch. Biochem. Biophys.*, **58**, 124–43.

Allen, J. F. (1975) *Nature (London)*, **256**, 599–600.

Almassy, R. J. & Dickerson, R. E. (1978) *Proc. Natl. Acad. Sci. USA*, **75**, 2674–8.

Annau, E., Banga, I., Blazsó, A., Bruckner, V., Laki, K., Straub, F. B. & Szent-Györgyi, A. (1936) *Hoppe-Seyler's Z. Physiol Chem.*, **244**, 105–16.

Aquila, H., Eiermann, W., Babel, W. & Klingenberg, M. (1978) *Eur. J. Biochem.*, **85**, 549–60.

Armond, P. A. & Arntzen, C. J. (1977) *Plant Physiol.*, **59**, 398–404.

Arnon, D. I. (1959) *Nature (London)*, **184**, 10–21.

Arnon, D. I., Whatley, F. R. & Allen, M. B. (1954) *J. Am. Chem. Soc.*, **76**, 6324–9.

Arntzen, C. J., Dilley, R. A., Peters, G. A. & Shaw, E. R. (1972) *Biochim. Biophys. Acta*, **256**, 85–107.

Atkinson, D. E. (1968) in *Metabolic Roles of Citrate* (Goodwin, T. W., ed.), pp. 23–40, Academic Press, New York & London.

Azzi, A., Chappell, J. B. & Robinson, B. H. (1967) *Biochem. Biophys. Res. Commun.*, **29**, 148–52.

Azzone, G. F. & Massari, S. (1973) *Biochim. Biophys. Acta*, **301**, 195–226.

Baddiley, J., Michelson, A. M. & Todd, A. R. (1949) *J. Chem. Soc.*, 582–6.

Bakeeva, L. E., Chentsov, Y. S. & Skulachev, V. P. (1978) *Biochim. Biophys. Acta*, **501**, 349–69.

Banerjee, R. K., Shertzer, H., Kanner, B. I. & Racker, E. (1977) *Biochem. Biophys. Res. Commun.*, **75**, 772–8.

Banga, I., Ochoa, S. & Peters, R. A. (1939) *Biochem. J.*, **33**, 1109–21.

Banks, B. E. C. & Vernon, C. A. (1970) *J. Theor. Biol.*, **29**, 301–26.

Barath, Z. & Kuntzel, H. (1972) *Proc. Natl. Acad. Sci. USA*, **69**, 1371–4.

Barrell, B. G., Bankier, A. T. & Drouin, J. (1979) *Nature (London)*, **282**, 189–94.

Bartley, W. & Davies, R. E. (1954) *Biochem. J.*, **57**, 37–49.

Bassham, J. A., Benson, A. A., Kay, L. D., Harris, A. Z., Wilson, A. T. & Calvin, M. (1954) *J. Am. Chem. Soc.*, **76**, 1760–70.

Bedbrook, J. R., Link, G., Coen, D. M., Bogorad, L. & Rich, A. (1978) *Proc. Natl. Acad. Sci. USA*, **75**, 3060–4.

Beinert, H., Palmer, G., Cremona, T. & Singer, T. P. (1965) *J. Biol. Chem.*, **240**, 475–80.

Belitzer, V. A. & Tsybakova, E. T. (1939) *Biokhimiya*, **4**, 516–534. Eng. Trans. in Kalckar, H. M. (1969) *Biological Phosphorylations* pp. 211–27, Prentice Hall, Englewood Cliffs, New Jersey, USA.

Bensley, R. R. & Hoer, N. L. (1934) *Anat. Rec.*, **60**, 449–55.

Benson, A. A., Bassham, J. A., Calvin, M., Goodale, T. C., Haas, V. A. & Stepka, W. (1950) *J. Am. Chem. Soc.*, **72**, 1710–8.

Bernard, C. (1878) *Leçons sur les Phénomènes de la Vie communs aux animaux et aux végétaux*. Baillière et Fils, Paris.

Bertazzoni, U., Scovassi, A. I. & Brun, G. M. (1977) *Eur. J. Biochem.*, **81**, 237–48.
Bishop, N. I. (1961) In: Ciba Foundation Symp. On *Quinones in Electron Transport*, (Wolstenholme, G. E. W. & O'Connor, C. M., eds.), pp. 385–409, Churchill, London.
Boardman, N. K. & Anderson, J. M. (1964) *Nature (London)*, **203**, 166–7.
Bohme, H. (1977) *Eur. J. Biochem.*, **72**, 283–9.
Bohme, H. & Cramer, W. A. (1972) *Biochemistry*, **11**, 1155–60.
Borisov, A. Yu. & Godik, V. I. (1973) *Biochim. Biophys. Acta*, **301**, 227–48.
Bosshard, H. R. & Zurrer, M. (1980) *J. Biol. Chem.*, **255**, 6694–9.
Boyer, P. D. (1975) *FEBS Lett.*, **58**, 1–6.
Boyer, P. D., Bieber, L. L., Mitchell, R. A. & Szabolcsi, G. (1966) *J. Biol. Chem.*, **241**, 5384–90.
Boyer, P. D., Cross, R. L. & Momsen, W. (1973) *Proc. Natl. Acad. Sci. USA*, **70**, 2837–9.
Boyer, P. D., Falcone, A. B. & Harrison, W. H. (1954) *Nature (London)*, **174**, 401–402.
Branton, D., Bullivant, S., Gilula, N. B., Karnovsky, M. J., Moor, H., Muhlethaler, K., Northcote, D. H., Packer, L., Satir, B., Satir, P., Speth, V., Staehlin, L. A., Steere, R. L. & Weinstein, R. S. (1975) *Science*, **190**, 54–56.
Branton, D. & Park, R. B. (1967) *J. Ultrastruct. Res.*, **19**, 283–303.
Brierley, G. P., Jurkowitz, M., Chavez, E. & Jung, D. W. (1977) *J. Biol Chem.*, **252**, 7932–9.
Brierley, G. P., Jurkowitz, M. & Jung, D. W. (1978) *Arch. Biochem. Biophys.*, **190**, 181–92.
Bril, C., Van der Horst, D. J., Poort, S. R., Thomas, J. B. (1969) *Biochim. Biophys. Acta*, **172**, 345–8.
Bryla, J. & Dzik, J. M. (1978) *Biochim. Biophys. Acta*, **504**, 15–25.
Bryla, J. & Harris, E. J. (1976) *FEBS Lett.*, **72**, 331–6.
Calvin, M. (1956) *J. Chem. Soc.*, 1895–915.
Caple, M. B., Chow, H. & Strouse, C. E. (1978) *J. Biol. Chem.*, **253**, 6730–7.
Carafoli, E. & Lehninger, A. L. (1971) *Biochem. J.*, **122**, 681–90.
Chance, B. (1965) *J. Biol. Chem.*, **240**, 2729–48.
Chance, B. & Hollunger, G. (1960) *Nature (London)*, **185**, 666–72.
Chance, B., Saronio, C. & Leigh, J. S. (1975) *J. Biol. Chem.*, **250**, 9226–37.
Chance, B., Smith, L. & Castor, L. (1953) *Biochim. Biophys. Acta*, **12**, 289–98.
Chance, B. & Williams, G. R. (1956) *Adv. Enzymol.*, **17**, 65–134.
Chance, B.,Wilson, D. F., Dutton, P. L. & Erecinska, M. (1970) *Proc. Natl. Acad. Sci. USA*, **66**, 1175–82.
Chappell, J. B. & Crofts, A. R. (1966) In: *Regulation of Metabolic Processes in Mitochondria* (Tager, J. M., Papa, S., Quagliariello, E. & Slater, E. C., eds.), pp. 293–316, Elsevier, Amsterdam.
Chappell, J. B. & Haarhoff, K. N. (1967) In: *Biochemistry of Mitochondria* (Slater, E. C., Kaniuga, Z. & Wojtczak, L. eds.), pp. 75–91, Academic Press & P. W. N.-Polish Scientific Publishers, London & Warsaw.
Chappell, J. B. & Robinson, B. H. (1968) In: *Metabolic Roles of Citrate* (Goodwin, T. W., ed.) pp. 123–33, Academic Press, London & New York.
Chua, N. H. & Schmidt, G. W. (1978) *Proc. Natl. Acad. Sci. USA*, **75**, 6110–14.
Claude, A. (1940) *Science*, **91**, 77–8.
Clayton, R. K. (1963) *Biochim. Biophys. Acta*, **75**, 312–23.
Cohn, M. (1953) *J. Biol. Chem.*, **201**, 735–50.
Cohn, M. & Drysdale, G. R. (1955) *J. Biol. Chem.*, **216**, 831–46.
Colowick, S. P., Kaplan, N. O., Neufeld, E. F. & Ciotti, M. M. (1952) *J. Biol. Chem.*, **195**, 95–105.
Crane, F. L., Widmer, C., Lester, R. L. & Hatefi, Y. (1959) *Biochim. Biophys. Acta*, **31**, 476–89.
Crompton, M., Hediger, M. & Carafoli, E. (1978) *Biochem. Biophys. Res. Commun.*, **80**, 540–6.
Crompton, M. & Heid, I. (1978) *Eur. J. Biochem.*, **91**, 599–608.

Crompton, M., Palmieri, F., Capano, M. & Quagliariello, E. (1974) *Biochem. J.*, **142**, 127–37.

Cross, R. L. & Kohlbrenner, W. E. (1978) *J. Biol. Chem.*, **253**, 4865–73.

Cunarro, J. & Weiner, M. W. (1975) *Biochim. Biophys. Acta*, **387**, 234–40.

Dakin, H. D. (1909) *J. Biol. Chem.*, **6**, 221–33.

Dancey, G. F. & Shapiro, B. M. (1976) *J. Biol. Chem.*, **251**, 5921–8.

Danielli, J. F. & Davson, H. (1935) *J. Cell. Comp. Physiol.*, **5**, 495–508.

Davenport, H. E., Hill, R. & Whatley, F. R. (1952) *Proc. Roy. Soc., Ser. B*, **139**, 346–58.

Davis, K. A. & Hatefi, Y. (1971) *Biochemistry*, **10**, 2509–16.

De Duve, C. (1973) *Science*, **182**, 85.

de Kouchkovsky, Y. & Joliot, P. (1967) *Photochem. Photobiol.*, **6**, 567–87.

De Luca, H. F. & Engstrom, G. W. (1961) *Proc. Natl. Acad. Sci. USA*, **47**, 1744–50.

de Vries, H., de Jonge, J. C., Schneller, J., Martin, R. P., Dirheimer, G. & Stahl, A. J. C. (1978) *Biochim. Biophys. Acta*, **520**, 419–27.

Dickerson, R. E., Takano, T., Eisenberg, D., Kallai, O. B., Samson, L., Cooper, A. & Margoliash, E. (1971) *J. Biol. Chem.*, **246**, 1511–35.

Doeg, K. A., Krueger, S. & Ziegler, D. M. (1960) *Biochim. Biophys. Acta*, **41**, 491–7.

Doering, G., Stiehl, H. & Witt, H. (1967) *Z. Naturforsch. B*, **22**, 639–44.

Dooijewaard, G., de Briun, G. J. M., van Dijk, P. J. & Slater, E. C. (1978) *Biochim. Biophys. Acta*, **501**, 458–69.

Downer, N. W., Robinson, N. C. & Capaldi, R. A. (1976) *Biochemistry*, **15**, 2930–6.

Drachev, L. A., Jasaitis, A. A., Kaulen, A. D., Kondrashin, A. A., Liberman, E. A., Nemecek, I. B., Ostroumov, S. A., Semenov, A. Y. & Skulachev, V. P. (1974) *Nature (London)*, **249**, 321–4.

Drysdale, G. R. & Lardy, H. R. (1953) *J. Biol. Chem.*, **202**, 119–36.

Duncan, H. M. & Mackler, B. (1966) *Biochemistry*, **5**, 45–56.

Dutton, H. J., Manning, W. M. & Duggar, B. M. (1943) *J. Phys. Chem.*, **47**, 308–13.

Duysens, L. N. M. (1954) *Nature (London)*, **173**, 692–3.

Duysens, L. N. M. & Amesz, J. (1959) *Plant Physiol.*, **34**, 210–13.

Duysens, L. N. M., Amesz, J. & Kamp, B. M. (1961) *Nature (London)*, **190**, 510–11.

Duysens, L. N. M. & Sweers, H. E. (1963) In: *Studies on Microalgae and Photosynthetic Bacteria* (Jap. Soc. Plant Physiol., eds.), pp. 353–72, Univ. of Tokyo Press, Tokyo.

Einbeck, H. (1919) *Biochem. Z*, **95**, 296.

Eisenbach, M., Cooper, S., Garty, H., Johnstone, R. M., Rottenberg, H. & Caplan, S. R. (1977) *Biochim. Biophys. Acta*, **465**, 599–613.

Elliott, W. B. & Haas, D. W. (1967) *Methods in Enzymol.*, **10**, 179–81.

Embden, G. & Zimmermann, M. (1927) *Z. Physiol. Chem.*, **167**, 137–40.

Emerson, R. & Arnold, W. (1932a) *J. Gen Physiol.*, **15**, 391–420.

Emerson, R. & Arnold, W. (1932b) *J. Gen. Physiol.*, **16**, 191–205.

Emerson, R. & Rabinowitch, E. (1960) *Plant Physiol.*, **35**, 477–85.

Erecinska, M., Veech, R. L. & Wilson, D. F. (1974) *Arch. Biochem. Biophys.*, **160**, 412–21.

Erecinska, M. & Wilson, D. F. (1978) *Arch. Biochem. Biophys.*, **188**, 1–14.

Estabrook, R. W. & Sactor, B. (1958) *J. Biol. Chem.*, **233**, 1014–19.

Evans, M. C. W., Reeves, S. G. & Cammack, R. (1974) *FEBS Lett.*, **49**, 111–14.

Feick, R. & Drews, G. (1978) *Biochim. Biophys. Acta*, **501**, 499–513.

Fenna, R. E. & Matthews, B. W. (1975) *Nature (London)*, **258**, 573–7.

Ferguson, S. J., Harris, D. A. & Radda, G. K. (1977) *Biochem. J.*, **162**, 351–7.

Fernandez-Moran, H. (1962) *Circulation*, **26**, 1039–65.

Fernandez-Moran, H., Oda, T., Blair, P. V. & Green, D. E. (1964) *J. Cell Biol.*, **22**, 63–100.

Fiske, C. H. & Subbarow, Y. (1929) *Science*, **70**, 381–2.

Fliege, R., Flugge, U., Werdan, K. & Heldt, H. W. (1978) *Biochim. Biophys. Acta*, **502**, 232–47.

Fonty, G., Goursot, R., Wilkie, D. & Bernardi, G. (1978) *J. Mol. Biol.*, **119**, 213–35.

Fox, C. F., Carter, J. R. & Kennedy, E. P. (1967) *Proc. Natl. Acad. Sci. USA*, **57**, 698–705.
Frenkel, A. (1954) *J. Am. Chem. Soc.*, **76**, 5568–9.
Friedman, S. & Fraenkel, G. (1955) *Arch. Biochem. Biophys.*, **59**, 491–501.
Gallerani, R., de Giorgi, C., de Benedetto, C. & Saccone, C. (1976) *Biochem. J.*, **157**, 295–300.
Gamble, J. G. & Lehninger, A. L. (1973) *J. Biol. Chem.*, **248**, 610–18.
Gervais, M., Groudinsky, O., Risler, Y. & Labeyrie, F. (1977) *Biochem. Biophys. Res. Commun.*, **77**, 1543–51.
Giaquinta, R. T., Ort, D. R. & Dilley, R. A. (1975) *Biochemistry*, **14**, 4392–6.
Gillum, A. M. & Clayton, D. A. (1978) *Proc. Natl. Acad. Sci. USA*, **75**, 677–81.
Gloe, A., Pfennig, N., Brockmann, H. Jr. & Trowitzsch, W. (1975) *Arch. Microbiol.*, **102**, 103–9.
Gomez-Puyou, A., de Gomez-Puyou, M., Becker, G. & Lehninger, A. L. (1972) *Biochem. Biophys. Res. Commun.*, **47**, 814–19.
Goodwin, T. W. (1976) In: *Chemistry and Biochemistry of Plant Pigments* (Goodwin, T. W., ed.) second edition, pp. 225–61, Academic Press, New York.
Govindjee, R., Rabinowitch, E. & Govindjee (1968) *Biochim. Biophys. Acta*, **162**, 539–44.
Graber, P. & Witt, H. T. (1976) *Biochim. Biophys. Acta*, **423**, 141–63.
Green, D. E. (1954) *Biological Reviews*, **29**, 330–66.
Green, D. E. & Baum, H. (1970) *Energy and the Mitochondrion*, Academic Press New York.
Green, D. E., Loomis, W. F. & Auerbach, V. H. (1948) *J. Biol. Chem.*, **172**, 389–403.
Green, D. E. & Perdue, J. F. (1966) *Proc. Natl. Acad. Sci. USA.*, **55**, 1295–302.
Gregg, C. T. (1967) *Methods in Enzymol.*, **10**, 181–5.
Griffiths, D. E., Cain, K. & Hyams, R. L. (1977) *Trans. Biochem. Soc.*, **5**, 205–7.
Gumaa, K. A., McLean, P. & Greenbaum, A. L. (1971) *Essays in Biochemistry*, **7**, 39–86.
Gutteridge, S., Winter, D. B., Bruynincky, W. J. & Mason, H. S. (1977) *Biochem. Biophys. Res. Commun.*, **78**, 945–51.
Guynn, R. W. & Veech, R. L. (1973) *J. Biol. Chem.*, **248**, 6966–72.
Hackenbrock, C. R. (1966) *J. Cell. Biol.*, **30**, 269–97.
Hahn, A. & Haarmann, W. (1928) *Z. Biol.*, **88**, 91–2.
Hale, G. & Perham, R. N. (1979) *Biochem. J.*, **177**, 129–37.
Halestrap, A. P. & Denton, R. M. (1974) *Biochem. J.*, **138**, 313–16.
Hallermayer, G. & Neupert, W. (1974) *Hoppe-Seyler's Z. Physiol. Chem.*, **355**, 279–88.
Harris, D. A. & Slater, E. C. (1975) *Biochim. Biophys. Acta*, **387**, 335–48.
Harris, R. A., Penniston, J. T., Asai, J. & Green, D. E. (1968) *Proc. Natl. Acad. Sci. USA*, **59**, 830–7.
Hatch, M. D. & Kagawa, T. (1976) *Arch. Biochem. Biophys.*, **175**, 39–53.
Hatch, M. D. & Slack, C. R. (1966) *Biochem. J.*, **101**, 103–11.
Hatefi, Y., Haavik, A. G., Fowler, L. R. & Griffiths, D. E. (1962a) *J. Biol. Chem.*, **237**, 2661–9.
Hatefi, Y., Haavik, A. G., Fowler, L. R. & Griffiths, D. E. (1962b) *J. Biol. Chem.*, **237**, 1676–80, 1681–5.
Hatefi, Y., Jurtshuk, P. & Haavik, A. G. (1961) *Biochim. Biophys. Acta*, **52**, 119–29.
Hauska, G., Reimer, S. & Trebst, A. (1974) *Biochim. Biophys. Acta*, **357**, 1–13.
Haxo, F. T. (1960) In: *Comparative Biochemistry of Photoreactive Pigments*, (Allen, M. B., ed.) pp. 339–60, Academic Press, New York.
Haxo, F. T. & Blinks, L. R. (1950) *J. Gen. Physiol.*, **33**, 389–422.
Heaton, G. M., Wagenvoord, R. J., Kemp, A & Nicholls, D. G. (1978) *Eur. J. Biochem.*, **82**, 515–21.
Heldt, H. W. & Rapley, L. (1970) *FEBS Lett.*, **7**, 139–42; **10**, 143–8.
Heldt, H. W. & Sauer, F. (1971) *Biochim. Biophys. Acta*, **234**, 83–91.
Henriques, F. & Park, R. B. (1976) *Arch. Biochem. Biophys.*, **176**, 472–8.
Henriques, F. & Park, R. B. (1978) *Biochem. Biophys. Res. Commun.*, **81**, 1113–18.
Heron, C., Corina, D. & Ragan, C. I. (1977) *FEBS Lett.*, **79**, 399–403.

Hill, G. C. (1976) *Biochim. Biophys. Acta*, **456**, 149–93.
Hill, K. E. & Wharton, D. C. (1978) *J. Biol. Chem.*, **253**, 489–95.
Hill, R. (1937) *Nature (London)*, **139**, 881–2.
Hill, R. (1939) *Proc. Roy. Soc. Ser. B*, **127**, 192–216.
Hill, R. (1954) *Nature (London)*, **174**, 501–3.
Hill, R. & Scarisbrick, R. (1940) *Proc. Roy. Soc. Ser. B*, **129**, 238–55.
Hill, R. & Scarisbrick, R. (1951) *New Phytol.*, **50**, 98–111.
Hinkle, P. C. & Horstman, L. L. (1971) *J. Biol. Chem.*, **246**, 6024–8.
Hogeboom, G. H., Claude, A. & Hotchkiss, R. D. (1946) *J. Biol. Chem.*, **165**, 615–29.
Hogeboom, G. H., Schneider, W. C. & Palade, G. E. (1948) *J. Biol. Chem.*, **172**, 619–36.
Höjeberg, B. & Rydström, J. (1977) *Biochem. Biophys. Res. Commun.*, **78**, 1183–90.
Huber, S. C. & Edwards, G. E. (1977) *Biochim. Biophys. Acta*, **462**, 583–602, 603–12.
Itagaki, E. & Hager, L. P. (1966) *J. Biol. Chem.*, **241**, 3687–95.
Jackson, J. B. & Crofts, A. R. (1968) *Eur. J. Biochem.*, **6**, 41–54.
Jagannathan, V. & Schweet, R. S. (1952) *J. Biol. Chem.*, **196**, 551–62.
Jeng, A. Y., Ryan, T. E. & Shamoo, A. E. (1978) *Proc. Natl. Acad. Sci. USA*, **75**, 2125–9.
Jensen, R. G. & Bassham, J. A. (1968) *Biochim. Biophys. Acta*, **153**, 227–34.
John, P. & Whatley, F. R. (1975) *Nature (London)*, **254**, 495–8.
Johnson, R. N. & Chappell, J. B. (1973) *Biochem. J.*, **134**, 769–74.
Johnson, R. N. & Hansford, R. G. (1977) *Biochem. J.*, **164**, 305–22.
Joliot, P. & Joliot, A. (1968) *Biochim. Biophys. Acta*, **153**, 625–34.
Joyard, J. & Douce, R. (1977) *Biochim. Biophys. Acta*, **486**, 273–85.
Judah, J. D. (1951) *Biochem. J.*, **49**, 271–85.
Julliard, J. H. & Gautheron, D. C. (1978) *Biochim. Biophys. Acta*, **503**, 223–37.
Junge, W. (1977) *Annu. Rev. Plant Physiol.*, **28**, 503–36.
Kaback, H. R. (1972) *Biochim. Biophys. Acta*, **265**, 367–416.
Kagawa, Y., Kandrach, A. & Racker, E. (1973) *J. Biol. Chem.*, **248**, 676–84.
Kagawa, Y. & Racker, E. (1966) *J. Biol. Chem.*, **241**, 2467–74.
Kasamatsu, H., Robberson, D. L. & Vinograd, J. (1971) *Proc. Natl. Acad. Sci. USA*, **68**, 2252–7.
Katoh, S., Suga, I, Shiratori, I. & Takamiya, A. (1961) *Arch. Biochem. Biophys.*, **94**, 136–41.
Katre, N. V. & Wilson, D. F. (1977) *Arch. Biochem. Biophys.*, **184**, 578–85.
Kayalar, C., Rosing, J. & Boyer, P. D. (1977) *J. Biol. Chem.*, **252**, 2486–91.
Keilin, D. (1925) *Proc. Roy. Soc. Ser. B*, **98**, 312–39.
Keilin, D. (1930) *Proc. Roy. Soc. Ser. B*, **106**, 418–44.
Keilin, D. (1966) *The History of Cell Respiration and Cytochrome.* Cambridge University Press.
Keilin, D. & Hartree, E. F. (1939) *Proc. Roy. Soc. Ser. B*, **127**, 167–91.
Keilin, D. & Hartree, E. F. (1955) *Nature (London)*, **176**, 200–6.
Kennedy, E. P. & Lehninger, A. L. (1949) *J. Biol. Chem.*, **179**, 957–72.
Kilpatrick, L. & Erecinska, M. (1977) *Biochim. Biophys. Acta*, **460**, 346–63.
Kita, K., Yamato, I. & Anraky, Y. (1978) *J. Biol. Chem.*, **253**, 8910–15.
Klein, R. M. (1970) *Ann. N.Y. Acad. Sci.*, **175**, 623–33.
Klingenberg, M. (1970) *Essays in Biochemistry*, **6**, 119–59.
Klingenberg, M. & Pfaff, E. (1966) In: *Regulation of Metabolic Processes in Mitochondria* (Tager, J. M., Papa, S., Quagliariello, E. & Slater, E. C., eds.), pp. 180–201, Elsevier, Amsterdam.
Klingenberg, M., Riccio, P. & Aquila, H. (1978) *Biochim. Biophys. Acta*, **503**, 193–210.
Knaff, D. B. & Arnon, D. I. (1969) *Proc. Natl. Acad. Sci. USA*, **63**, 963–9.
Knoop, F. (1905) *Beitr. Chem. Physiol. Pathol.*, **6**, 150.
Knox, R. S. (1969) *Biophys. J.*, **9**, 1351–62.
Kok, B. (1956) *Biochim. Biophys. Acta*, **22**, 399–401.
Kok, B. (1957) *Nature (London)*, **179**, 583–4.
Kok, B. (1961) *Biochim. Biophys. Acta*, **48**, 527–33.

Kok, B., Forbush, B. & McGloin, M. (1970) *Photochem. Photobiol.*, **11**, 457–75.

Kolodner, R. & Tewari, K. K. (1975) *Biochim. Biophys. Acta*, **402**, 372–90.

Korb, H. & Neupert, W. (1978) *Eur. J. Biochem.*, **91**, 609–20.

Korkes, S. A., del Campillo, A., Gunsalus, I. C. & Ochoa, S. (1951) *J. Biol. Chem.*, **193**, 721–34.

Kornberg, H. L. & Elsden, S. R. (1961) *Adv. Enzymol.*, **23**, 401–70.

Kornberg, H. L. & Krebs, H. A. (1957) *Nature (London)*, **179**, 988–91.

Kraayenhof, R. (1969) *Biochim. Biophys. Acta*, **180**, 213–15.

Krab, K. & Wikstrom, M. (1978) *Biochim. Biophys. Acta*, **504**, 200–4.

Krahling, J. B., Gee, R., Murphy, P. A., Kirk, J. R. & Tolbert, N. E. (1978) *Biochem. Biophys. Res. Commun.*, **82**, 136–41.

Kramer, R. & Klingenberg, M. (1977) *FEBS Lett.*, **82**, 363–7.

Krebs, H. A. (1943) *Adv. Enzymol.*, **3**, 191–252.

Krebs, H. A. & Eggleston, L. V. (1940) *Biochem. J.*, **34**, 442–59.

Krebs, H. A. & Johnson, W. A. (1937) *Enzymologia*, **4**, 148–56.

Kroon, A. M., Pepe, G., Bakker, H., Holtrop, M., Bollen, J. E., van Bruggen, E. F. J., Cantatore, P., Terpstra, P. & Saccone, C. (1977) *Biochim. Biophys. Acta*, **478**, 128–45.

Kunau, W. & Dommes, P. (1978) *Eur. J. Biochem.*, **91**, 533–44.

Kundig, W., Ghosh, S. & Roseman, S. (1964) *Proc. Natl. Acad. Sci. USA*, **52**, 1067–74.

Kuntzel, H. & Noll, H. (1967) *Nature (London)*, **215**, 1340–5.

Kurup, C. K. R. & Brodie, A. F. (1966) *J. Biol. Chem.*, **241**, 4016–22.

Land, J. M. & Clark, J. B. (1974) *FEBS Lett.*, **44**, 348–51.

LaNoue, K., Mizani, S. M. & Klingenberg, M. (1978) *J. Biol. Chem.*, **253**, 191–8.

Lauquin, G. J. M., Brandolin, G., Lunardi, J. & Vignais, P. V. (1978) *Biochim. Biophys. Acta*, **501**, 10–19.

Lehner, K. & Heldt, H. W. (1978) *Biochim. Biophys. Acta*, **501**, 531–44.

Lehninger, A. L. (1949) *J. Biol. Chem.*, **178**, 625–44.

Lehninger, A. L. (1951) *J. Biol. Chem.*, **190**, 345–59.

Lehninger, A. L. (1974) *Proc. Natl. Acad. Sci. USA*, **71**, 1520–4.

Leiser, M. & Gromet-Elhanan, Z. (1977) *Archiv. Biochem. Biophys.*, **178**, 79–88.

Leloir, L. F. & Munoz, J. M. (1939) *Biochem, J.*, **33**, 734–46.

Lemberg, R. & Barrett, J. (1973) *Cytochromes.* Academic Press, New York.

Lemberg, R. & Mansley, G. E. (1966) *Biochim. Biophys. Acta*, **118**, 19–35.

Leung, K. H. & Hinkle, P. C. (1975) *J. Biol. Chem.*, **250**, 8467–71.

Lewis, F. S., Rutman, R. J. & Avadhani, N. G. (1976) *Biochemistry*, **15**, 3362–6, 3367–72.

Liberman, E. A., Topaly, V. P., Tsofina, L. M., Jasaitis, A. A. & Skulachev, V. P. (1969) *Nature (London)*, **222**, 1076–8.

Lichtenthaler, H. K. & Park, R. B. (1963) *Nature (London)*, **198**, 1070–2.

Lieberman, J. R., Bose, S. & Arntzen, C. J. (1978) *Biochim. Biophys. Acta*, **502**, 417–29.

Lin, L. H. & Beattie, D. S. (1978) *J. Biol. Chem.*, **253**, 2412–18.

Linn, T. C., Pettit, F. H. & Reed, L. J. (1969) *Proc. Natl. Acad. Sci. USA*, **62**, 234–41.

Lipmann, F. (1945) *J. Biol. Chem.*, **160**, 173–90.

Lipmann, F. (1946) In: *Currents in Biochemical Research*, (Green, T. E. ed.), pp. 137–48, Interscience, New York.

Lipmann, F. (1948) *Harvey Society Lectures*, pp. 99–123, C. C. Thomas, Springfield, Illinois, USA.

Lipmann, F., Kaplan, N. O., Novelli, G. D., Tuttle, L. C. & Guirard, B. M. (1947) *J. Biol. Chem.*, **167**, 867–70.

Lohmann, K. (1929) *Naturwiss.*, **17**, 624–5.

Lorimer, G. H. (1978) *Eur. J. Biochem.*, **89**, 43–50.

Losada, M., Whatley, F. R. & Arnon, D. I. (1961) *Nature (London)*, **190**, 606–10.

Low, H. & Vallin, I. (1963) *Biochim. Biophys. Acta*, **69**, 361–74.

Luck, D. J. L. & Reich, E. (1964) *Proc. Natl. Acad. Sci. USA*, **52**, 931–8.

Lundegardh, H. (1962) *Physiol. Plant.*, **15**, 390–8.

Lynen, F. & Ochoa, S. (1953) *Biochim. Biophys. Acta*, **12**, 299–314.

Maccecchini, M., Rudin, Y., Blobel, G. & Schatz, G. (1979) *Proc. Natl. Acad. Sci. USA*, **76**, 343–7.

MacDonald, R. E., Lanyi, J. K. & Greene, R. V. (1977) *Proc. Natl. Acad. Sci. USA*, **74**, 3167–70.

McGivan, J. D. & Chappell, J. B. (1975) *FEBS Lett.*, **52**, 1–7.

McLean, J. R., Cohn, G. L., Brandt, I. K. & Simpson, M. V. (1958) *J. Biol. Chem.*, **233**, 657–63.

MacMunn, C. A. (1887) *J. Physiol.*, **8**, 51–65.

Mahler, H. R., Sarkar, N. K. & Vernon, L. P. (1952) *J. Biol. Chem.*, **199**, 585–97.

Malkin, R. & Bearden, A. J. (1971) *Proc. Natl. Acad. Sci. USA*, **68**, 16–19.

Malnoe, P. & Rochaix, J. D. (1978) *Molec. Gen. Genet.*, **166**, 269–75.

Malviya, A. N., Parsa, B., Yodaiken, R. E. & Elliott, W. B. (1968) *Biochim. Biophys. Acta*, **162**, 195–209.

Marra, E., Doonan, S., Saccone, C. & Quagliariello, E. (1978) *Eur. J. Biochem.*, **83**, 427–35.

Marres, C. A. M. & Slater, E. C. (1977) *Biochem. Biophys. Acta*, **462**, 531–48.

Martius, C. (1938) *Hoppe Seyler's Z. Physiol. Chem.*, **247**, 104–10.

Mason, T. L. & Schatz, G. (1973) *J. Biol. Chem.*, **248**, 1355–60.

Meijer, A. J., Gimpel, J. A., Deleeuw, A., Tager, J. M. & Williamson, J. R. (1975) *J. Biol. Chem.*, **250**, 7728–38.

Meijer, A. J. & van Dam, K. (1974) *Biochim. Biophys. Acta*, **346**, 213–44.

Menke, W. (1962) *Annu. Rev. Plant Physiol.*, **13**, 27–44.

Meyer, J. (1977) *Arch. Biochem. Biophys.*, **178**, 387–95.

Mimuro, M. & Fujita, Y. (1978) *Biochim. Biophys. Acta*, **504**, 406–12.

Mitchell, P. (1961) *Nature (London)*, **191**, 144–8.

Mitchell, P. (1975) *FEBS Lett.*, **59**, 137–9.

Mitchell, P. (1976) *Biochem. Soc. Trans.*, **4**, 399–430.

Mitchell, P. & Moyle, J. (1967) *Biochem. J.*, **105**, 1147–62.

Molisch, H. (1925) *Z. Botanik*, **17**, 577–93.

Monod, J. (1956) In: *Enzymes, Units of Biological Structures and Function* (Gaebler, O. H. ed.), pp. 7–28 Academic Press, New York.

Moore, A. L., Jackson, C., Halliwell, B., Dench, J. E. & Hall, D. O. (1977) *Biochem. Biophys. Res. Commun.*, **78**, 483–91.

Moore, C. & Pressman, B. C. (1964) *Biochem. Biophys. Res. Commun.*, **15**, 562–7.

Moorman, A. F. M., van Ommen, G. B. & Grivell, L. A. (1978) *Molec. Gen. Genet.*, **160**, 13–24.

Moreau, F., Dupont, J. & Lance, C. (1974) *Biochim. Biophys. Acta*, **345**, 294–304.

Morimoto, R., Merten, S., Lewin, A., Martin, N. C. & Rabinowitz, M. (1978) *Molec. Gen. Genet.*, **163**, 241–55.

Morton, R. A. (1958) *Nature (London)*, **182**, 1764–7.

Mowbray, J. (1974) *FEBS Lett.*, **44**, 344–7.

Moyle, J. & Mitchell, P. (1973a) *FEBS Lett.*, **30**, 317–20.

Moyle, J. & Mitchell, P. (1973b) *Biochem. J.*, **132**, 571–85.

Moyle, J. & Mitchell, P. (1977) *FEBS Lett.*, **77**, 136–40.

Moyle, J. & Mitchell, P. (1978) *FEBS Lett.*, **90**, 361–5.

Munn, E. A. (1974) *The Structure of Mitochondria*. Academic Press, London.

Nakatani, H. Y. & Barber, J. (1977) *Biochim. Biophys. Acta*, **461**, 510–12.

Nass, M. M. K. & Nass, S. (1963) *J. Cell Biol.*, **19**, 593–611, 613–29.

Nass, M. M. K., Nass, S. & Afzelius, B. A. (1965) *J. Exptl. Cell Res.*, **37**, 516–39.

Neumann, J. & Jagendorf, A. T. (1964) *Archiv. Biochem. Biophys.*, **107**, 109–19.

Nicholls, D. G. (1974) *Eur. J. Biochem.*, **50**, 305–15.

Nishimura, M. (1963) *Biochim. Biophys. Acta*, **66**, 17–21.

O'Brien, T. W. & Kalf, G. F. (1967) *J. Biol. Chem.*, **242**, 2172–9, 2180–5.

Ochoa, S. (1943) *J. Biol. Chem.*, **151**, 493–505.

Ochoa, S., Mehler, A. & Kornberg, A. (1947) *J. Biol. Chem.*, **167**, 871–2.

Oelze, J. (1976) *Biochim. Biophys. Acta*, **436**, 95–100.

Oesterhelt, D. & Stoeckenius, W. (1971) *Nature New Biol. (London)*, **233**, 149–52.

Oesterhelt, D. & Stoeckenius, W. (1973) *Proc. Natl. Acad. Sci. USA*, **70**, 2853–7.

Ogston, A. G. (1948) *Nature (London)*, **162**, 963.

Ohnishi, T., Leigh, J. S., Ragan, C. I. & Racker, E. (1974) *Biochem. Biophys. Res. Commun.*, **56**, 775–82.

Ohnishi, T., Lim, J., Winter, D. B. & King, T. E. (1976) *J. Biol. Chem.*, **251**, 2105–9.

Okamoto, H., Sone, N., Hirata, H., Yoshida, M. & Kagawa, Y. (1977) *J. Biol. Chem.*, **252**, 6125–31.

Oleszeko, S. & Moudrianakis, E. N. (1974) *J. Cell Biol.*, **63**, 936–48.

Ovchinnikov, Yu. A. (1972) In: *Mitochondria: Biogenesis and Bioenergetics. Biomembranes: Molecular Arrangements and Transport Mechanisms*, FEBS symposia, **28**, pp. 279–306 North Holland, Amsterdam.

Packer, L. (1960–62) *J. Biol. Chem.*, **235**, 242–9; **236**, 214–20; **237**, 1327–31.

Packer, L. (1973) In: *Mechanisms in Bioenergetics* (Azzone, G. F., Ernster, L., Papa, S., Quagliariello, E. & Siliprandi, N., eds.), pp. 33–52 Academic Press, New York.

Packer, L. & Konishi, T. (1978) In: *Energetics and Structure of Halophilic Microorganisms* (Caplan, S. R. & Ginzburg, M. eds.), pp. 143–63, Elsevier–North Holland, Amsterdam.

Padmanaban, G., Hendler, F., Patzer, J., Ryan, R. & Rabinowitz, M. (1975) *Proc. Natl. Acad. Sci. USA*, **72**, 4293–7.

Palade, G. E. (1952) *Anat. Rec.*, **114**, 427–51.

Palmer, J. M. (1979) *Biochem. Soc. Trans.*, **7**, 246–52.

Pande, S. V. (1975) *Proc. Natl. Acad. Sci. USA*, **72**, 883–7.

Pande, S. V. & Parvin, R. (1978) *J. Biol. Chem.*, **253**, 1565–73.

Papa, S. (1976) *Biochim. Biophys. Acta*, **456**, 39–84.

Park, R. B. & Biggins, J. (1964) *Science*, **144**, 1009–11.

Park, R. B. & Pfeifhofer, A. O. (1969) *J. Cell Science*, **5**, 299–311.

Park, R. B. & Sane, P. V. (1971) *Annu. Rev. Plant Physiol.*, **22**, 395–430.

Parson, W. W. & Cogdell, R. J. (1975) *Biochim. Biophys. Acta*, **416**, 105–49.

Parsons, D. F., Williams, G. R. & Chance, B. (1966) *Ann. NY Acad. Sci.*, **137**, 643–66.

Penniston, J. T., Harris, R. A., Asai, J. & Green, D. E. (1968) *Proc. Natl. Acad. Sci. USA*, **59**, 624–31.

Perlman, S., Abelson, H. T. & Penman, S. (1973) *Proc. Natl. Acad. Sci. USA*, **70**, 350–3.

Petrack, B. & Lipmann, F. (1961) In: *Light and Life* (McElroy, W. D. & Glass, B., eds.), pp. 621–30, John Hopkins, Baltimore.

Petty, K. M., Jackson, J. B. & Dutton, P. L. (1977) *FEBS Lett.*, **84**, 299–303.

Pfaff, E., Klingenberg, M. & Heldt, H. W. (1965) *Biochim. Biophys. Acta*, **104**, 312–15.

Poincelot, R. P. (1973) *Arch. Biochem. Biophys.*, **159**, 134–42.

Portis, A. R., Chon, C. J., Mosbach, A. & Heldt, H. W. (1977) *Biochim. Biophys. Acta*, **461**, 313–25.

Portis, A. R. & Heldt, H. W. (1976) *Biochim. Biophys. Acta*, **449**, 434–46.

Poyton, R. O. & Schatz, G. (1975) *J. Biol. Chem.*, **250**, 752–61.

Priestley, J. (1772) *Phil. Trans. Roy. Soc.*, **62**, 147–252.

Prince, R. C. & Dutton, P. L. (1977) *Biochim. Biophys. Acta*, **462**, 731–47.

Prince, R. C., Tiede, D. M., Thornber, J. P. & Dutton, P. L. (1977) *Biochem. Biophys. Acta*, **462**, 467–90.

Pullman, M. E., Penefsky, H. S., Datta, A. & Racker, E. (1960) *J. Biol. Chem.*, **235**, 3322–9.

Quastel, J. H. & Wooldridge, W. R. (1928) *Biochem. J.*, **22**, 689–702.

Racker, E. (1970) In: *Membranes of Mitochondria and Chloroplasts* (Racker, E. ed.), pp. 127–71, Van Nostrand Reinhold, New York.

Radmer, R. & Kok, B. (1975) *Annu. Rev. Biochem.*, **44**, 409–33.

Raff, R. A. & Mahler, H. R. (1972) *Science*, **177**, 575–82.

Ragan, C. I. & Racker, E. (1973) *J. Biol. Chem.*, **248**, 6876–84.
Ragan, C. I. & Hinkle, P. C. (1975) *J. Biol. Chem.*, **250**, 8472–6.
Randall, D. D., Rubin, P. M. & Fenko, M. (1977) *Biochim. Biophys. Acta*, **485**, 336–49.
Rao, P. V. & Keister, D. L. (1978) *Biochem. Biophys. Res. Commun.*, **84**, 465–73.
Rathnam, C. K. M. & Edwards, G. E. (1975) *Arch. Biochem. Biophys.*, **171**, 214–25.
Reed, D. W. & Clayton, R. K. (1968) *Biochem. Biophys. Res. Commun.*, **30**, 471–5.
Reed, L. J., DeBusk, B. G., Gunsalus, I. C. & Hornberger, C. S. (1951) *Science*, **114**, 93–4.
Reed, L. J., Pettit, F. H., Eley, M. H., Hamilton, L., Collins, J. H. & Oliver, R. M. (1975) *Proc. Natl. Acad. Sci. USA*, **72**, 3068–72.
Reeves, S. G. & Hall, D. O. (1978) *Biochim. Biophys. Acta*, **463**, 275–98.
Reynafarje, B. & Lehninger, A. L. (1978) *J. Biol. Chem.*, **253**, 6331–4.
Reynolds, J. A. & Stoeckenius, W. (1977) *Proc. Natl. Acad. Sci. USA*, **74**, 2803–4.
Rieske, J. S. (1976) *Biochim. Biophys. Acta*, **456**, 195–247.
Rieske, J. S., Zaugg, W. S. & Hansen, R. E. (1964) *J. Biol. Chem.*, **239**, 3023–30.
Rifkin, M. R., Wood, D. D. & Luck, D. J. L. (1967) *Proc. Natl. Acad. Sci. USA*, **58**, 1025–32.
Ringler, R. L., Minakami, S. & Singer, T. P. (1963) *J. Biol. Chem.*, **238**, 801–10.
Romijn, J. C. & Amesz, J. (1977) *Biochim. Biophys. Acta*, **461**, 327–38.
Rosing, J. & Slater, E. C. (1972) *Biochim. Biophys. Acta*, **267**, 257–90.
Ross, E. & Schatz, G. (1976) *J. Biol. Chem.*, **251**, 1991–6.
Ross, R. T. & Calvin, M. (1967) *Biophys. J.*, **7**, 595–614.
Rossi, C. S. & Lehninger, A. L. (1963) *Biochem. Z.*, **338**, 698–713.
Ruben, S., Randall, M., Kamen, M. & Hyde, J. L. (1941) *J. Am. Chem. Soc.*, **63**, 877–9.
Rubin, M. S. & Tzagoloff, A. (1973) *J. Biol. Chem.*, **248**, 4275–95.
Ryrie, I. J. & Jagendorf, A. T. (1971) *J. Biol. Chem.*, **246**, 3771–4.
Safer, B. & Williamson, J. R. (1973) *J. Biol. Chem.*, **248**, 2570–9.
Sager, R. & Ishida, M. R. (1963) *Proc. Natl. Acad. Sci. USA*, **50**, 725–30.
Saha, S., Ouitrakul, R., Izawa, S. & Good, N. E. (1971) *J. Biol. Chem.*, **246**, 3204–9.
Salemme, F. R. (1977) *Annu. Rev. Biochem.*, **46**, 299–330.
Sanders, J. P. M., Heyting, C., Verbeet, M. P., Meijlink, F. C. P. W. & Borst, P. (1977) *Molec. Gen. Genet.*, **157**, 239–61.
Sane, P. V., Goodchild, D. J. & Park, R. B. (1970) *Biochim. Biophys. Acta*, **216**, 162–78.
San Pietro, A. & Lang, H. M. (1958) *J. Biol. Chem.*, **231**, 211–29.
Sato, N., Ohnishi, T. & Chance, B. (1972) *Biochim. Biophys. Acta*, **275**, 288–97.
Sauer, K., Mathis, P., Acker, S. & van Best J. A. (1978) *Biochim. Biophys. Acta*, **503**, 120–34.
Schachman, H. K., Pardee, A. B. & Stanier, R. Y. (1952) *Arch. Biochem. Biophys.*, **38**, 245–60.
Schatz, G., Haslbrunner, E. & Tuppy, H. (1964) *Biochem. Biophys. Res. Commun.*, **15**, 127–32.
Schliephake, W., Junge, W. & Witt, H. T. (1968) *Naturforsch. B*, **23**, 1571–8.
Schurmann, P. & Wolosiuk, R. A. (1978) *Biochim. Biophys. Acta*, **522**, 130–8.
Seeley, G. R. (1973) *J. Theoret. Biol.*, **40**, 173–87.
Senior, A. E. (1973) *Biochim. Biophys. Acta*, **301**, 195–226.
Shertzer, H. G. & Racker, E. (1976) *J. Biol. Chem.*, **251**, 2446–52.
Singer, S. J. & Nicolson, G. L. (1972) *Science*, 720–31.
Singer, T. P. & Kearney, E. B. (1954) *Biochim. Biophys. Acta*, **15**, 151–3.
Singer, T. P., Kearney, E. B. & Kenney, W. C. (1973) *Adv. Enzymol.*, **37**, 189–272.
Sirevag, R. (1974) *Archiv. Microbiol.*, **98**, 3–18.
Sjöstrand, F. S. (1953) *Nature (London)*, **171**, 30.
Sjöstrand, F. S. (1978) *J. Ultrastruct. Res.*, **64**, 217–45.
Slater, E. C. (1949) *Nature (London)*, **163**, 532.
Slater, E. C. (1953a) *Nature (London)*, **172**, 975–8.
Slater, E. C. (1953b) *Annu. Rev. Biochem.*, **22**, 17–56.
Smirnov, B. P. (1960) *Biokhimiya*, **25**, 545–55.

Smith, D. D., Selman, B. R., Voegeli, K. K., Johnson, G. & Dilley, R. A. (1977) *Biochim. Biophys. Acta*, **459**, 468–82.

Sottocasa, G. L., Kuylenstierna, B., Ernster, L. & Bergstrand, A. (1967) *J. Cell. Biol.*, **32**, 415–38.

Sottocasa, G., Sandri, G., Panfili, E., de Bernard, B., Gazzotti, P., Vasington, F. & Carafoli, E. (1972) *Biochem. Biophys. Res. Commun.*, **47**, 808–13.

Spanner, D. C. (1964) *Introduction to Thermodynamics*, pp. 213–27, Academic Press, London.

Stanley, H. K. & Tubbs, P. K. (1975) *Biochem. J.*, **150**, 77–88.

Stern, K. G. (1939) *Cold Spring Harbor Symp. Quant. Biol.*, **7**, 312–22.

Stiehl, H. H. & Witt, H. T. (1969) *Z. Naturforsch. B*, **246**, 1588–98.

Stiles, J. W. & Crane, F. L. (1966) *Biochim. Biophys. Acta*, **126**, 179–81.

Stocking, C. R. (1971) *Methods Enzymol.*, **23**, 221–8.

Stoeckenius, W. & Kunau, W. F. (1968) *J. Cell Biol.*, **38**, 337–57.

Stouthamer, A. H. (1977) In: *Microbial Energetics*, Symp. 27 Soc. Gen. Microbiol. (Haddock, B. A. & Hamilton, W. A. eds.) pp. 285–315, Cambridge University Press.

Stouthamer, A. H. & Bettenhaussen, C. (1973) *Biochim. Biophys. Acta*, **301**, 53–70.

Sugden, P. H. & Randle, P. J. (1978) *Biochem. J.*, **173**, 659–68.

Swanson, R., Trus, B. L., Mandel, N., Mandel, G., Kallai, O. B. & Dickerson, R. E. (1977) *J. Biol. Chem.*, **252**, 759–75.

Tagawa, K. & Arnon, D. I. (1962) *Nature (London)*, **195**, 537–43.

Takano, T., Trus, B. L., Mandel, N., Mandel, G., Kallai, O. B., Swanson, R. & Dickerson, R. E. (1977) *J. Biol. Chem.*, **252**, 776–85.

Thiers, R. E. & Vallee, B. L. (1957) *J. Biol. Chem.*, **226**, 911–20.

Tischler, M., Friedrichs, D., Coll, K. & Williamson, J. R. (1977) *Arch. Biochem. Biophys.*, **184**, 222–36.

Tischler, M. E., Pachence, J., Williamson, J. R. & LaNoue, K. F. (1976) *Arch. Biochem. Biophys.*, **173**, 448–62.

Toeniessen, E. & Brinkmann, E. (1930) *Hoppe–Seyler's Z. Physiol. Chem.*, **187**, 137–59.

Tottmar, S. O. C. & Ragan, C. I. (1971) *Biochem. J.*, **124**, 853–65.

Trebst, A. V., Tsujimoto, H. Y. & Arnon, D. I. (1958) *Nature (London)*, **182**, 351–5.

Tyree, B. & Webster, D. A. (1978) *J. Biol. Chem.*, **253**, 7635–7.

Tzagoloff, A. & Meagher, P. (1971) *J. Biol. Chem.*, **246**, 7328–36.

Tzagoloff, A. & Meagher, P. (1972) *J. Biol. Chem.*, **247**, 594–603.

Utter, M. F., Barden, R. E. & Taylor, B. L. (1975) *Adv. Enzymol.*, **42**, 1–72.

Vambutas, V. K. & Racker, E. (1965) *J. Biol. Chem.*, **240**, 2660–7.

van Beseoouw, A. & Wintermans, J. F. G. M. (1978) *Biochim. Biophys. Acta*, **529**, 44–53.

van Niel, C. B. (1941) *Adv. Enzymol.*, **1**, 263–328.

Vasington, F. D. & Murphy, J. V. (1962) *J. Biol. Chem.*, **237**, 2670–7.

Vercesi, A., Reynafarje, B. & Lehninger, A. L. (1978) *J. Biol. Chem.*, **253**, 6379–85.

Vishniac, W. & Ochoa, S. (1951) *Nature (London)*, **167**, 946–8.

Vogel, O. (1977) *Biochim. Biophys. Res. Commun.*, **74**, 1235–41.

Wagner, G., Hartmann, R. & Oesterhelt, D. (1978) *Eur. J. Biochem.*, **89**, 169–79.

Warburg, O. (1913) *Pfuger's Arch. Physiol.*, **154**, 599–617.

Warburg, O. (1949) *Heavy Metal Prosthetic Groups and Enzyme Action*. Oxford University Press.

Warburg O. & Christian, W. (1932) *Naturwiss.*, **20**, 688.

Watson, J. A. & Lowenstein, J. M. (1970) *J. Biol. Chem.*, **245**, 5993–6002.

Weier, T., Stocking, C. R., Thomas, W. W. & Drever, H. (1963) *J. Ultrastruct. Res.*, **8**, 122–43.

Weinhouse, S., Medes, G. & Floyd, N. F. (1944) *J. Biol. Chem.*, **155**, 143–51.

Weiss, H. (1976) *Biochim. Biophys. Acta*, **456**, 291–313.

Weitzman, P. D. J. & Danson, M. J. (1976) *Curr. Topics in Cell Regulation*, **10**, 161–204.

Werkheiser, W. C. & Bartley, W. (1957) *Biochem. J.*, **66**, 79–91.

West, I. C. & Mitchell, P. (1974) *FEBS Lett.*, **40**, 1–4. \

Whatley, F. R., Tagawa, K. & Arnon, D. I. (1963) *Proc. Natl. Acad. Sci. USA*, **49**, 266–70.

Whereat, A. F., Orishimo, M. W. & Nelson, J. (1969) *J. Biol. Chem.*, **244**, 6498–506.

Wieland, H. (1932) *On the Mechanism of Oxidation.* Yale University Press.

Wigglesworth, J. M., Packer, L. & Branton, D. (1970) *Biochim. Biophys. Acta*, **205**, 125–35.

Williams–Smith, D. L., Heathcote, P., Shira, C. K. & Evans, M. C. W. (1978) *Biochem. J.*, **170**, 365–71.

Wilson, A. T. & Calvin, M. (1955) *J. Am. Chem. Soc.*, **77**, 5948–57.

Wilson, D. F., Stubbs, M., Veech, R. L., Erecinska, M. & Krebs, H. A. (1974) *Biochem. J.*, **140**, 57–64.

Winget, D. G., Kanner, N. & Racker, E. (1977) *Biochim. Biophys. Acta*, **460**, 490–9.

Wintersberger, E. (1966) In: *Regulation of Metabolic Processes in Mitochondria* (Tager, J. M., Papa, S., Quagliariello, E. & Slater, E. C. eds.), pp. 439–53, Elsevier, Amsterdam.

Wood, H. G. & Werkman, C. H. (1940) *Biochem. J.*, **32**, 1262–71.

Yakushiji, E. & Okunuki, K. (1940) *Proc. Imp. Acad (Tokyo)*, **16**, 299–302.

Yang, T. & Jurtshuk, P. (1978) *Biochim. Biophys. Acta*, **502**, 543–8.

Yaoi, H. & Tamiya, H. (1928) *Proc. Imp. Acad. (Tokyo)*, **4**, 436–9.

Zannoni, D., Melandri, B. A. & Baccarini-Melandri, A. (1976) *Biochim. Biophys. Acta*, **423**, 413–30.

Index

THAMES POLYTECHNIC LIBRARY